IS SCIENCE PROGRESSIVE?

SYNTHESE LIBRARY

STUDIES IN EPISTEMOLOGY,

LOGIC, METHODOLOGY, AND PHILOSOPHY OF SCIENCE

VOLUME 177

ILKKA NIINILUOTO

Department of Philosophy, University of Helsinki

IS SCIENCE PROGRESSIVE?

D. REIDEL PUBLISHING COMPANY

A MEMBER OF THE KLUWER ACADEMIC PUBLISHERS GROUP

DORDRECHT / BOSTON / LANCASTER

Library of Congress Cataloging in Publication Data

CIP

Niiniluoto, Ilkka.
Is science progressive?
(Synthese library; v. 177)
Essays written between 1978 and 1983, some translated from the Finnish.
Includes index.
1. Science–Philosophy–Addresses, essays, lectures. 2. Realism–Addresses, essays, lectures. I. Title.
Q175.3.N54 1984 501 84-15915
ISBN 90-277-1835-0

Published by D. Reidel Publishing Company,
P.O. Box 17, 3300 AA Dordrecht, Holland.

Sold and distributed in the U.S.A. and Canada
by Kluwer Academic Publishers,
190 Old Derby Street, Hingham, MA 02043, U.S.A.

In all other countries, sold and distributed
by Kluwer Academic Publishers Group,
P.O. Box 322, 3300 AH Dordrecht, Holland.

Printed in The Netherlands

To Ritva

CONTENTS

PREFACE

This collection brings together several essays which have been written between the years 1975 and 1983. During that period I have been occupied with the attempt to find a satisfactory explicate for the notion of truthlikeness or verisimilitude. The technical results of this search have partly appeared elsewhere, and I am also working on a systematic presentation of them in a companion volume to this book: *Truthlikeness* (forthcoming hopefully in 1985). The essays collected in this book are less formal and more philosophical: they all explore various aspects of the idea that progress in science is associated with an increase in the truthlikeness of its results. Even though they do not exhaust the problem area of scientific change, together they constitute a step in the direction which I find most promising in the defence of critical scientific realism.

★

Chapter 1 appeared originally in Finnish as the opening article of a new journal *Tiede 2000* (no. 1/1980) – a Finnish counterpart to journals such as *Science* and *Scientific American*. This explains its programmatic character. It tries to give a compact answer to the question 'What is science?', and serves therefore as an introduction to the problem area of the later chapters.

Chapter 2 is a revised translation of my inaugural lecture for the chair of Theoretical Philosophy in the University of Helsinki on April 8, 1981. It appeared in Finnish in *Parnasso* 31(1981), pp. 278–282. Like Chapter 1, it displays a programme in trying to show how philosophy can be a 'scientific' and 'progressive' enterprise in spite of its differences with factual empirical sciences. It also contains my reply to Richard Rorty's challenge of analytical philosophy.

Chapter 3 is the oldest of these essays. It was written in the summer of 1975 at the same time when I prepared my first paper on truthlikeness for the 5th International Congress of Logic, Methodology, and Philosophy of Science in London, Ontario. It shows that many questions about scientific change which became fashionable in the 1960's and 1970's were discussed on a sophisticated level already in the 19th century by William Whewell and Charles Peirce. At the same time, it serves as a critical evaluation of Karl Popper's falsificationism. The paper appeared first in *Ajatus* **37**(1978), pp. 272–327. It is reprinted here without changes, but some new notes (numbered as 2a, 5a, etc.) are added.

Chapter 4 is a paper read in a symposium on 'Evolution', organized by Tutkijaliitto in Helsinki, November 20–22, 1980. It has appeared in the Proceedings of the symposium in 1982. The paper deals with Peirce's, Popper's, and Stephen Toulmin's attempts to use evolutionary concepts and analogies in the analysis of the growth of knowledge.

Chapter 5 is perhaps the central article in this collection, since it deals directly with the notion of scientific progress. To answer the factual question, Is science progressive?, we first have to consider the conceptual question, What is meant by progress in science?, and the methodological question, How is scientific progress appraised? The chapter outlines a defence – against Thomas Kuhn, among others – of what I call the 'realist' view: science makes progress insofar as it gains true or highly truthlike information about reality. It appeared in *Synthese* **45** (1980), pp. 427–464.

Chapter 6 is my contribution to the Second Joint International Congress of History and Philosophy of Science, Pisa, 1978, and it has appeared in the Proceedings in 1981 (Hintikka, Gruender, and Agazzi, Eds., *Synthese Library* vol. 145). It gives an overall introduction and evaluation of the 'structuralist' approach to theories and theory-change of Joseph Sneed and Wolfgang Stegmüller, and therefore it is more technical than the other chapters. Some new notes are added in order to make the presentation up-to-date.

Chapter 7 is written for the Fourth Joint International Congress of History and Philosophy of Science, Blacksburg, 1982, and its abridged version will appear in the Proceedings of the Congress. It continues directly the argument of Chapter 5 by defending realism against recent criticisms due to Larry Laudan and Hilary Putnam. It also gives more information about my programme for defining degrees of truthlikeness for various kinds of scientific statements and for estimating these degrees on the basis of some evidence.

Chapter 8 is a revised translation of my paper for a series of lectures on the history of mathematics, organized in the University of Helsinki in 1980. It appeared in Finnish in 1982 in the Proceedings of this series. The paper is an introduction to some philosophical and methodological issues about the history of mathematics, and it relates the nature of progress in mathematics to the general issues of scientific change.

Chapter 9 is partly based on my paper for a symposium on 'Scientific Progress and the Social Sciences', organized in the University of Tampere on April 1980. It was read at Cornell University on November 1982. It argues that the realist view of progress as approach to the truth is applicable also to the descriptive social sciences. It also compares the ontological and epistemological position of scientific realism with the views of some continental philosophical traditions (Kantianism, Marxism, phenomenology, hermeneutics).

Chapter 10 is based on an unpublished lecture on models of scientific change and their relevance to science policy, presented in the Academy of Finland, on March 1, 1978, and on my paper 'Comments on the Finalization Thesis', presented as a comment to Wolfgang van den Daele's lecture in an International Seminar on Science Studies, January 1977, Espoo. By criticising the Finalization Thesis, I outline a view of applied sciences as 'design sciences' which attempt to establish so-called technical norms.

Chapter 11 is a case-study which evaluates the claim that management science (or operations research) is facing a 'crisis' in Kuhn's sense. It is based on a lecture read in the Finnish Society of Operations Research on March 5, 1981, and in Gent on December 1982. It turns out that the Kuhnian model of science fits surprisingly well with OR. This fact, I argue, has important consequences to the problem-solving accounts of scientific progress.

Chapter 12 is partly based on my paper 'Tekniikan filosofiasta' which appeared in Finnish in 1982 and on my paper for the 17th World Congress of Philosophy, Montreal, 1983. It tries to show that science and technology have important differences in their aims and patterns of development. It also re-evaluates the realist conception of progress with respect to the technological sciences.

*

In writing these essays, I have enjoyed the stimulating atmosphere at the Department of Philosophy, University of Helsinki. The leisure needed for the completion of these papers and for collecting them in a book was partly provided in 1982 by the grant for Senior Scholars from the Academy of Finland. For practical help in working with this volume, I am grateful to Mrs. Auli Kaipainen and Mr. Hannu Simpura.

I dedicate this book to my wife who has supported all the aspects of progress that I have made in preparing this book.

Ilkka Niiniluoto

CHAPTER 1

THE NATURE OF SCIENCE

»One of the most wonderful features of reasoning» is its tendency »to correct itself, and the more so the more wisely its plan is laid». These words from the 1898 essay 'Methods for Attaining the Truth' summarize Charles Peirce's vision of science as a »self-corrective process»[1] – which still is one of the most pertinent attempts to characterize science and the scientific method. It helps us to clarify the sense in which features such as objectivity, critical attitude, autonomy, and progress belong to the nature of science.

1. THE DEMARCATION PROBLEM

How can one distinguish science from that which is not science? Following Karl Popper's suggestion in *Logik der Forschung* (1934), this question concerning the difference between science and non-science is usually called the *problem of demarcation.*[2]

For Popper, demarcation is the central problem within the philosophy of science. The overall aim of the philosophy of science, we might say, is to give us a coherent and well-argued conception of what science is or what it should be. While working scientists often remain content with an institutional characterization of science – science is simply what they have learned to do through their scientific education[3] – philosophers try to formulate and bring forward alternative views about the nature of science and submit them to open and critical discussion.

There are several reasons why the philosophical enterprise of finding the characteristics of science is worthwhile and important. First, prevailing conceptions of science have an influence on the actual practice of the scientists – even when they have been adopted unconsciously. By making such implicit conceptions of science explicit, philosophers may improve the self-understanding of the scientists.

Secondly, historians of science have to use some criteria to determine which persons and works belong to the history of science – as opposed to the history of art, literature, religion, magic, technology, etc. Did the ancient Egyptians practice science? Were the mediaeval alchemists scientists or not? To answer questions like these, we need knowledge about the solutions to the demarcation problem among earlier cultures and consideration of its presently suggested solutions.

Thirdly, sociologists of science need a criterion for distinguishing scientific communities from other – e.g. religious or political communities.

Fourthly, some criterion for distinguishing science from non-science is needed by teachers in universities, editors of scientific journals, and administrative panels of science policy who have to be able to recognize some work, study or plan as 'scientific'.

Fifthly, a solution of the demarcation problem is presupposed by the attempts to build a *scientific world view*, i.e., a synthesis of the results of various special sciences. Criteria are needed for judging what claims may rationally be accepted as elements of our scientific image of the reality.[4] In particular, a boundary line has to be drawn which separates science from *pseudosciences* – systematically organized disciplines which claim to be 'sciences' but fail to satisfy the critical standards of scientific inquiry.[5]

The demarcation between science and *non*-science is primarily a matter of conceptual clarity and does not have an evaluative function. There are numerous human activities which are not – and do not pretend to be – 'scientific' (e.g. art, religion, sport, gardening, dancing) but still are valuable by their own standards. However, the demarcation between science and pseudoscience has the normative function of separating scientific and *un*-scientific activities from each other.

Prima facie it seems that the problem of demarcation has to be distinguished from the *problem of evaluation* concerning the quality of scientific works. First we separate science and non-science, and then within science we recognize good and bad science. On the other hand, it may be thought that these two problems are in fact interrelated: a work is *good* science if it satisfies very clearly the standards of being scientific at all. As we shall see below, some attempts to characterize science in terms of its progressive nature have the consequence that the terms 'science' and 'good science' become virtually equivalent.

2. Definitions of Science

One difficulty with the demarcation problem is the fact that there are several types of entities which may be called 'scientific': men and women, groups and institutions, concepts and statements, thoughts and inferences, arguments and experiments, methods and instruments, articles and works, theories and world views.

The term 'science' may thus refer to the scientific *institution* (scientific workers, organization of research), the research *process*, the scientific *method* or scientific *knowledge*. These four aspects of science are usually defined relative to each other in a circular way. Science as an institution is that part of the society which produces scientific knowledge by maintaining scientific research. Scientific research means systematic and institutionalized pursuit of new knowledge by using scientific methods. Scientific methods in turn are those tools for producing scientific knowledge which have been accepted by the scientific community. Scientific knowledge consists of the results of research which uses scientific methods.[6]

These statements are reasonable – but they fail to define science as long as none of the four elements is characterized independently of the others.

One important tradition is to try to define science by means of its *object.* For Plato, genuine knowledge (*episteme*) is always directed at the invariant realm of ideas. The paradigmatic Platonic science is thus geometry which gives rational and provable knowledge about ideal geometrical objects.

A variant of this approach is represented by Norman Campbell in his *What is Science?* (1921). He starts by saying that »science consists in the study of the external world of nature», but rejects this definition and replaces it with another: »Science is the study of those judgments concerning which universal agreement can be obtained» (p. 27). The problem with this kind of definitions is that they usually are at the same time too strong (Campbell has to exclude historiography and economics from science) and too weak (how can we know now that 'universal agreement' concerning astrological generalizations will never be obtained in the future?).[7]

Another possibility is to try to define science by the scientific attitude or by the scientific method. The closing paragraph of David Hume's *Enquiry* (1748) is a famous expression of this view:

> »When we run over libraries, persuaded of these principles, what havoc must we make? If we take in our hand any volume; of divinity or school metaphysics, for instance; let us ask, *Does it contain any abstract reasoning concerning quantity or number*? No. *Does it contain any experimental reasoning concerning matters of fact and existence*? No. Commit it then to the flames: for it can contain nothing but sophistry and illusion.»

Similarly, John Kemeny (1959) defines science and its method by acceptable standards of inference (»the cycle of induction, deduction, and verification») and by the undogmatic attitude of scientists (»eternal search for improvement of theories which are only tentatively held») (p. 176).

The problem with these kinds of definitions is the fact that the methods of science are historically developing. Moreover, our understanding of the nature of scientific method is also continuously changing. For these reasons, the characterization of science by methods which are now found to be acceptable is questionable.

Paul Feyerabend defends, in *Against Method* (1975), »methodological anarchism». He claims that there are no general methodological rules for science which we should not sometimes violate in the name of scientific progress. Feyerabend concludes that in fact the problem of demarcation has no solution: science cannot be distinguished from politics, art, myths and fairy tales. Recently he has compared the status of astrology to physics or medicine (cf. Feyerabend, 1979).

Science is a living institution which develops and changes through time. It is doubtful whether the »essence» of the scientific enterprise can be expressed in some abstract definition.[8] This does not mean that anyone of us can use the term 'science' as he or she pleases. There are certain general features – all associated with the »self-corrective» nature of scientific inquiry – which seem to be inseparably connected with any reasonable conception of science. These features include objectivity, critical attitude, autonomy, and progress.

3. Objectivity

Scientific inquiry is based upon the interaction between the investigator (subject) and the object of research. The aim of this interaction is to acquire new information about the object. Science is *subjective* insofar as its results are influenced by personal motives, likings, wishes, beliefs and presuppositions; science is *objective* insofar as its results are based upon the effects of the object of research to the knowing subject and as its results correspond to the properties of the object.

History of science shows clearly that the scientific knowledge of each age has been more or less subjective, bound to temporally changing patterns of thought. Each age has »its own truth», as it is often said. This subjectivity of the investigators is balanced in science by the general principle of *publicity*: only the outcomes of a critical discussion within the scientific community may be tentatively accepted as the results of inquiry. In this sense, the real 'subjects' of scientific knowledge are the scientific communities rather than individual scientists. However, it has to be acknowledged that, due to 'sectarianism', even the claims accepted by actual scientific communities may be 'subjective'.

In order to be scientific, inquiry has to be objective at least in two senses. First, the object of investigation has to be real in Peirce's sense, i.e., its characters should be »independent of what anybody may think them to be».[9] Secondly, the object should be allowed to influence the formation of the result of inquiry, and this influence should be intersubjectively recognizable. For example, the radiation emitted by radio stars is intersubjectively observable, but no weight of evidence is possessed by 'information' based upon non-repeatable experiments, secret documents, private intuitions or instincts of the investigators, 'messages' obtained through 'higher inspiration' or divine revelation.

4. Critical Attitude

Francis Bacon required in *Novum Organum* (1620) that the investigators who »interpret the nature» should purify their minds of all prejudice (»idols») and proceed in science through gradual and cautious generalizations. Even though Bacon's treatise contains healthy criticism against faith in authorities and against unclear thinking, its conception of the critical functions of the scientific method does not correspond to the actual practice of science.

First, the scientists in fact always act on the basis of some initial background assumptions which they do not question – at least in the short run. Thomas Kuhn calls such assumptions, when they are shared by a scientific community, a »paradigm» or a »disciplinary matrix». In this view, the picture of »pure» research which is free from all initial assumptions is a myth: »normal science» is rather a matter of conscious articulation and employment of a given paradigm.[10]

Secondly, the proposal of new hypotheses in science need not be cautious: science may make progress through bold guesses which contain new and even improbable ideas. The critical nature of science is founded on the strict requirements that are operative in the testing and justification of hypotheses. (Cf. Chapter 3.)

Popper's solution to the demarcation problem is the proposal that only those hypotheses which in principle are falsifiable are scientific. In other words, a hypothesis is scientific if and only if it is possible to say which 'basis sentences' (i.e., intersubjectively agreed observational statements) would falsify it. For example, Einstein's general theory of relativity is scientific, because it predicted that the light rays will curve in a gravitational field – and would have been refuted if this prediction had not been verified in the observations concerning the 1919 eclipse. Theories which always can be saved from falsification by 'immunization strategies' are not scientific in Popper's sense.[11] Popper's demarcation criterion is too tight, however, since it excludes from science not only the claim 'There are brownies' but also the unrestricted existential claims 'There are electrons' or 'There are planets'.[12]

Popper's criterion of falsifiability may be weakened to a requirement of *testability*[13]: scientific hypotheses should be intersubjectively testable, i.e., it should be possible to derive – deductively or inductively – from them (and suitable additional assumptions) consequences the validity of which can be publicly controlled. Hypotheses which do not pass such tests have to be rejected or corrected. This requirement of testability can be extended also to the paradigms, even though they are given up by the scientific community very slowly – not earlier than they have gradually lost their credibility with the increase of contrary evidence and with the emergence of more able competitors.

Science is self-corrective in the sense that no authorities can claim for themselves a permanent position. All assumptions and claims made in science may in principle be questioned and refuted by publicly presented evidence. In this sense, science is more critical than any other method for attaining knowledge.

The tenacity of scientific communities in the change of paradigms – Kuhnian normal science – and repression of new ideas (sometimes combined with methods of oppression to silence the supporters of a new way of thinking) have to some extent obscured the critical nature of scientific approach. There are groups who pretend to be »critical» against the »dominating sciences» but base their own thinking upon uncritical wishes that there are short cuts to the truth which pass the way of science.

5. Autonomy

Nicholas Rescher (1978) has emphasized that the idea of the self-corrective nature of science is associated with the principle of autonomy: science is *self*-corrective in the sense that the procedure for the revision of the results of science belongs exclusively to the scientific community itself, and should not be interfered with by any extrascientific groups.

The testing and appraisal of scientific statements should be based upon criteria which are concerned with their epistemic status. Any appeal to the fact that the truth or the falsity of a hypothesis would be desirable e.g. for political, religious or moral reasons is inappropriate within science. The community of investigators ceases to be a scientific community if it gives up – or is forced to give up – this principle of autonomy.[14]

6. Progress

The assumption that all so-called 'scientific knowledge' is strictly speaking true is an illusion. The growth of science does not simply mean the accumulation of new truths over and above the system of old truths. The development of science includes 'revolutionary' breaks in which the basic concepts and theories have to be re-evaluated.

The self-corrective nature of science meant for Peirce that science is able to locate and eliminate errors: in tests of scientific hypotheses it is very probable that a false hypothesis will be rejected (cf. Chapters 4 and 5 below). In this view, the progress of science consists essentially in the replacement of earlier theories with new results which are true or at least closer to truth than the old ones. In other words, science makes progress by approaching to the truth.[15]

Imre Lakatos (1970) has taken the step of proposing progress as the demarcation criterion for science: instead of evaluting the 'scientific' nature of single statements or theories, Lakatos focuses on the success of long-term scientific research programmes. In the Lakatosian view, pseudo-sciences are unscientific not because they contain 'metaphysical' or unfalsifiable assumptions, but rather because they are unable to support progressive research programmes. Instead, they constitute research programmes which are degenerative, i.e., their 'hard core' assumptions together with the 'positive heuristics' are unable to develop new theories which make successful new predictions. Paul Thagard (1977) has argued that Lakatos's criterion has to be complemented by some additional requirements: astrology is pseudoscientific because it is not progressive and its supporters do not make any serious efforts to solve the problems associated with their theory and do not make any detailed comparisons of their program with rival progressive theories.

7. Science and Social Progress

Bertrand Russell opens his *The Scientific Outlook* (1931) with the following remark:

> »To say that we live in an age of science is a commonplace, but like most commonplaces it is only partially true. From the point of view of our predecessors, if they could view our society, we should, no doubt, appear to be very scientific, but from the point of view of our successors it is probable that the exact opposite would seem to be the case.» (p. vii)

Russell predicts that science – scientific thinking and the techniques that science develops for the manipulation of nature, man, and his social milieu – will gain an increasingly central position in society. However, he makes the important addition:

> »If, therefore, a scientific civilization is to be a good civilization it is necessary that increase in knowledge should be accompanied by increase in wisdom. I mean by wisdom a right conception of the ends of life. This is something which science itself does not provide. Increase of science by itself, therefore, is not enough to guarantee any genuine progress, though it provides one of the ingredients which progress requires.» (p. x.)

It is well-known that, evaluated in quantitative terms, the output of science has for the last 200 years grown exponentially with the doubling time of 10–15 years. Our century with its wars and constantly growing military armament has not yet shown signs of a corresponding »increase in wisdom». Science in itself, contrary to the belief of Bacon, does not remove the miseries of human life. Even if science as a cognitive enterprise is a self-corrective process, it alone cannot guarantee a balanced development within the growth of human knowledge and human wisdom.

Notes

1 See Peirce's *Collected Papers*, vol. 5, para. 575. (Below we shall use the standard way of giving this reference by '*CP* 5. 575'.) Cf. also Chapters 3, 4, and 5 below.

2 Popper formulates this problem as that of finding »a criterion which would enable us to distinguish between the empirical sciences on the one hand, and mathematics and logic as well as 'metaphysical' systems on the other» (Popper, 1959, p. 34). In accordance with the wide range of the German word *Wissenschaft*, Popper thus intends the term 'science' to cover all systematic empirical disciplines, not only natural science. It is problematic, however, that he wants here to place non-empirical disciplines (i.e., mathematics and logic) on the same side of the demarcation line as meta-

physics. This is not in harmony with Popper's proposal of calling the demarcation problem »Kant's problem» (*ibid.*), since logic and mathematics were *Wissenschaften* for Kant (see e.g. the Preface to the second edition of *The Critique of Pure Reason*) and he even tried to demonstrate »how metaphysics is possible as science» (see *Prolegomena*). Popper has later told that his original idea in 1919 was to demarcate science from pseudoscience (exemplified, in his view, by Marx and Freud) by designating the »critical attitude» as the »true scientific attitude» (cf. Popper, 1974, pp. 29–33, 62–63). From this perspective, however, we should again expect to find logic, mathematics – and even philosophy itself (cf. Chapter 2 below) – on the side of science.

3 For a suggestive picture of science as a tradition-bound activity where the criteria of science – or 'good' science' – are learned by imitating the earlier paradigmatic achievements, see Kuhn (1962).

4 A world view, in this sense, is a collection of factual statements about the world. It is scientific, if its all elements have been obtained by scientific methods, and non-scientific otherwise. The concept of scientific world view presupposes a *cognitivist* analysis of science in the sense of Levi (1967). Even though Popper often claims that scientific hypotheses are accepted only for further testing, he sometimes expresses a position which could be called fallibilist cognitivism: »The always changing, always inconclusive results of the critical discussion constitute what may be called 'the science of the day'» (Popper, 1974a, p. 68). A fully developed *Weltanschauung* should contain, besides a world view in the above sense, epistemological and ethical principles.

5 For an amusing introduction to pseudosciences, see Gardner (1957). See also Grim (1982). One of the less amusing recent pseudosciences is 'creationism' which claims to base a 'scientific' account of creation upon the Genesis. Ironically enough, some creationists defend their attack on Darwinism by appealing to Popper's claim that Darwinism is a non-testable »metaphysical research programme» (Popper, 1974a, pp. 133–143).

6 For example, the widely used OECD definitions characterize research as »systematic pursuit of new scientific knowledge» – without giving any indication of what scientific knowledge might be.

7 One problem for the definition of science is the following: all systematically organized bodies of factual information do not represent science – as is shown e.g. by the Phone Catalogue of Helsinki. The question whether a singular event or fact is relevant to science seems to depend on the available scientific theories.

8 Laudan (1982) is skeptical over the possibility of making a sharp distinction between science and non-science: he suggests that creationism should be treated as »bad science» rather than »pseudoscience» (see also Laudan, 1983). In my view, this conclusion is too strong: there may be clear examples of science and pseudoscience even if the borderline between these areas is vague.

9 See *CP* 5.405. Cf. Peirce's well-known distinction between the methods of tenacity, authority, intuition and science in 'The Fixation of Belief' (*CP* 5.358–387).

10 For Kuhnian normal science, see also Chapter 6 below.

11 Popper's most important applications of his criterion – Freud, Marx, Darwinism – have all been disputed.

12 Popper emphasizes that spatio-temporally restricted existential statements (such as 'There are rabbits in this room now') may be falsifiable. He adds that whenever an unrestricted existential statement »appears to belong to empirical science, it will in fact do so *not on its own account,* but *by virtue of being a consequence of a corroborated falsifiable theory*» (Popper, 1974b, p. 1038). Since universal statements do not entail any existential ones, Popper admits here that scientific theories contain (restricted) existential claims. If these statements are 'scientific', it seems to me a little strange to insist that the unrestricted existential claims which are entailed by them both »belong to empirical science» and are »metaphysical». It is also inconvenient to combine this view with Popper's practice of thinking about scientific theories as

deductively closed sets of sentences.

13 Some logical empiricists proposed the principle of testability as a criterion for distinguishing 'empirically meaningful' and 'meaningless' statements. In my view, it is more plausible to apply this criterion to Popper's problem of demarcation.

14 The problem of autonomy is discussed also in Chapter 10 below.

15 For an analysis of this idea, see Chapter 5 below.

Bibliography

Campbell, N., *What is Science?*, Methuen, London, 1921. (Dover, New York, 1952.)

Feyerabend, P., *Against Method: Outline of an Anarchistic Theory of Knowledge,* New Left Books, London, 1975.

Feyerabend, P., 'Dialogue on Method', in G. Radnitzky and G. Andersson (eds.), *The Structure and Development of Science,* D. Reidel, Dordrecht, 1979, pp. 63–131.

Gardner, M., *Fads and Fallacies in the Name of Science. The Curious Theories of Modern Pseudoscientists and the Strange, Amusing and Alarming Cults that Surround Them. A Study in Human Gulligibility,* Dover, New York, 1957.

Grim, P. (ed.), *Philosophy of Science and the Occult,* State University of New York Press, Albany, 1982.

Kemeny, J., *A Philosopher Looks at Science,* Van Nostrand, New York, 1959.

Kuhn, T.S., *The Structure of Scientific Revolutions,* The University of Chicago Press, Chicago, 1962. (2nd ed. 1970.)

Laudan, L., 'Commentary: Science at the Bar – Causes for Concern', *Science, Technology & Human Values* 7 (1982), pp. 16–19.

Laudan, L., 'The Demise of the Demarcation Problem', in R.J. Cohen and L. Laudan (eds.), *Physics, Philosophy and Psychoanalysis,* D. Reidel, Dordrecht, 1983, pp. 111–127.

Levi, I., *Gambling with Truth,* Alfred A. Knopf, New York, 1967. (2nd ed., The M.I.T. Press, Cambridge, Mass., 1973.)

Peirce, C.S., *Collected Papers* (ed. by C. Hartshorne and P. Weiss), vols. 1–6, Harvard University Press, Cambridge, Mass., 1931–35.

Popper, K.R., *The Logic of Scientific Discovery,* Hutchinson, London, 1959.

Popper, K.R., 'Autobiography', in P.A. Schilpp (ed.), *The Philosophy of Karl Popper,* Part I, Open Court, La Salle, 1974, pp. 1–181. (a)

Popper, K.R., 'Replies to My Critics', in P.A. Schilpp (ed.), *The Philosophy of Karl Popper,* Part II, Open Court, La Salle, 1974, pp. 961–1197. (b)

Rescher, N., *Scientific Progress,* Blackwell, Oxford, 1978.

Russell, B., *The Scientific Outlook,* London, 1931.

Thagard, P., 'Why Astrology is a Pseudoscience', in P.D. Asquith and I. Hacking (eds.), *PSA 1978,* vol. 1, Philosophy of Science Association, East Lansing, 1978, pp. 223–234.

CHAPTER 2

HOW IS PHILOSOPHY POSSIBLE AS A SCIENCE?

Philosophy is in many ways an extraordinary discipline. It differs from all other 'special sciences' through its history, methods, object and results. The position of philosophy is unique within the system of sciences and more generally within the whole human culture. Immanuel Kant's famous question – how is philosophy possible as a science? – is a permanently topical theme on which all researchers, teachers, and practitioners of philosophy have to find a personal standpoint.[1]

1. PHILOSOPHY AS AN ACADEMIC DISCIPLINE

Even though it is a matter of dispute when the history of science begins and which science is the oldest one, it is in any case literally true that philosophy is the oldest of 'academic' subjects. In the mediaeval universities philosophy flourished – in spite of the fact that this 'mother of sciences' was subordinated as the 'maiden of theology' (*ancillae theologiae*). In modern universities, philosophy has maintained an important position as an independent academic discipline.

When the first university was founded in Finland in 1640, its teachers included one professor in jurisprudence, one in medicine, three in theology, and six in the Faculty of Philosophy.[2] The fields represented by these six professors were Greek and Hebrew, eloquence, mathematics, physics, »moral and history» (practical philosophy), and »logic and poetry» (theoretical philosophy). Nicolaus Laurentii Nycopensis[3], the first *professor logices et poëseos* in the Academy of Turku, was relieved in 1650, by his own wish, from the task of reading the poetic scribbles of the students, and the title of his chair was changed to »logic and metaphysics». This corresponds to the present statutes concerning the chair of theoretical philosophy in the University of Helsinki: its field consists of the history of philosophy, logic, epistemology, and »systematic philosophy» (i.e., metaphysics and ontology).

Contemporary philosophy, in Finland as well as elsewhere, has all the marks of a special discipline: an established position in the universities, own journals and scientific societies, congresses and international unions. In this sense, philosophers constitute an independent 'scientific community' just as the investigators within other fields. The peculiar features of philosophy become visible, however, when we consider its object and methods of research.

2. Objects of Philosophy

One of the tasks of philosophy is to consider questions which are not – or at least not yet – studied by any special science. It tries to find shape, structure, and organization within such fields of reality where systematic inquiry has not yet gained a foothold. In this way it has for 2500 years served as a 'stem' or 'stalk' from which new special sciences have emerged. This process is still continuing – and is just now visible in logic and the theory of language. The historical task of philosophy is thus to act as a forerunner of other sciences and to extend rational thinking to previously untouched fields. Philosophy stands on the borderline between that-which-is-already-understood and that-which-is-not-yet-understood and makes scouting expeditions over this border.

Philosophy goes beyond all special fields of inquiry also in the sense that it tries to build upon their results systematic world views. The sum of the results of the special sciences at a certain moment of time does not yet constitute a fullblown *Weltanschauung,* since to tie them together one needs a philosophical outlook about the nature of reality, the preconditions for attaining knowledge and the mission of man in the world. This endeavour is visible not only in the finished 'great' philosophical systems but also in the everlasting strive for deeper and more complete understanding of reality.

The critical function of philosophy consists in the analysis and evaluation of the presuppositions, contents, and consequences of all forms of human culture. These forms of culture include, among others, language and conceptual frameworks, ideas and religions, science and art, even philosophy itself. In this way, a number of branches of systematic philosophy are defined in terms of their object: philosophy of language, social philosophy, philosophy of religion, philosophy of science, philosophy of technology, philosophy of art, etc. Relative to the system of sciences this means that the basic concepts, theories and methods of any scientific discipline may become the objects of philosophical investigation.

The field of philosophy is endless. There are good grounds for thinking that human knowledge will grow for ever without reaching an upper bound – without becoming complete (cf. Chapter 5 below). Beyond the border of that-which-is-already-understood there will always be room for the expeditions of philosophical thinking. Moreover, the task of systematizing the knowledge which is already gained becomes more and more demanding. Further, new material for critical thinking emerges all the time. The fear that philosophy will become empty of problems when new special sciences keep parting from it is groundless already for the reason that the same process generates new special areas within philosophy: when philosophy of nature developed to natural science, a new philosophical field – the philosophy of natural science – was created. The same pattern repeats itself in other branches of science: social philosophy – social science – the philosophy of social science; philosophy of language – linguistics – philosophy of linguistics.

3. Methods of Philosophy

Some branches of philosophy employ methods from other sciences: logicians use exact mathematical tools of concept formation, historians of philosophy base their work on the methods of textual criticism and interpretation. The attempts to establish new philosophical results usually apply a way of thinking which can be described by the words *problematization, explication*, and *argumentation.*

Philosophical thinking does not start from problems but rather from the awakening of problems. The starting point may be any idea, concerning reality or our relation to reality, which has not earlier been found problematic. With critical doubt as his weapon a philosopher tries to reveal some questionable or unclear assumptions involved with this idea. In this way, he creates a new problem situation. A classical expression of this approach is St. Augustine's remark that he knows what time means when no one asks it, but does not know it any more when someone asks him.

The second task is to formulate putative answers to philosophical problems. This process of explication involves reformulation of questions, analysis and definition of concepts, presentation of new ideas and points, and formulation of philosophical theses. In some cases, the result may be that the original problem dissolves, i.e., it will turn out to be a pseudo-problem created by careless abuse of words.

The third task is to compare and to criticize the putative answers by presenting arguments to support or to refute them. Relevant questions in this connection are e.g. the following: Is the proposed solution consistent, clear, and compatible with other philosophical theses? Does it really solve the original problem? Does it have interesting consequences for other similar problems? The presented arguments may give rise to new philosophical problems, to attempts at their solution, and so on. This process gives internal continuity to philosophical discussion – which is comparable to development of other scientific disciplines.

Socratic dialogues in Plato's works give vivid illustrations of the use of philosophical methods. As a more recent example, one may mention the discussion about the growth of scientific knowledge. According to the traditional view, science grows cumulatively, i.e., by adding new well-founded pieces of knowledge upon the earlier ones. This view was effectively problematized two decades ago by Karl Popper, Thomas Kuhn, and Paul Feyerabend who brought attention to incontinuities within the development of science, due to the corrigible and often inaccurate character of scientific knowledge and to changes in conceptual and theoretical frameworks. In 1960's a number of explications for scientific change emerged, among them Popper's theory of verisimilitude, Kuhn's distinction between cumulative normal science and scientific revolutions, and Lakatos's methodology of scientific research programmes. Argumentation about the merits and demerits of these proposals is still lively continuing. It received a new 'push' when Popper's notion of verisimilitude turned out to be inapplicable for the comparison between false theories. This new problematization lead to new attempts to explicate the notion of verisimilitude which now are debated among philosophers.[4]

4. KANT AND RORTY

In Preface to the second edition of *Kritik der reinen Vernunft* (1787) Kant, Professor of Logic and Metaphysics at the University of Königsberg, presented his evaluation of the development of science. According to Kant, logic reached its completion already in the hands of Aristotle and has not since then made any advances. Mathematics found the »sure path of science» already in the ancient Greece with the discovery of geometrical proofs, and physics »entered the highway of science» in the new age with the experimental method. Unfortunately, Kant adds, metaphysics »has not yet had the good fortune to attain to the sure scientific method» which would guarantee its progress. In his *Prolegomena zu einer jeden künftigen Metaphysik, die als Wissenschaft wird auftreten können* (1783) Kant had already described the situation as follows:

> »It seems almost ridiculous, while every other science is continually advancing, that in this, which pretends to be wisdom incarnate, for whose oracle everyone inquiries, we should constantly move round the same spot, without gaining a single step.» (Kant, 1950, p. 4.)

In Kant's view, the main reason for the nonprogressive character of metaphysics was its dogmaticism, i.e., human reason had advanced in speculations without first attempting to critically evaluate its own abilities. The task for a critical philosophy, Kant argued, is to examine the possibility and the limits of human knowledge – and the time for scientific metaphysics comes only afterwards.

It has been said that Kant was the most 'professional' of all professors of philosophy. It is not an exaggeration to say that his influence has been decisive for the fact that systematically cultivated critical analysis – theory of knowledge with its various applications – has received a central position within modern philosophy. Even though Wilhelm Windelband's idea of dividing the whole history of philosophy to three ages – pre-Kantian, Kantian and post-Kantian[5] – may seem a little overdone, the influence of the old Königsbergian professor is visible in all systematic trends of philosophy in our time (such as analytic philosophy, pragmatism, phenomenology, and marxism).

Kant is a natural target when the whole idea of systematic philosophy is questioned. In his important evaluation of modern philosophy, *Philos-*

ophy and the Mirror of Nature (1980), Richard Rorty criticizes the Kantian idea that philosophy could be a science which studies the foundations of other fields of culture. Rorty suggests that the theory of knowledge should be abandoned and replaced by *edifying* philosophy which substitutes »*Bildung* (education, self-formation)» for knowledge as the goal of thinking (p. 359) and aims at continuing »the conversation of mankind» rather than at »discovering truth». This edifying philosophy – whose 'heroes' for Rorty are Ludwig Wittgenstein, Martin Heidegger and John Dewey – does not try to establish any systems or eternal truths. Rather it may help to prevent philosophy »from attaining the secure path of science» (p. 372) and converting it »into a boring academic specialty» (p. 385).[6]

5. Philosophy and Science

The question whether philosophy could or should be 'scientific' is naturally relative to our conception of science (cf. Chapter 1 above). Therefore, this question will appear in new light when our views about the nature of science change. Kant and many of his successors who have tried to make philosophy a 'rigorous science' clearly have had in mind the cumulative model of scientific growth as a criterion for science. While this model may, with some reservations, be applicable to logic and mathematics (cf. Chapter 8 below), it is not generally valid for other special sciences.

Kant was of course consistent when he required the same kind of growth from philosophy as from other sciences. This sort of consistency has not always been easy to follow – at least according to Nowakowa and Nowak (1978). They argue that while the »positivists» (logical empiricists) have supported the mistaken cumulative model of scientific growth and the marxists the correct view of dialectical and self-corrective development, paradoxically positivist philosophy itself has progressed in accordance with the marxist model and marxist philosophy in accordance with the positivist model.

A reply both to Kant and Rorty is, in my view, contained already in the above remarks on the method of philosophy as problematization, explication and argumentation. It is incompatible with the critical nature of philosophy to bind it with the cumulative model of growth – and it is not needed, either, for philosophy to qualify as 'scientific'. In other words, Kant's hope and Rorty's fear that philosophy would reach the sure path of science are equally unwarranted.

It follows that there is no need to make a sharp distinction between 'systematic' and 'edifying' philosophy in Rorty's sense: philosophy, as described in this chapter, does not fit with Rorty's description of 'systematic' or 'edifying' philosophy.[7] Nevertheless, this philosophy is in some sense both 'systematic' and 'edifying' at the same time – these words describe two interrelated aspects of philosophical activity which at their best support each other. In particular, one way of keeping up the 'conversation of mankind' is to practice philosophy in a systematic fashion.

6. PROGRESS IN PHILOSOPHY

Philosophy is an activity which is based upon critical thinking and argumentation. It does not consist of final 'knowledge' but rather of continuous strive for the clarification, organization, and re-evaluation of concepts and thoughts. Just like other scientific thinking, this activity may be appraised for its novelty, clarity, precision, systematicity, comprehensiveness, depth, and fruitfulness. These criteria are public and objective, and they make it possible to evaluate works across the – often artificial – boundaries between philosophical 'schools'.

The popular view of philosophy as the contemplation of eternally unsolved questions thus turns out to be a mistake which results from the tendency to search for the achievements of philosophy in a wrong place. Philosophy can find results which have a permanent value, but they mainly concern new fruitful ways of posing questions, structuring problems to subproblems, clarifying conceptual distinctions, proposing new ideas or discoveries, and presenting arguments. In this sense, philosophy is a progressive discipline.

The progressive nature of philosophy can be defended also by noting that sometimes philosophers are able to find final negative results: for example, they may show that some intuitive conception or some set of assumptions is logically inconsistent.[8] Moreover, in some cases they may be able to demonstrate that consequence Q logically follows from premise P. Other philosophers may of course challenge the conclusion Q by attacking the assumption P, since no dogmatic presuppositions are acceptable within philosophy. However, even in this case it may be claimed that the conditional assertion 'If P then Q' has been established as a permanent result of philosophical inquiry.[9]

Even though philosophy has reached some results of this kind, perhaps most philosophical arguments are looser, not logically conclusive, so that they cannot be transformed to demonstrated conditional statements. The re-evaluation and clarification of such arguments is a task where the history of philosophy and systematic philosophy meet each other.

One of such arguments is contained in Rorty's thesis for the impossibility or undesirability of epistemology in the post-Kantian sense. This argument relies essentially on certain results of modern analytic philosophy. Slightly simplifying the matter, it has the form: *if* Quine is right in his criticism of the analytic-synthetic distinction, and Sellars is right in his criticism of the 'myth of the given', and Davidson is right in his criticism of the notion of conceptual framework, etc., *then* systematic epistemology is impossible.[10]

Rorty's argument is paradoxical, however. The reason for this is the fact that the clarification of the foundations of philosophy is a *philosophical* task. Theses about the aims or the possibility of philosophy are themselves philosophical (epistemological) claims. Therefore, the tighter connection Rorty is able to find between the premises and the conclusion of his argument, i.e., the more convincing argument he finds for the impossibility of epistemology, the better epistemological result he has achieved.

The better argument for the impossibility of epistemology is found, the better proof this is of the possibility of epistemology. It follows from this that a proof of the impossibility of epistemology is impossible.

Notes

1 Actually Kant was questioning the possibility of »metaphysics» (as well as mathematics and »pure natural science»), but the question may be generalized to the whole of philosophy (cf. below).

2 This institution was founded in Turku but continues its existence as the University of Helsinki.

3 Nycopensis is an ancestor of the leading contemporary Finnish philosopher G.H. von Wright.

4 Cf. Chapter 5 below.

5 See Windelband (1907).

6 For comments on Rorty, see Hacking (1980) and Kim (1980). For Rorty's earlier discussion of progress in philosophy, see Rorty (1967).

7 It does not strive for »universal commensuration» within a »permanent framework for inquiry» (pp. 368, 380), but still it aims at results rather than mere »conversation». In my view, the idea of making progress in science by approach to the truth does not presuppose universal neutral frameworks of discourse (cf. Chapter 5 below).

8 Such reductive arguments have always been used in philosophy, but their validity and presuppositions have also continued to be a matter of dispute (cf. Zeno's paradox).

9 This strategy corresponds to the approach that has been called 'if-thenism' in the philosophy of mathematics (cf. Chapter 8 below).

10 This argument is mainly developed in Chapter IV of Rorty (1980). (See also *ibid.*, p. 10.) Rorty's treatment of systematic philosophy in Chapter VIII is a generalization of this argument. For him, the difference between analytic and other sorts of philosophy is »relatively unimportant», since analytic philosophy is »one more variant of Kantian philosophy» (p. 8). However, Rorty does not claim that he has given a general proof of the impossibility of all kinds of systematic philosophy: »Perhaps a new form of systematic philosophy will be found which has nothing whatever to do with epistemology but which nevertheless makes normal philosophical inquiry possible» (p. 394).

Bibliography

Hacking, I., 'Is the End in Sight for Epistemology?', *Journal of Philosophy* **77** (1980), pp. 579–588.

Kant, I., *Critique of Pure Reason*, G. Bell and Sons, London, 1930.

Kant, I., *Prolegomena to Any Future Metaphysics*, The Bobbs-Merrill Company, Indianapolis, 1950.
Kim, J., 'Rorty on the Possibility of Philosophy', *Journal of Philosophy* **77** (1980), pp. 588–597.
Nowakova, I. and Nowak, L., 'Marxism and Positivism: The Idea of a Scientific Philosophy', *Poznan Studies in the Philosophy of the Sciences and the Humanities* **4** (1978), pp. 210–232.
Rorty, R., 'Introduction: Metaphilosophical Difficulties of Linguistic Philosophy', in R. Rorty (ed.), *The Linguistic Turn: Recent Essays in Philosophical Method,* The University of Chicago Press, Chicago, 1967.
Rorty, R., *Philosophy and the Mirror of Nature*, Blackwell, Oxford, 1980.
Windelband, W., *Die Geschichte der neueren Philosophie* I–II, Vierte Auflage, von Breitkopf & Härtel, Leipzig, 1907.

Chapter 3

NOTES ON POPPER AS FOLLOWER OF WHEWELL AND PEIRCE

1. Introduction

History of methodology has become, in our age, an intensively studied subject which is hoped to make significant contributions to our understanding of both the history and the philosophy of science. Because of the repercussions to currently debated issues, the nineteenth-century methodology seems to be a particularly rewarding topic for a philosopher of science.[1] It is no accident that, at the same time when philosophers of science have been searching for new directions after the rejection of the central theses of logical empiricism, there has been a notable revival of interest in philosophers who in various ways were opposed to the influential traditions of French positivism (Comte) and British empiricism (Herschel, Mill, Jevons). Among these thinkers we find 'Kantians' from Kant himself to J.F. Fries, William Whewell, and Ernst Apelt, 'realists' like Bernard Bolzano and Charles Sanders Peirce, 'hermeneutic' historians like Wilhelm Dilthey, and working scientists like Karl Marx.[2]

It is clear that this revival is not only based upon historical interest about the later influence of these thinkers – for example, Butts notes that William Whewell (1794–1866) had no philosophical followers and that Charles Peirce (1839–1914) was the only »among creative philosophers in the late nineteenth and early twentieth centuries» who knew intimately Whewell's work and appreciated it more than Mill's theory of scientific method (Butts, 1968, p. vii).[2a] The most natural explanation for the Whewell renaissance in the 1960's is the fact that in many respects Whewell can be regarded, to a much greater extent than his most distinguished opponent John Stuart Mill, as one of our contemporaries.

Whewell was systematically occupied with the same problems as the critics of logical empiricism one hundred years later. His philosophy of science seems strikingly modern to us essentially for the same reason as the English translation of Karl Popper's *Logik der Forschung* (1934) did in 1959. Popper's *LSD*[3] shows that already in the heydays of logical empiricism he represented ideas and emphasized problems which later in the sixties have become subjects of intensive philosophical discussion. As we shall see later, however, these problems were largely the same as the standard problems of nineteenth-century methodologists. Anyone who sets out to systematically compare Popper's methodology with the nineteenth-century conceptions of science will find out that many ideas of Whewell's and Peirce's, for example, are surprisingly similar to the epistemology and philosophy of science which Sir Karl has taught for many years in the conviction that it is new and revolutionary.[4] (A similar observation on the relation between Berkeley and Mach was made by Popper in his, 'A Note on Berkeley as Precursor of Mach and Einstein'; see Popper, 1963, p. 171.) Rather than as the first and foremost representative of a new epoch, Popper can thus be regarded as an upholder of an important nineteenth-century tradition in methodology[5,5a] – to the extent, indeed, that Laurens Laudan has not unjustifiably claimed Popper to be »probably closer to the nineteenth-century methodological tradition than is any other living philosopher» (Laudan, 1973a, p. 306).

Observations and discoveries of this sort are potentially very important for the history of the philosophy of science. For example, both Berkeley and Mach were puzzled by the same problem, viz. the foundations of Newton's mechanics, and their responses turned out to be very similar. Did Berkeley have any – direct or indirect – influence upon Mach's thinking? If Mach developed his ideas largely independently, how is the similarity of their criticism of Newton to be explained? Answers to these questions may help the historian to redescribe the development of the philosophy of physics. Corresponding questions may be raised about the relations between Whewell, Peirce, and Popper; and it is of particular interest to try to find out what common sources of inspiration they have had.

A philosopher of science, on the other hand, is not primarily interested in the historical problem about one philosopher's influence upon another. So far as he is intrigued by the same problems as his predecessors, he will direct his attention to the way in which they have formulated their views and to the arguments they have presented, and he will attempt to assess the merits of these formulations and arguments in relation to each other and against his own views. It may very well happen that earlier formulations of the same idea are found to be better than more recent ones. Moreover, the 'anticipations' of modern conceptions are seldom complete or accurate, and a more careful analysis of their content may bring unnoticed variants of current doctrines into his consideration. There exist many ways in which historical studies may give us a new perspective from which to reconsider current issues in the philosophy of science.

It has been suggested that the recent interest in »tracking down writers who found Popper's theory first» shows »the general acceptance of his views», since »there is perhaps less interest in discovering historical precedents of people's mistaken views» than for »ideas which appear to be substantially correct» (see Settle, 1974, pp. 742–743). Popper's essay on Berkeley and Mach serves nicely as a counter-example to this thesis, and this paper is hoped to be another one. The above thesis ignores the fact the weaknesses of Popper's philosophical and methodological thought. Instead of repeating, again and again, tables showing the differences between Popper and the logical empiricists[6], we should test his methodological theory against its tradition and see whether it solves all the problems which its predecessors solved and those which they failed to solve.

In the Popper volume in *The Library of Living Philosophers,* Peter Medawar presents a very illuminating account of the attempts by English and Scottish philosophers to formulate the hypothetico-deductive conception of science in the nineteenth century. In the same volume, Eugene Freeman and Henryk Skolimowski systematically compare the philosophies of Peirce and Popper.[7] It is somewhat disappointing that both of these contributions represent a more historically than critically orientated approach. Thus, in spite of their philosophical insight and careful work, Freeman and Skolimowski are contended with a presentation of a summary of similarities and differences between Peirce and Popper; Sir Karl's only response to this summary is an expression of gratitude »for the many enlightening passages from Peirce, which show to what degree Peirce anticipated some of my central ideas» (Popper, 1974b, p. 1072). As we shall see below, this summary is neither complete nor entirely reliable. Further, it gives rise to many interesting questions which are not discussed by Freeman and Skolimowski – they acknowledge Popper's remark that he was not familiar with Peirce's ideas until the early fifties[8], but they do not go on to propose any conjectures about Peirce's indirect influence on Popper or about sources which are common to them. Apart from these historical questions, they seem uncritically (that is, without any arguments) to value Popper's falsificationism higher than Peirce's fallibilism – as something in which one of Peirce's anticipations has »reached full flower» (Freeman and Skolimowski, 1974, p. 479). Precisely the opposite view, that Popper has betrayed fallibilism in the original Peircean sense, has been defended by Isaac Levi; see Levi, 1967a, p. 387.

In this paper, some observations on Popper's philosophy are made in the light of Whewell's and Peirce's theories of science. I have tried my best to treat Whewell and Peirce, too, as living philosophers. This paper is only partly intended as an exercise in the history of methodology, for I have attempted to *use* the history of methodology in doing philosophy of science. A short paper like this cannot do justice to the wealth of ideas of three original philosophers. In a more complete account, more attention should be paid to the 'metaphysical' background assumptions underlying Whewell's, Peirce's, and Popper's methodological views. In choosing the points of comparison, I have been selective without aiming at completeness, with the focus on topics related to the dynamics of scientific knowledge – to its discovery, evolution, growth, and epistemic status.

2. The Growth of Knowledge

Peirce and Popper express in the following way their views about the fundamental task of philosophy:

My philosophy may be described as the attempt of a physicist to make such conjecture as to the constitution of the universe as the methods of science may permit, with the aid of all that has been done by previous philosophers. (Peirce, *CP*, 1.7.)

I, however, believe that there is at least one philosophical problem in which all thinking men are interested. It is the problem of cosmology: *the problem of understanding the world – including ourselves, and our knowledge, as part of the world.* All science is cosmology, I believe, and for me the interest of philosophy, no less than of science, lies solely in the contributions which it has made to it. (Popper, *LSD*, p. 15.)

They share – with Kant (cf. *LSD*, p. 13) – the view that philosophy deals with genuine 'cosmological' problems which cannot be dissolved as linguistic puzzles. In a similar way, Whewell, in 1849, complained that men of his time »often seem inclined to dispose of philosophical questions» as »questions merely of words» (Whewell, 1860, p. 243).

The distinction and opposition of the subjective and objective elements of our knowledge – of Thoughts and Things, Ideas and Sensations, Theory and Facts – constitutes for Whewell the 'Fundamental Antithesis of Philosophy'. Whewell's philosophical work has the general aim of laying into open man's intellectual progress, consisting in the 'idealization of facts' which takes place within the development of science. Peirce attacks the problems of philosophy by developing a new dynamic theory of the 'fixation of belief' and a new 'pragmatic' theory of meaning which teaches us 'how to make our ideas clear' (*CP*, 5.358–410). Popper treats the problem of knowledge as a part of the general problem of cosmology, and he propounds the thesis that the problem of the *growth* of knowledge is the central problem of epistemology and that it can be best studied by studying the growth of scientific knowledge (*LSD*, p. 15).

The problem of the growth of knowledge has sometimes been called 'Popper's problem'. This name is not telling at least for two reasons. In the first place, the sceptical problem of induction is called 'Hume's problem' partly for the reason that it could not even be formulated until certain transformations in the sciences had taken place (cf. Hacking, 1975, pp. 176–185). The growth of knowledge, on the other hand, was recognized as a problem as soon as systematic science had emerged – and especially after the scientific revolution in the sixteenth century. Secondly, Popper was by no means the first important thinker who gave this problem a crucial place in his philosophical system. In this respect, he was preceded by many philosophers, Whewell and Peirce among them, even though it is clear that Einstein's revolution in physics offered him an opportunity of of giving a new emphasis upon this old problem.

The most suggestive formulation of the problem of the growth of knowledge in the nineteenth century was given by William Whewell. After having finished his *History of the Inductive Sciences* (1837), the first full-scale account of the history of the physical sciences, Whewell told the »moral to the story»[9] in his *Philosophy of the Inductive Sciences, Founded upon their History* (1840). Without pretending to propose an entirely new problem, he remarks, in the Preface to the second edition, that his generation is in better intellectual position than Francis Bacon to develop a philosophy of science:

> Bacon only divined how sciences might be constructed; we can trace, in their history, how their construction has taken place. However sagacious were his conjectures, it may be expected that they will be further illustrated by facts which we know to have really occurred... It cannot be an unfit task, at the present day, to endeavour to extract from the actual past progress of science, the elements of a more effectual and substantial Method of Discovery. (Whewell, 1847, p. vi.)

Whewell, Peirce, and Popper are all trying to solve the problem of the growth of knowledge; they all treat it as the key problem in the philosophy of science; and in spite of important differences, their answers to this problem share a striking family resemblance.

3. The Tree of Knowledge

In his Herbert Spencer Lecture of 1961, 'Evolution and the Tree of Knowledge', Popper contrasts an 'evolutionary tree' with the growing 'tree of human knowledge'. The former grows, like a 'family tree', up from a common stem into more and more branches, while the latter grows »almost in the opposite direction». In contrast with the increasing specialization and differentiation which dominates the growth of our tools, instruments, and applied knowledge, the growth of 'fundamental research' is dominated by a »tendency towards increasing integration towards unified theories» (Popper, 1972, pp. 262–263).

Popper attributes this theory of the *integrative growth* of scientific knowledge to Herbert Spencer. However, Spencer was only a youth of 20 years when William Whewell, in 1840, presented this very theory as the main 'moral' drawn from his detailed studies in the history of science. According to Whewell, integrative growth is the principal feature of the progress in science. Science progresses from lower-level empirical generalizations to more and more inclusive theories which incorporate past results. At each step of scientific discovery, the 'facts' are 'colligated', or bound together, by a new conception; this process Whewell calls – in a non-standard manner – induction. The most dramatic and convincing sign of progress is, in Whewell's view, the *consilience of inductions*, which takes place when 'inductive' steps from different classes of facts lead to the same results, that is, when two separate generalizations or laws are found to be consequences or special cases of the same comprehensive theory. Whewell and Popper thus use the same terms to describe the morphology of scientific progress:

> This gradation of truth, successively included in other truths, may be conveniently represented by *Tables* resembling the genealogical tables by which the derivation of descendants from a common ancestor is exhibited; except that it is proper in this case to invert the form of the Table, and to make it converge to unity downwards instead of upwards, since it has for its purpose to express, not the derivation of many from one, but the collection of one truth from many things. (Whewell, 1847, p. 76.)

> Assuming the same upward direction of time, we should have to represent the tree of knowledge as springing from countless roots which grow up into the air rather than down, and which ultimately, high up, tend to unite into one common stem. (Popper, 1972, pp. 262–263.)

Structurally speaking, Popper's tree of knowledge is nothing more than Whewell's Inductive Table (or 'Tree of Science'): the former is obtained by turning the latter upside down.[10]

In the Whewell-Mill controversy, the question was raised whether successive theories may be incompatible with each other. Whewell held the view the later generalizations 'included', without 'negativating', the earlier generalizations in the same branch of an Inductive Table. Mill, on the other hand, accepted this idea only for different 'descriptions', but claimed that different 'explanations' may not all be true. It might seem, therefore, that Mill is here making, against Whewell, the point which Duhem and Popper have later emphasized – that successive explanatory theories may strictly speaking contradict each other (cf. Duhem, 1954, p. 193, and Popper, 1972, pp. 197–202). Whewell's position is more complex, however, and it is not clear at all whether he intended to claim that the successive theories always are compatible in the strict sense. The 'consistency' of the Cartesian vortex theory and the Newtonian theory of gravitation meant, for Whewell, that in the process of making the former »consistent with facts» it will be modified by successive 'transformations' until it becomes indistinguishable from the latter (see Whewell, 1860, pp. 251–252, 492–503). And while Mill argued that of several different explanations »only one can be true in any degree, and... the other... must be altogether false» (see *ibid.*, p. 251), Whewell thought that each of these explanations may be »to a certain extent true» (*ibid.*, p. 252). As Whewell here assumes that successive theories at least partially share a common factual content, his view is not far from a 'principle of correspondence' which claims that »a new theory should contain the old one approximately» (Popper, 1972, p. 202).

4. Appropriate Conceptions

Whewell made one important qualification to his 'Tree of Science': he noted that the merging of two branches of the tree is »not merely a juxtaposition of materials, by which the new proposition contains all that its component parts contained», for in induction

> besides mere collection of particulars, there is always a *new conception*, a principle of connexion and unity, supplied by the mind, and superinduced upon the particulars. (Whewell, 1847, p. 77.)

In his Aphorism XV concerning science, Whewell states that

> an Induction is not the mere *sum* of the Facts which are colligated. The Facts are not only brought together, but seen in a new point of view. (Whewell, 1847, p. 469.)

Peirce says that he agrees with Whewell in that »progress in science depends upon the observation of the right facts by minds *furnished with appropriate ideas*» (*CP*, 6.604). In his 'first requirement for the growth of knowledge' Popper says that

> the new theory should proceed from some *simple, new, and powerful, unifying idea* about some connection or relation... between hitherto unconnected things... or facts... or new 'theoretical entities'. (Popper, 1963, p. 241.)

Whewell is thus very well aware of the fact that conceptual change plays an important role within the development of science.[11] At each step towards a more comprehensive theory the previous results, 'facts', are bound together by a new unifying conception. This inductive step involves, besides generalization, a reinterpretation of 'facts' in the light of a theory or a new conceptual scheme. Science does not only *describe* observable phenomena and regularities among them, as the 'Sensational School' and its followers like Comte and Mill have claimed, but it also tries to explain and interpret them.

Whewell learned from Kant's (and Plato's) philosophy that the two poles of the 'Fundamental Antithesis', viz. Ideas and Facts, are inseparable. As Facts cannot be separated from Ideas, he argued, the descriptive view of science (Comte, Mill) is not valid. Whewell believed that the progress of knowledge gradually brings into light 'necessary truths' which are first known *a posteriori* and then become known *a priori* (Whewell, 1860, p. 354). Necessary truths are, for Whewell, »those of which we cannot distinctly conceive the contrary» (*ibid.*, p. 463) after our Ideas have reached the sufficient degree of development (cf. *ibid.*, p. 524). Necessary truths include, besides the truths of geometry and arithmetic, also physical principles like 'All matter is heavy'. This rather curious doctrine of necessary truths which »flow from our ideas», but are learned by means of »experiment and observation» (*ibid.*, p. 477), is perhaps the most questionable element of Whewell's philosophy.[12] Twentieth-century 'Kantians', who have learned the same lesson as Whewell, have usually given more emphasis upon our relative freedom in selecting and even overthrowing the conceptual schemes that are employed in the interpretation of observations.

Apart from this difference, Whewell's view is surprisingly modern. He argues against the now familiar 'theoretical – observational' dichotomy by saying that »to report facts for scientific purposes without some aid from theory, is not only useless, but impossible» (*ibid.*, p. 155).

Indeed not only is an existing theory requisite to supply the observer with instruments and methods, but without theory he cannot even describe his observations. (*Ibid.*, p. 154.)

Whewell's criticism of Mill repeats itself in Popper's insistence, against the logical empiricists, that observations are »interpretations in the light of theories» (Popper, *LSD*, p. 30n). Kant's critical philosophy is the common source of this view, both in Whewell and in Popper (cf. Popper, 1963, p. 191; Popper, 1974a, pp. 46, 65; and Popper, 1974b, p. 1033).

5. Evolutionary Approach to Knowledge

Whewell's dynamic conception of science reflects his influences from, and reactions to, the German philosophy of Kant, Fichte, Schelling, and Hegel (see Whewell, 1860, pp. 307–314), which he combined with sound knowledge about the physical sciences and their development. In his youth, Peirce was deeply engaged in the study of Kant's philosophy (for Peirce's Kantianism, see Apel, 1974), and his later cosmological thoughts were influenced by Schelling and Hegel – cf. his ideas that matter is »mere specialized and partially deadened mind» (*CP*, 6.102) and that »continuity is an indispensable element of reality» (*CP*, 5.436).

Besides these common influences, Peirce's philosophy has a new characteristic feature which is lacking in Whewell's generation, viz. biological conceptualizations in epistemology.[12a] Charles Darwin's *The Origin of Species* was published in 1859, and it exercised a strong influence upon the young Peirce who graduated from Harvard in the same year.[13] Peirce did not value Herbert Spencer's evolutionary philosophy high, and he disapproved the attempts of Chauncey Wright, one of his friends and older contemporaries, to combine Mill's nominalism with Darwin's evolutionism (*CP*, 5. 12). Peirce developed his own 'thoroughgoing evolutionism' which forms a central part of his systematic philosophy. For example, in addition to 'tychism' (chance) and 'synechism' (continuity), one of Peirce's cosmological categories is 'agapism', evolutionary love. In view of his evolutionary cosmology, he regarded all the regularities of nature and of mind as a sort of habits which are still evolving products of growth (*CP*, 6.102), with the important implications that the universe is indeterministic and the laws of nature are not stable but do in fact wiggle.[14]

Peirce wrote in 1896 that »the evolutionary theory in general throws great light upon history and especially upon the history of science» (*CP*, 1.103), and he proposes to view the progress of science from an evolutionary perspective. A Darwinian mode of evolution might, according to him, consist in »slight fortuitous modification» of a judgement at each of its recalls to the mind (*CP*, 1.107). A Lamarckian mode of evolution might consist in our efforts to make our opinion to represent the known facts »by perpetually modifying our opinion» when more and more observations are collected (*CP*, 1.108). Peirce argues that the first sort of evolution does not exist in science which is »controlled and exact»; and he adds that the second sort of evolution does not have »a prominent place in science». Yet, Peirce notes, there is another sort of Darwinian evolution which takes place in science:

We are studying over phenomena of which we have been unable to acquire any satisfactory account. Various tentative explanations recur to our minds from time to time, and at each occurrence are modified by omission, insertion, or change in the point of view, in an almost fortuitous way. Finally one of these takes such an aspect that we are led to dismiss it as impossible. Then all the energy of thought which had previously gone to the consideration of that becomes distributed among the other explanations, until finally one of them becomes strengthened in our minds. (CP, 1.107.)

He seems to think, however, that the discovery and the choice of scientific hypotheses is usually more systematic than this 'natural selection' of the explanations recurring to our minds. He continues with the remark that science mainly progresses by leaps rather than by minute changes:

It advances by leaps; and the impulse for each leap is either some new observational resource, or some novel way of reasoning about the observations. Such novel way of reasoning might, perhaps, be considered as a new observational means, since it draws attention to relations between facts which would previously have been passed by unperceived. (*CP*, 1.109.)

It is not difficult to find a Kantian (or Whewellian; cf. Section 4) tone in Peirce's conclusion that science advances by leaps which involve seeing old observations in a new way. Peirce's general theory of belief, mostly developed in his articles of 1877–1878 (see *CP*, 5.358–410), has a more evolutionary flavour, however. He does not refer to 'Darwinian evolution' in this connection; but a very similar account of the 'natural selection' of hypotheses has been developed by Popper, who calls it a »largely Darwinian theory of the growth of knowledge» (Popper, 1972, p. 261).

Peirce learnt from Alexander Bain the definition of belief as »that upon which a man is prepared to act» (*CP*, 5.12). On the basis of this dispositional analysis of belief, he describes the acquisition of belief as follows:

(E) → belief → habit → action → surprise →
→ doubt → inquiry → belief →

Belief establishes in our nature a habit or a rule of action; acting on a belief may lead us to an unexpected and disappointing result; this surprise irritates our doubt until a settlement of opinion, a new belief, is attained through inquiry.

Popper introduced in 1966 (even if the idea is older) the following schema for the growth of knowledge:

(P) $P_1 \to TT \to EE \to P_2$

(see Popper, 1972, pp. 119, 126, 164, 243, 287). Here P_1 is a problem from which we start; the first step in problem-solving is the proposal of an imaginative conjectural solution *TT* (the 'tentative theory'); the next step is the attempted 'error elimination' *EE* by severe critical examination; and finally P_2 represents the new 'problem-situation' which emerges from the attempt to solve problem P_1.

As theories of the growth of belief (or 'subjective knowledge', as Popper calls it) there is no essential difference between Peirce's schema (E) and Popper's schema (P): in contrast with the 'classical' or 'pre-Darwinian' epistemology, Peirce has clearly recognized the dispositional and 'expectational' nature of 'subjective knowledge' (cf. Popper, 1972, pp. 66, 71, 145). Moreover, Peirce and Popper share the following fundamental assumption:

> All human knowledge, up to the highest flights of science, is but the development of our inborn instincts. (*CP*, 2.754.)

> All acquired knowledge, all learning, consists of the modification (possibly the rejection) of some form of knowledge, or disposition, which was there previously, and in the last instance of inborn dispositions. (Popper, 1972, p. 71.)

This assumption naturally invites us to consider the relations of schemas (E) and (P) to the higher forms of human knowledge, i.e., to scientific knowledge.

6. Cognitivism and Pragmatism

In his article 'Mr. Mill's Logic', in 1849, Whewell warns against confusing »knowledge with practical tendencies» – he claims that a sharp distinction should be made between »practical habits, practical skill, instincts, and the like» which »appear in action, and in action only» and human knowledge which involves »general propositions» and »general words or phrases». Science, he argues, cannot result from mere instinct or animal tendencies to action, because »instinct by its nature is not conscious and general»:

> For though action may be modified by habit, and habit by experience, in animals as well as in men, such experience, so long as it retains that merely practical form, is no part of the materials of science. (Whewell, 1860, p. 242.)

In a similar way, Popper argues that the 'commonsense theory of knowledge' fails to see the significant difference between dispositional 'subjective knowledge' and »knowledge in the objective sense, human knowledge, which consists of linguistically formulated expectations submitted to critical discussion» (Popper, 1972, p. 66). Human knowledge depends upon »linguistically formulated theories» which belong to 'world 3' (*ibid.*, p. 74); the growth of subjective knowledge can be analysed only in terms of »the third-world objects in which it is anchored» (*ibid.*, p. 164).[15]

Even if Whewell and Popper agree on a distinction, they disagree on its consequences. Whewell argues that the way in which man *learns* from experience is quite different from that of an animal (see Whewell, 1860,

p. 243). He makes this point so sharply[16] that one could not expect him to be impressed by Popper's claim: »From the amoeba to Einstein, the growth of knowledge is always the same» (Popper, 1972, p. 261) and representable by the scheme (P) of trial and error, the main difference being that »Einstein consciously seeks for error elimination» (*ibid.*, pp. 24–25).

Peirce relates his schema (E) to science by regarding the scientific method as the correct method of *inquiry*. He rejects the *a priori* method and the methods of tenacity and authority; the correct method of inquiry should be of the kind »by which our beliefs may be determined by nothing human, but by some external permanency» (*CP*, 5.384). The scientific method is precisely of this sort – its 'fundamental hypothesis' is that »there are Real things, whose characteristics are entirely independent of our opinions about them» (*CP*, 5.384). In a similar way as Popper, Peirce thus emphasizes the objective and co-operative nature of the scientific method.

Peirce's schema (E) 'locates' the scientific activity – the inquiry – between doubt and belief. It thus outlines the general setting or the framework in which Peirce proposes to consider science: in one direction, it suggests that science is ultimately nothing but an outgrowth of our natural instincts of breeding and feeding (*CP*, 6.500); in another direction, it shows how new scientific problems result from action based upon the earlier conclusions of inquiry. Schema (E) does not tell anything specific about the structure of inquiry, which in Peirce basically consists of the 'abduction' from a surprising fact to a hypothesis (explanation) and of the subsequent testing and eventual rejection of this hypothesis (see Sections 7 and 8). If explanatory theories are correlated with beliefs, the method of science has some structural similarities to schema (E), even if the crucial element of (E), the *acting upon a belief*, is quite different from the 'corresponding' element of the scientific method, viz. the *testing of a hypothesis.*[17]

If we compare Peirce's view with Popper's, we note that Popper thinks that his schema (P) really reflects, not only the process of adaptation used by living organisms, but also the evolution of objective knowledge. Schema (P) illustrates his thesis that science does not start from observations, but »should be visualized as *progressing from problems to problems*» (Popper, 1963, p. 222). He explains this idea more carefully as follows:

> Admittedly, our expectations, and thus our theories, may precede, historically, even our problems. *Yet science starts only with problems.* Problems crop up especially when we are disappointed in our expectations, or when our theories involve us in difficulties, in contradictions; and these may arise either within a theory, or between two different theories, or as the result of a clash between our theories and our observations. Moreover, it is only through a problem that we become conscious of holding a theory. It is the problem which challenges us to learn; to advance our knowledge; to experiment; and to observe. (Popper, 1963, p. 222.)

These remarks about the generation of problems seems to reduce Popper's position remarkably close to Peirce's schema (E); we might describe the above account by the following pattern:

(S) → theory → expectation → disappointment → problem → theory →

There is an important difference of emphasis in Peirce and Popper, however. The 'expectations' and 'disappointments' in schema (S) are typically 'theoretical' problems which concern our failure to be successful in our explanations. They are results of clashes between theories or between theories and observation, not between theories and practical action. Schema (S) thus places science into the context of cognitive and intellectual problems, while Peirce's schema (E) places it into the context of practical problems which concern rational human deliberation and action. In Popper's terms, we might say that (S) exhibits the second-world part of scientific activity in the third-world context of objective theories and problems, while (E) represents this second-world activity in the first-world context of man's real-life environment.[18]

A distinction is often made between the cognitive problem, 'What do we know?', and the practical problem, 'What should we do?'. The former is about knowledge, the latter about rational action. Whewell, Peirce, and Popper are all *cognitivists* in the sense that they regard the search of knowledge, or true information about the reality, as the primary aim of science. They all are opposed to *behaviouralism* which claims that there is no scientific knowledge at all, in the sense of tentatively accepted results of scientific inquiry, but all the results of science should be viewed as mere recommendations for action in practical decision making situations.[19] In particular, Peirce's pragmatism should not be confused with the behaviouralist absorption of the cognitive problem to the practical one. Peirce treats the cognitive problem as the primary one (cf. *CP*, 1.55, 1.76), but still his schema (E) reminds us of this real-life fact: scientific knowledge has the character that it is, in each day, being acted upon. It is *trustworthy* in varying degrees; and this is not a merely accidental feature of science, but a feature which philosophers of science should take seriously. Many cognitivists have, indeed, tried to relate their theory of scientific knowledge to the fact that men do rely, in their practical activity, upon the results of scientific inquiry. As Levi has recently emphasized, this is precisely the spirit of Peirce's 'fallibilist' epistemology (cf. Section 11): in spite of the corrigibility of all scientific knowledge, it may be rational to act upon an uncertain conjecture if the goal of the action is so valuable that the risk of failure is worth taking (see Levi, 1967a, p. 387; and also Levi, 1971).[19a]

Popper's attitude towards the practical problem has been quite different: for a long time, he simply ignored it. His followers have been eager to warn against the confusion of science with technology (Agassi) and to argue that the problem of trustworthiness is »not a problem in philosophy of (strict) science», but »a problem of technology and life» (Settle, 1974, p. 703). This difference to Peirce is reflected in Popper's schemas (P) and

(S), which – in contrast with Peirce's schema (E) – are exclusively concerned with 'negative arguments', with criticism and severe testing, not with acting upon beliefs. Criticism is, for Popper, the »artificially intensified selection pressure» which constitutes the »struggle for survival of hypotheses» (Popper, 1974b, p. 1024). One of his favourite ideas is that the critical method »consists in letting our hypotheses die in our stead» (Popper, 1972, p. 248). Only recently he has given any serious consideration to the practical problem of rational action (see Popper, 1972, pp. 21–23). He correctly points out that »even practical problems are theoretical» or »imbued with theories» (Popper, 1974a, pp. 106–107) – an insight which is already contained in Peirce's schema (E) – but he naively claims that these problems reduce to his solution of the cognitive problem: »We should», he says, »*prefer* as basis for action the best-tested theory» (Popper, 1972, p. 22). At the same time, he argues that »from a rational point of view, we should not 'rely' on any theory», for there »can be no good reasons» for expecting that our choice will be a successful one. We shall, he says, »base our theoretical predictions as well as practical decisions» upon the hypothesis which has a higher degree of corroboration than its competitors (*ibid.*, p. 83), even if this degree »says nothing whatever about future performance, or about the 'reliability' of a theory» (*ibid.*, p. 18). He further clarifies his view by saying that our decision to do something should be based upon the best-tested scientific theories in our possession for the reason that otherwise our decision would »collapse under criticism» which freely makes use of such theories (Popper, 1974b, p. 1025).

There is always a cognitive element in rational action, and it is our responsibility to take into account the relevant scientific evidence in making our decisions. What Popper forgets here, however, is the value component in rational action, that is, the practical utilities involved in deliberation.[20] If we decide not to act upon a theory h which is the best-tested one, because of the serious and risky consequences that might result, who can blame our choice? What is needed here is a method of balancing and weighting the strength of our evidence against the risks or the values of the consequences of contemplated acts. In this respect, Popper's account is superseded by the Bayesian analysis of rational decision making under uncertainty.[20a]

7. The Logic of Discovery

The nature of scientific discovery has been, and still is, one of the most debated subjects among the theorists of scientific inference, and a great number of different accounts of it have been given since the days of Plato. Aristotle's term for the process of arriving at the 'first principles' of science was *epagoge*, which is usually translated as 'induction'. For Aristotle, these 'first principles' are *definitions*, so that his account of induction is different from the more familiar 'enumerative' and 'eliminative' sorts of induction. The *inductivist* view of inductive generalization as 'the' method of discovery of scientific laws was developed already in the Antiquity and in the Middle Ages, probably as a misinterpretation of *Analytica Priora* II, 23. In spite of the doubts of Sextus Empiricus (2nd century), Francis Bacon, and David Hume, it gained new popularity among the eighteenth-century Newtonians (Thomas Reid[21]) – as a misinterpretation, again, of Newton's method of analysis – and among the nineteenth-century positivists and empiricists (Comte, Mill). 'Eliminative' variants of the inductivist view were represented by Robert Grosseteste's method of resolution (13th century), Francis Bacon's theory of gradual and successive induction starting from experience and observation, and John Stuart Mill's methods of inquiry. Representatives of the rationalist school, on the other hand, thought that the basic principles of science are obtainable *a priori* – by reason or 'intellectual intuition'. The Renaissance Aristotelians in Padua argued that the inference from observable effects to knowledge about causes takes place through nondemonstrative (but deductive) syllogisms. Galileo and Newton regarded the discovery of laws as analogous to the method of *analysis* of the Greek geometers.[22] Finally, the hypothetico-deductive conception of scientific inference allows the discovery of scientific hypotheses to take place in almost any conceivable way whatsoever as long as the results are deductively testable. In contrast with all the other views mentioned above, it assumes that the process of discovery is quite irrelevant to the problems which are studied in the theory of scientific inference.

Inductivism was attacked in the nineteenth century from many directions. One of the most well-argued and careful attacks was William Whewell's theory of induction as a process which – like Aristotle's induction – in inseparable from scientific concept formation. In spite of his acceptance of Bacon's recommendation of »a continuous advance from observation, by limited steps, through successive gradations of generality» (Whewell, 1860, p. 131), he argued that every step in scientific discovery involves the suggestion and the appropriation of a new conception (see Section 4). Against James Spedding, one of the editors of *The Works of Francis Bacon*, Whewell argued that science does not progress in the way that Bacon had claimed, viz. by making and accumulating a great collection of facts from which philosophers might then extract scientific truths by the application of a right method (*ibid.*, p. 153). Such methods of procedure, he remarks,

> do not belong to the *Epoch of Discovery*, but to the Period of *verification* and *application* of the discovery which follows. (*Ibid.*, p. 154.)

In a similar way, the chemist Justus Liebig argued, in 1863, that no scientist ever could follow the methods of Baconian induction, with the further conclusion that there does not exist any fool-proof method for scientific discovery (see Laudan, 1968, p. 33).

Whewell also argued that Bacon did not »justly appreciate the sagacity, the inventive genius, which all discovery requires» (*ibid.*, p. 138), and claimed that 'happy guessing' plays a crucial role within scientific discovery:

The Conceptions by which Facts are bound together are suggested by the sagacity of discoverers. This sagacity cannot be taught. It commonly succeeds by guessing; and this success seems to consist in framing several *tentative hypotheses* and selecting the right one. But a supply of appropriate hypotheses cannot be constructed by rule, nor without inventive talent. (Whewell, 1847, pp. 467–468, Aphorism VIII concerning science.)

In a similar way, Peirce emphasized the indispensability of imagination in science:

There is, after all, nothing but imagination that can ever supply him an inkling of truth. He can stare stupidly at phenomena; but in the absence of imagination they will not connect themselves together in any rational way... It is not too much to say that next after the passion to learn there is no quality so indispensable to the successful prosecution of science as imagination. (Peirce, *CP*, 1.46)

Popper's arguments against induction continue the nineteenth-century criticism of inductivism – Liebig and Duhem are mentioned in *LSD*, p. 32n, but Whewell and Peirce are not.[23] Popper makes a sharp distinction – typical of the hypothetico-deductive account – between »the process of conceiving a new idea, and the methods and results of examining it logically», where the initial stage of discovery is not susceptible of logical analysis (*LSD*, p. 31).

There is no such thing as a logical method of having new ideas, or a logical reconstruction of this process. My view may be expressed by saying that every discovery contains 'an irrational element', or 'a creative intuition', in Bergson's sense. (Popper, *LSD*, p. 32.)

The principal mistake of the inductivists, Popper argues, is that they have confused psychological problems (that is, the reconstruction of the steps involved in discovery) with logical ones (that is, the reconstruction of the subsequent deductive tests to which hypotheses are subjected after their discovery).[24]

Freeman and Skolimowski (1974, p. 511) suggest that there is a 'logic of discovery' which both Peirce and Popper ascribe to. This claim is problematic for the reason that – in spite of the name 'The Logic of Scientific Discovery' that Popper gave to the English translation of his *Logik der Forschung*[25] – Popper seems to deny the existence of any logical reconstruction of the process of discovery in science. Peirce, on the other hand, developed a theory of *abduction* (also called *retroduction* and *hypothetic inference*) which gives an account of the inferential step in »the first starting of a hypothesis and the entertaining of it, whether as a simple interrogation or with any degree of confidence» (*CP*, 6.525). While induction, in Peirce's later writings, is »the operation of testing a hypothesis by experiment» (*CP*, 6.526), abduction consists in »studying facts and devising a theory to explain them» (*CP*, 5.145). He also makes it very clear that abduction is a sort of inference which may at least partly be reconstructed in logical terms:

Long before I first classed abduction as an inference it was recognized by logicians that the operation of adopting an explanatory hypothesis – which is just what abduction is – was subject to certain conditions. Namely, the hypothesis cannot be admitted, even as a hypothesis, unless it be supposed that it would account for the facts or some of them. (*CP*, 5.189.)

Thus Peirce's theory of abduction is, in a sense, a 'logic of discovery'. In view of this, N. R. Hanson (1961) has regarded Popper as one of the main opponents of the Peircean approach to scientific discovery – a view which seems to be precisely the opposite to Freeman's and Skolimowski's thesis. It is argued below, however, that in this respect Peirce and Popper have more in common than Hanson assumes, while there still exist in their views some fundamental dissimilarities which Freeman and Skolimowski fail to notice.

In 1903, Peirce (*CP*, 5.189) gave the following schema for abductive inference:

The surprising fact, *C*, is observed;
(P) But if *A* were true, *C* would be a matter of course.
Hence, there is reason to suspect that A is true.

In this inference, an explanatory hypothesis is »abductively conjectured» from a surprising fact *C* and from the premise that *A* explains *C*. In brief, abduction is an inference to an explanation. However, in 'A Theory of Probable Inference' (1883), Peirce gave a more specific account of abduction (hypothetic inference) as accomplishing the discovery of *causes* which are the middle terms of explanatory syllogisms. While an ordinary syllogism (in figure *Barbara*) proceeds from Rule and Case to Result:

Rule. Every *M* is a *P*,
Case. *S* is an *M*;
Result. *S* is a *P*,

induction proceeds from Case and Result to Rule, and hypothetic inference from Rule and Result to Case (*CP*, 2.712). Abduction, in this sense, is an inference from the explanandum and the generalization employed in the explanans to the *initial condition* of an explanatory argument.

The idea that there is an inference from effects to causes, or from facts to their explanation, is traditional (cf. *Analytica Posteriora* I, 13). Peirce's discussion of such an inference reminds of Aristotle's account of the hitting by guess upon the middle term of an explanatory syllogism (*Analytica Posteriora* I, 34), a passage which Whewell quotes with approval in his *Philosophy of Discovery* (1860, p. 21). The requirement that a hypothesis should explain the known facts was explicitly formulated by Thomas Hobbes and Robert Boyle (see Laudan, 1966, p. 95), and it goes back to the idea of 'saving the appearances'. This requirement of explanation was emphasized by Whewell, too. Peirce has the merit of being more explicit than his predecessors in his account of the structure of valid explanations. (In note 4, we made the remark that Peirce was one of the nineteenth-century philosophers who gave a clear formulation to the 'Popper-Hempel-theory' of explanation.) The main novelties of his theory are the account of 'statistical syllogisms' which contain Rules of the form 'The proportion r of the M's are P's' and the extension of the above presented ideas about induction and abduction to the case of probabilistic reasoning.

Freeman and Skolimowski (1974), pp. 511–512, complain that Peirce claims both abduction and induction to be ways to new knowledge. Even if Peirce's terminology varies from time to time, this is no »incongruity» in his philosophy. It is true that Peirce once says that the *discovery of laws* is accomplished by induction and the *discovery of causes* by hypothetic inference (*CP*, 2.713). In this context, 'induction' means the same as 'statistical generalization', i.e., inference from a sample to a population. Peirce explains very carefully that the inductive reasoning from a sample to a population M presupposes that the attribute P to be detected in the sample should be 'predesignated', i.e., agreed upon in advance (*CP*, 2.736). Induction thus presupposes a reference class M and an initial choice (or discovery) of a concept 'P'; it only starts after the *form* of the statistical law. 'The proportion x of the M's are P's', has been agreed upon. In this sense, induction amounts to the estimation of the unknown population parameter x on the basis of a subsequently drawn sample from M. As the result of this induction – 'The proportion .27 of the M's are P's', say – is not usually known or even thought of before the sample is drawn, there is a very natural sense in which one may say that this statistical law was 'discovered by induction'.

The inductivists thought that the way in which hypotheses are arrived at may be sufficient for justifying them. The hypothetico-deductive theorists denied precisely this point and claimed that the process of discovery is entirely irrelevant to the subsequent evaluation of a hypothesis.[26] Whewell, Peirce, and Popper share the view that there are criteria for the evaluation of hypotheses before they are tested – a view which goes beyond the simple hypothetico-deductive model, as Salmon (1966), p. 115, correctly notes. If we wish, such criteria may be said to constitute a 'logic of discovery'.[27] Thus, Hanson (1961) suggests that there is 'logic of discovery' which concerns *reasons for suggesting a hypothesis* in the first place, instead of reasons for its acceptance. But this way of speaking about discovery would be misleading. For example, in Peirce's schema (P) for abduction, the conclusion is that it is reasonable to suspect that a hypothesis A is true. This schema cannot describe the discovery of hypothesis A, since A occurs already in one of the premises of (P). It rather expresses a principle for the *before-trial evaluation* (or *a priori* appraisal) of A. As a rule for the 'logic of before-trial evaluation', Peirce's idea might be expressed by saying that hypotheses with real explanatory power, i.e., hypotheses which explain known facts, are worthy of further consideration. In this form, Peirce's schema (P) formulates a principle which Popper certainly would accept – even though Popper mistakenly has suggested that the hypothesis with the greatest explanatory power relative to some preexisting set of problems is always the same as the hypothesis with the greatest potential (overall) explanatory power (or information content).[28]

Peirce's and Popper's views may be reconciled to the extent that they agree on the existence of a logic of before-trial evaluation of hypotheses. However, they seem to disagree upon the nature of this logic. For Popper, the main element of this logic is the information content of a hypothesis (see Section 10). But in contrast with Popper, Peirce clearly includes among the before-trial criteria of evaluation the path along which the hypothesis has in fact been reached.[29] He concludes the account of Kepler's discoveries as follows:

> Thus, never modifying his theory capriciously, but always with a sound and rational motive for just the modification he makes, it follows that when he finally reaches a modification – of most striking simplicity and rationality – which exactly satisfies the observations, *it stands upon a totally different logical footing from what it would if it had been struck out at random, or the reader knows not how, and had been found to satisfy the observations.* Kepler shows his keen logical sense in detailing the whole process by which he finally arrives at the true orbit. This is the greatest piece of Retroductive reasoning ever performed. (*CP*, 1.74, my italics.)

Peirce's discussion about Kepler's achievement suggests an even more drastic difference between his and Popper's views on scientific discovery. Peirce's schema (P) can be modified so that it is applied to *types* of hypotheses instead of single hypotheses (cf. Hanson, 1961). The conclusion is then of the following form: 'It is reasonable to suspect that some hypothesis of kind *K* is true'. The second premise of (P) can then be weakened to the assumption that it is reasonable to suspect that some hypothesis of kind *K* explains the surprising fact *C*. Thus, the following modification of schema (P) is obtained:

The surprising fact, *C*, is observed.
(Q) It is reasonable to suspect that some hypothesis of kind *K* explains *C*.
Hence, there is reason to suspect that some hypothesis of kind *K* is true.

If the scientist has first reasoned to the conclusion of schema (Q), and a particular hypothesis *A* of kind *K* occurs to him where *A* has not been suggested or thought of before, then we may say that schema (Q) shows a part of the 'logic' involved in the discovery of *A*. The next step, then, is to see, whether *A is* a hypothesis which explains *C*, i.e., whether *A* in fact satisfies the second premise of schema (P). If it does, schema (P) suggests that *A* is worthy of further consideration; otherwise it will be dropped, and another hypothesis of kind *K* is sought for.[30]

This account of scientific discovery fits very well to Peirce's description of Kepler's work. Peirce represents Kepler's path to the discovery of of the law of Mars's motion as a series of 'retroductions', starting from inference to such vaguely stated principles as 'the sun must have something to do with causing the planets to move in their orbits' and 'a general description of Mars's motion should refer to the sun as a fixed point'. Kepler's more specific hypotheses were of the type indicated in these principles; but they turned out to be unsatisfactory for the reason that they did not explain all the known facts (i.e., Tycho Brahe's earlier observations on the position of Mars at different times), even if they led him closer to the discovery of his final hypothesis of elliptical orbit.

It is remarkable that this account is not representable in the standard terms of Popperian methodology.[31] Popper usually writes as if it were always fairly easy to find a great number of rival explanations – i.e., explanatory theories which 'solve all the problems' which their predecessor solved and those which it failed to solve – so that the problem of the before-trial evaluation is that of finding the most *testworthy* among them (see, for example, Popper, 1972, pp. 13–15). Kepler's labourious process of discovery serves well to remind us of the extraordinary efforts and difficulties that the hitting upon *even one* satisfactory explanation may involve – not to mention the hitting upon two or more such explanations.[32] It also reminds us that it is not always immediately clear whether a hypothesis really explains the known facts: finding this out may require very elaborate reasoning and calculation. Kepler rejected his intermediate conjectures, because they did not account for the known facts; he did not submit them to new tests at all. These conjectures were not testworthy in Popper's sense, since they were not even explanations of the known facts. Peirce's account of them shows, however, that there were good reasons for proposing them and for studying whether they are explanations of Brahe's observations. Moreover, part of the evaluation of these reasons proceded the discovery of these conjectures (cf. schema (Q)).

The consideration of a new hypothesis A cannot neatly be described as involving three separate and subsequent initial steps: (i) the discovery of A (ii) the evaluation of the testworthiness of A, and (iii) the testing of A. Before evaluating the testworthiness of rival explanations, hypotheses have to be proposed as worthy of being candidates for explanations. Step (ii) is thus rather complex, and it may get mixed with step (i) in ways which philosophers of science should attempt to analyze. There may, after all, be more 'logic' in the scientific discovery than Popper has ever ventured to suggest.[32a]

8. Boldness and Severity

Whewell remarks in his *Philosophy of Discovery* that one of the great »intellectual lessons» of Plato is the insight that »in order to understand the external world, we must indeed observe carefully, but we must also guess boldly» (Whewell, 1860, p. 36). Popper has learnt this lesson – to the extent, indeed, that boldness has become the most crucial element in his methodology. Boldness is required, Popper thinks, at each step of scientific progress: it is not only needed for the purpose of making progress possible, but also for making our pursuit scientific. By being bold in proposing new theories, it will be possible to submit them to severe tests and thereby to learn from our mistakes:

If we have made this our task, then there is no more rational procedure than the method of trial and error – of conjecture and refutation: of boldly proposing theories; of trying our best to show that these are erroneous; and of accepting them tentatively if our critical efforts are unsuccessful. (Popper, 1963, p. 51.)

A very clear and suggestive formulation of the 'Popperian' idea of science as progressing through *severe tests of bold conjectures* was given already by William Whewell:

A facility in devising hypotheses, therefore, is so far from being a fault in the intellectual character of a discoverer, that it is, in truth, a faculty indispensable to his task. It is, for his purpose, much better that he should be too ready in contriving, too eager in pursuing systems which promise to introduce law and order among a mass of unarranged facts, than that he should be barren of such inventions and hopeless of such success. (Whewell, 1847, p. 54.)

To try wrong guesses is, with most persons, the only way to hit upon right ones. The character of the true philosopher is, not that he never conjectures hazardously, but that his conjectures are clearly conceived, and brought into rigid contacts with facts. (Whewell, 1847, p. 55.)

Again, in his Aphorisms IX and X concerning science, Whewell requires that the discoverer should »rigorously» compare his tentative hypotheses with facts and »resolutely» reject those which do not »bear the test» (Whewell, 1847, p. 468). He also recognizes the principle of 'learning from mistakes' by noting that hypotheses involving a portion of incompleteness and error may often be of service to science:

But all who discover truths, must have reasoned upon many errours to obtain each truth; every accepted doctrine must have been one chosen out of many candidates. (Whewell, 1847, p. 56.)

A tentative scientific hypothesis which explains some surprising facts should be subsequently tested by experiment. Whewell, Peirce and Popper agree on the requirement – which was well-known in the nineteenth century – that scientific hypotheses should be testable. An important addition to this principle is the further requirement that it should be possible to test the hypothesis against evidence which is not included in the fact it was designed to explain:

The hypothesis which we accept ought to explain phenomena which we have observed. But they ought to do more than this: our hypotheses ought to *foretel* phenomena which have not yet been observed; – at least of the same kind as those which the hypothesis was invented to explain. (Whewell, 1847, pp. 62–63.)

For, secondly, we require that the new theory should be *independently testable.* That is to say, apart from explaining all the *explicanda* which the new theory was designed to explain, it must have new and testable consequences (preferably consequences of a *new kind*); it must lead to the prediction of phenomena which have not so far been observed. (Popper, 1963, p. 241.)

In the above quotation, which is his 'second requirement for the growth of knowledge', Popper emphasises the value of predictions of a *new kind.* Precisely the same idea was expressed already by Whewell:

We have here spoken of the prediction of facts *of the same kind* as those from which

> our rule was collected. But the evidence in favour of our induction is of a much higher and more forcible character when it enables us to explain and determine cases of a *kind different* from those which were contemplated in the formation of our hypothesis. (Whewell, 1847, p. 65.)

Popper says that this requirement is indispensable, since »it is always possible to produce a theory to fit any given set of explicanda» (Popper, 1963, pp. 241–242). Peirce was very well aware of this problem, which he illustrated by amusing numerical examples and by noting that the so called Bode's law »was shattered by the first discovery of a true planet after its enunciation» (*CP*, 2.739). By sufficient ingenuity, he remarks, we might always find a characteristic which is common to all the members of our (finite) sample. If this characteristic has not been agreed upon in advance (cf. Section 7), then this discovery serves only »to suggest a question, and ought not to create any belief» (*CP*, 2.738), so that for the subsequent testing of the hypothesis suggested by this discovery requires the examination of new instances – in other words, the hypothesis should be 'independently testable'.

Peirce defines the severity of a test in the following way:

> We all know that as soon as a hypothesis has been settled upon as preferable to others, the next business in order is to commence deducing from it whatever experiential predictions are extremest and most unlikely among those deducible from it, in order to subject them to the test of experiment. (*CP*, 7.182.)

In other words, if e is a test statement for a hypothesis h, in the light of background assumptions b, then Peirce suggests that we should choose e such that $(h \,\&\, b) \vdash e$ and $P(e/b)$ is as small as possible. (Here $P(e/b)$ is the relative probability of e given b.) In modern theory of semantic information, the amount of information that e transmits about h, relative to b, is usually defined to be proportional to the difference

(1) $P(e/h \,\&\, b) - P(e/b)$,

which shows that Peirce's strategy in testing is a reasonable one.[33] A normalization of difference (1) has been given by Popper as a measure of the severity of the test e, interpreted as supporting evidence of the theory h, given the background knowledge b (Popper, 1963, p. 391).

A number of interesting similarities in Whewell's, Peirce's, and Popper's views about 'boldness' and 'severity' have been noted above. But when we come to a closer study of Popper's more detailed development of these ideas, we find a perplexing mixture of genuine insight and streched exaggeration. One of the basic ideas which Popper develops in his *LSD* is the contention that 'bold', 'testable', 'falsifiable', 'improbable', 'having empirical content' and 'scientific' are at least extensionally equivalent attributes of hypotheses in the sense that their measures covary with each other. 'Refutable', 'corroborable', 'informative' and 'having explanatory power' are later added to this list (cf. *LSD*, p. 270; Popper, 1963, pp. 217, 391); and with some qualifications, Popper even claims that verisimilitude (see Section 11) can be measured by content (cf. Popper, 1972,

p. 53). Popper views the scientist as a revolutionary who is heroically engaged in finding the boldest explanations and is sincerely attempting to overthrow the best theories, never being too cautious in his proposals and never trying to justify any hypothesis. 'Permanent revolution!' is his battle-cry in science.[34]

As a factual statement about the development of science the Popperian revolution in permanence would be mistaken, as we can recognize, for example, by reading Whewell' account of the conflict between the Cartesians and the Newtonians in his 'On the Transformation of Hypotheses in the History of Science'. In the same spirit as Thomas Kuhn in *The Structure of Scientific Revolutions* more than one hundred years later, Whewell points out that before a defeated hypothesis disappears »with the old generation» it is transformed and modified »so as to have all palpable falsities squeezed out of it, and subsiduary provisions added», which makes it possible for its defenders to »maintain a sort of verbal consistency» until the original hypothesis »become inextricably confused, or breaks down under the weight of the auxiliary hypotheses thus fastened upon it» (Whewell, 1860, pp. 492–503). Sometimes a policy of 'immunizing' the best of our theories against results which appear to refute it may be reasonable for a while; and in spite of his claims to the contrary (Popper, 1972, p. 30), Popper does not have much insight to the value of the periods of 'normal science' between 'scientific revolutions' (cf. Kuhn, 1970b). A remark pointing towards this direction was made already by Whewell, who found it very conceivable that

> a boldness in the invention and application of hypotheses which was propitious to science in its infancy, may be one of the greatest dangers of its more mature period... (Whewell, 1860, p. 15.)

Even if the motivation of this view may lie in his questionable doctrine of necessary truth (see Section 4), Whewell's basic point here is worth remembering: there are different stages of development and different degrees of maturity in science, and different degrees of boldness may be required on the part of scientists in these different stages.

9. Falsifiability

Popper equates boldness and falsifiability: the more a hypothesis says, the more informative it is, the more potential falsifiers it has, the more falsifiable and improbable it is. Since scientists are seeking informative and true theories, he argues, they should have the highest respect for falsifiable theories. To learn about the truth, they should consciously look for errors in their theories, seek counter-examples to them, and make sincere attempts at their refutation.

Freeman and Skolimowski (1974), p. 511, claim that there is one element missing from Peirce's arguments about science to make them »entirely Popperian», viz. an account of the way in which »various hypotheses have been *refuted* to give rise to new hypotheses». This pronouncement is doubtful for three reasons. In the first place, we have already seen in Section 5 that 'refutation' – overthrow of theories which have let us to false expectations – plays an important role within Peirce's general schema for the growth of knowledge. Secondly, the fact that Peirce pays less attention to the refutation of hypotheses than Popper is naturally explainable by his much greater concern with probabilistic hypotheses, which are not falsifiable (cf. *LSD*, p. 190). Thirdly, there is an important 'anti-Popperian' element in Peirce's views about falsifiability.

The following quotation shows that Peirce (in 1896) had a keen awareness of the methodological importance of falsifiability:

> It is a great mistake to suppose that the mind of the active scientist is filled with propositions which, if not proved beyond all reasonable cavil, are at least extremely probable. On the contrary, he entertains hypotheses which are almost wildly incredible, and treats them with respect for the time being. Why does he do this? Simply because any scientific proposition whatever is always liable to be refuted and dropped at short notice. A hypothesis is something which looks as if it might be true and were true, and which is capable of verification or refutation by comparison with facts. The best hypothesis, in the sense of the one most recommending itself to the inquirer, is the one which can be the most readily refuted if it is false. (*CP*, 1.120.)

Freeman and Skolimowski (1974), p. 512, quote the last sentence and ask, whether we could have »anything more Popperian than this». So far as the quotation goes, it is indeed not implausible to regard this passage as one of the most striking anticipations of Popper in Peirce's writings. But already the next sentences point to differences in Peirce's and Popper's views:

> This far outweighs the trifling merit of being likely. For after all, what is a *likely* hypothesis? It is one which falls in with our preconceived ideas. But these may be wrong. Their errors are just what the scientific man is out gunning for more particularly. But if a hypothesis can quickly and easily be cleared away so as to go towards *leaving the field free for the main struggle,* this is an immense advantage. (*CP*, 1.120, my italics.)

Peirce thinks that, after the most easily falsifiable hypotheses have been eliminated, there follows »the main struggle», obviously with hypotheses which are less easily falsifiable. Popper, on the other hand, regards scientists as dealing with more and more falsifiable hypotheses – in other words, he would deny that science ever enters at all the 'main struggle' which Peirce speaks about.

In his reply to Medawar's discussion of the nineteenth-century hypothetico-deductivists, Popper claims that his ideas of (i) degrees of testability and of (ii) identifying these degrees with degrees of empirical content have not been anticipated (Popper, 1974b, p. 1036). However, it is very clear from the above quotation that Peirce made the anticipation (i), but probably not the anticipation (ii). There seems to be a difference between Peirce's 'most readily refutable' and Popper's 'falsifiable' hypotheses. There is, in fact, a very natural sense in which a hypothesis with a relatively low information content may happen to be easily refutable by the scientists: some 'potential falsifiers' of a hypothesis may be more accessible to us than others.

Peirce makes also the important remark that the scientists are particularly interested in 'gunning for' and refuting hypotheses which are likely relative to their background assumptions. This idea, even though it is a simple one, suggests an argument (cf. also Chalmers, 1973) which disproves perhaps the most central thesis of the Popperian falsificationism:

> I can therefore gladly admit that falsificationists like myself much prefer an attempt to solve an interesting problem by a bold conjecture, *even (and especially) if it soon turns out to be false,* to any recital of a sequence of irrelevant truisms. We prefer this because we believe that this is the way in which we can learn from our mistakes; and that in finding that our conjecture was false, we shall have learnt much about the truth, and shall have got nearer to the truth. (Popper, 1963, p. 231.)

Popper claims here that the most efficient way of approaching to the truth is by means of the falsification of bold conjectures.[35] This contradicts the idea that falsification should be directed towards cautious hypotheses which are close to our prejudice and background assumptions. That the latter view is correct, and hence Popper's thesis is mistaken, can easily be shown by applying Popper's own measure of information content:

$$\text{cont}(h) = 1 - P(h),$$

where $P(h)$ is the (logical) probability of h. (We might as well considered content relative to background knowledge, but this would not change the argument.) Suppose that we tentatively adopt a hypothesis h. If h happens to be true, by adopting it we 'gain' its information content cont(h). If h turns out to be false and we refute it, we gain the content of $\sim h$ (i.e., the negation of h), which is

$$\text{cont}(\sim h) = 1 - P(\sim h) = 1 - (1 - P(h)) = P(h).$$

If h is bold in Popper's sense, then cont(h) is high and $P(h)$ is low. Thus, we may gain much information by adopting a bold hypothesis. But when h is bold, our gain in refuting h – which is measured by $P(h)$ – is low. In contradiction to Popper's thesis, we see that *the bolder a hypothesis we falsify, the less information we gain.* Further, *the more probable a hypothesis we falsify, the more information we gain.* Finally, *the bolder a hypothesis we adopt, the more information we may gain.*

Popper's attempt to combine boldness (as a search for highly informative theories) and severity (as a negative approach to testing) is mistaken. There is nothing wrong, of course, in bold conjectures and sincere refutations as such, but is not advantageous to strive at both of them at the same time.

10. Fallibilism

'Fallibilism' is a term which Popper sometimes uses to characterize his epistemological position (see Popper, 1963, pp. 228, 238). I conjecture that he has learnt this word from Peirce, who was the first to introduce it into philosophy. Fallibilism is a doctrine which claims that human knowledge – and scientific knowledge in particular – is fallible or liable to error, so that »people cannot attain absolute certainty concerning questions of fact» (Peirce, *CP*, 1.149). Freeman and Skolimowski (1974), p. 514, find the term 'fallibilism' very inapt, since they think that it implies a view of science as »foredoomed by fate always to err». However, like other dispositions, the human propensity to make mistakes may have a nonuniversal degree varying from time to time – so that one may distinguish between different, weaker and stronger, variants of fallibilism.

The concept of knowledge is often used, in a presystematic sense, as designating the propositional content of the tentatively accepted results of inquiry (at some particular time t). Let us call such results of inquiry *knowledge*$_1$ (at time t). To be interesting at all, this notion presupposes that some form of cognitivism is valid, and behaviouralism is false (cf. Section 6). Both Peirce and Popper assume the existence of knowledge$_1$, even though they very strongly emphasize its temporary nature. Peirce says that science »does not consist so much in *knowing*, nor even in 'organized knowledge'», but rather in the pursuit of finding out (*CP*, 1.44). The 'established truths' of science are nothing more than »propositions into which the economy of endeavour prescribes that, for the time being, further inquiry shall cease» (*CP*, 5.589). Similarly, Popper says that the man of science is made by »his persistent and recklessly critical *quest* for truth», not by »his *possession* of knowledge» (*LSD*, p. 281). Still, a hypothesis which does better in the face of criticism than its competitors may, »temporarily and tentatively, be accepted as part of current scientific teaching» (Popper, 1972, p. 261).

Since Plato, philosophers have usually defined the concept of knowledge with a success condition, i.e., so that only true propositions may be known. According to this idea, let us define *knowledge*$_2$ as *true knowledge*$_1$. Scepticism may then be characterized as the doctrine that knowledge$_2$ is impossible – or more particularly, that it is not possible for us to obtain scientific knowledge$_2$.[36] The opponents of scepticism have traditionally endeavoured to show, not only that knowledge$_2$ is possible, but also that we can obtain 'absolutely certain' or 'final' knowledge$_2$ whose truth is immediately evident to our reason or to our senses or else is completely justifiable. Knowledge in this strong sense – which we may call *knowledge*$_3$ – is assumed to be *incorrigible*, that is, incompatible with all the states of affairs which would imply its falsity. More precisely, if we know$_3$ that h, then we know$_3$ that k for all k such that $\sim k$ is incompatible with our knowing$_3$ that h. Knowledge$_3$ therefore satisfies the principle 'knowing implies knowing that one knows'.[37] In contrast with knowledge$_2$, knowledge$_3$ is not falsifiable in principle, since there is conclusive evidence for its truth.

From Plato to Descartes and Kant, and from Aristotle to Bacon and Mill, the rationalists and the empiricists have dreamt of science as a body of known$_3$ factual truths. Ingenious arguments have been presented to show that there are absolutely reliable sources of $knowledge_3$; or that there exists effective criteria of truth; or that $knowledge_3$ is obtainable by combining the works of our reason or of our senses with the correct method of inquiry. These arguments were sharply criticized in the eighteenth and nineteenth centuries (see Laudan, 1973a), even if they had many influential supporters. For example, William Whewell argued that, in the progress of science, not only is the amount of $knowledge_2$ increased, but also an ever increasing portion of $knowledge_2$ becomes to be recognized as $knowledge_3$ (cf. Section 4). This Whewellian version of infallibilism was treated – not unjustifiably – with suspect by his empiricist opponents. Arguments were also presented to show that the methods of hypothesis, enumerative induction, and eliminative induction do not lead us to $knowledge_3$ even from $known_3$ premises. But if there is no basis of $known_3$ truths, how is scientific $knowledge_2$ possible at all? Is there any alternative to the ideal of demonstrative science – but scepticism and the irrationality of science? Fallibilism can be regarded as a systematic attempt to give an answer to this question.

All fallibilists deny the existence of (factual) scientific $knowledge_3$. It is consistent with this position to maintain the possibility of $knowledge_2$ in science. In other words, it may be claimed that at least in some cases we may obtain true $knowledge_1$ in science, but it is still impossible to give conclusive evidence for its truth. In this view, scientific $knowledge_1$ may be true as well as false, and it is always corrigible in principle. This view may be called *fallibilism in the weak sense.*[38]

Fallibilism in the strong sense denies not only $knowledge_3$ but also $knowledge_2$. It claims that all scientific $knowledge_1$ is strictly speaking false, but in contrast with scepticism it still admits the possibility of making *progress towards the truth.* True $knowledge_1$ is, in this view, the limit of inquiry which we can approach and approximate, even if it cannot be reached in a finite number of steps.

Between the weak and strong forms of fallibilism there are a number of intermediate positions which represent fallibilism in different degrees.[39] For example, a weak fallibilist may claim that most items of scientific $knowledge_1$ are true, or that most of them are false, or even that probably all of them obtained so far are false. A strong fallibilist may claim that most items of scientific $knowledge_1$ are already fairly close to the truth, or that most of them are still very far from it. It is also possible, without any contradiction[40], to combine ideas representing the weak and the strong forms of fallibilism, as we can see by studying Peirce's and Popper's epistemologies.

According to Peirce, fallibilism says that »people cannot attain absolute certainty concerning question of fact», with the important qualification that it neither affirms nor denies that »men cannot attain a sure knowledge of the creations of their own minds» (*CP*, 1.149).[41] Peirce regards assertions in an absolute shape as one of the obstacles to the advance of knowledge.

»That we can be sure of nothing in science is an ancient truth», taught already by the Academy of Plato (*CP*, 1.137). Infallibility in scientific matters seems to Peirce »irresistibly comical»; and he continues with the following remark:

> In those sciences of measurement which are the least subject to error – metrology, geodesy, and metrical astronomy – no man of self-respect ever now states his result, without affixing to it its *probable error*; and if this practice is not followed in other sciences it is because in those the probable errors are too vast to be estimated. (*CP*, 1.9.)

Other obstacles to the advance of knowledge are the assumption that »this or that law or truth has found its last and perfect formulation» and the claim that »this, that, or the other element of science is basic, ultimate, independent of aught else, and utterly inexplicable» (*CP*, 1.139–140; cf. Popper's arguments against the 'essentialist' doctrine that science aims at ultimate explanation; see Popper, 1963, pp. 103–107). Still another obstacle which Peirce mentions is the assumption that »this, that, and the other never can be known» (*CP*, 1.138). He says:

> There is nothing, then, to prevent our knowing outward things as they really are, and it is most likely that we do thus know them in numberless cases, although we can never be absolutely certain of doing so in any special case. (*CP*, 5.311.)

This passage gives a very clear formulation of the *weak* form of fallibilism. The same point is made in Peirce's distinction between 'perfect' and 'sure' knowledge which corresponds to the distinction between knowledge$_2$ and knowledge$_3$:

> Suppose our opinion with reference to a given question to be quite settled, so that inquiry, no matter how far pushed, has no surprises for us on this point. Then we may be said to have attained *perfect knowledge* about that question. (*CP*, 4.62.)
>
> Perhaps we have already attained to perfect knowledge about a number of questions; but we cannot have an unshakeable opinion that we have attained such perfect knowledge about any given question. That would be not only perfectly to know, but perfectly to know that we do perfectly know, which is what is called *sure knowledge.* (*CP*, 4.63)

Popper attributes to Xenophanes the view that »we search for truth, but may not know when we have found it» (Popper, 1963, p. 226). All scientific theories must be regarded as conjectural: »even if we hit upon a true theory, we shall as a rule be merely guessing, and it may well be impossible for us to know that it *is* true» (*ibid.*, p. 225). Popper is more prone than Peirce to use phrases like »most of our theories are false anyway» (*ibid.*, p. 28) and »we may seek for truth, for objective truth, though more often than not we may miss it by a wide margin» (*ibid.*, p. 16). But when he was accused of making truth just an illusion, he replied that »some of our hypotheses may well be true» (Popper, 1972, p. 30). Popper's position, then, represents a somewhat more radical form of weak fallibilism than Peirce's.

During the last one hundred years, much work on induction has been based upon the assumption that (i) rational estimates can be made of the degree in which the available evidence supports the claim that some particular item of scientific knowledge_1 is true, and (ii) these degrees have something to do with the epistemic (i.e., logical or subjective) probabilities of hypotheses relative to evidence. Popper's arguments against this program – which he often misinterprets and misunderstands – will not be discussed in this paper (see, for example, Niiniluoto, 1973). It is more interesting here to note that Peirce did not accept these assumptions, (i) and (ii), even if they are entirely compatible with fallibilism in the weak sense (cf. *CP*, 2.663 and 5.169). Induction lends, Peirce argues, »no definite probability to its conclusion». »It is nonsense to talk of the probability of a law, as if we could pick universes out of a grab-bag and find in what proportion of them the law held good» (*CP*, 2.780). Peirce's probabilistic theory of induction is non-Bayesian in the sense that he does not attribute probabilities to sentences or propositions, i.e., to the conclusions of inductive inference, but instead to modes of inference. These 'truth frequencies' are essentially identical with long run operating characteristics of modern statistics a la Neyman and Pearson. As both Peirce and Popper are opponents of the Bayesian school of induction, it is quite disappointing to find Popper claiming that one of Peirce's mistakes was his attempt to »give a *probabilistic* theory of induction» (Popper, 1974b, p. 1032). Popper explains here that he uses the word 'induction' to stand for »the myth that the *repetition* of something» provides a »rational basis for the acceptance of hypotheses». But, we may ask, who would seriously even consider the acceptance and the rejection of statistical hypotheses – which was Peirce's major concern – without some kind of repetition of instances?

Both the Peircean and the Popperian versions of fallibilism contain important elements from the strong forms of fallibilist views. To these belongs the idea of truth as the limit of inquiry, which is discussed in the next section.

11. Approach to the Truth

Laudan (1973a) has shown that the idea of science as a self-corrective operation which in the long run eliminates false hypotheses and leads us closer to the truth was developed in the middle of the eighteenth century by David Hartley, Georges Le Sage, and Joseph Priestley. Hartley saw an analogy between the method of hypothesis in science and the iterative methods of approximating roots of equations in mathematics. Among the nineteenth-century thinkers Laudan finds this idea in Herschel, Comte, Whewell, Peirce, Bernard, T.H. Huxley, and Duhem; and in the next century in Reichenbach and Popper. Peirce was not »the founder and first promulgator of the view that the methods of scientific inference are self-corrective», as it has been customary to assume, but he is »the crucial, logical and historical link», between nineteenth- and twentieth-century discussions of this idea (Laudan, 1973a, pp. 275–276).

Besides the endorsement of the self-correcting thesis in *methodology*, which Laudan has lucidly described, there has simultaneously existed a *metaphysical* idea of progress towards the truth. This tradition, which Laudan ignores in his paper, starts with German idealists, and it has later excercised influence upon marxist philosophers, especially Engels and Lenin.[42] Whewell notes that »Hegel traces the manifestation of the identity of the idea and fact in the *progress* of human knowledge» (Whewell, 1860, p. 309), and in some measure approaches his doctrine that »the only way in which we can approach to truth is gradually and successively», advancing »from the perception to the idea; from the fact to the theory; from the apprehension of truths as actual to the apprehension of them as necessary» – i.e., advancing from knowledge$_2$ to knowledge$_3$ (*ibid.*, p. 307). Not only Whewell, but also Peirce was influenced by this 'metaphysical' tradition. It is unfortunate that Laudan has not paid any attention to Peirce's general theory of truth and reality, because it seems that only from this perspective one can do justice to Peirce's approach to the self-correcting thesis.

The idea of 'approach to the truth' can be interpreted in several different ways. For example, a cumulative theory of scientific progress might assert that when scientific knowledge$_1$ advances the amount of knowledge$_2$ in it increases monotonously. In the interpretation which is relevant to the self-correcting thesis, it is recognized that scientific knowledge$_1$ does not only grow, but its items will be *replaced* by new laws and theories. A succession of theories $T_1, T_2, \ldots T_n, \ldots$ may then approach to the truth in two different ways: (i) theories $T_1, T_2, \ldots, T_{k-1}$ are false, but all of T_n, for $n \geqslant k$, are contained in a true theory T, and (ii) all of theories $T_1, T_2 \ldots$ are false, but each T_1 is closer to the truth than its predecessors. In other words, the true theory will be reached either (i) in a finite number of steps, or (ii) at best as an ideal limit only.[43] The possibility of successions corresponding to the case (i) is maintained in the weak form of fallibilism, while fallibilism in the strong sense assumes all of them to exemplify the case (ii).[44]

It was noted in the preceding section that both Peirce and Popper are weak fallibilists in the sense that they both admit the possibility of reaching true theories in finite number of steps (see also *CP*, 7.220). Popper thinks that most of our theories are false and that there are cases where we have »to work with theories which are *at best* approximations» (Popper, 1963, p. 235). Consequently, he has treated successions of type (ii) as a paradigm case of scientific progress, and he has – so far unsuccessfully (cf. Miller, 1974, 1975, and Niiniluoto, 1977) – attempted to define a notion (and a measure) of *verisimilitude* which would combine the ideas of approximation to the truth and of information content. The motivation for this – so far uncompleted – methodological program is the assumption that

> while we can never have sufficiently good arguments in the empirical sciences for claiming that we have actually reached the truth, we can have strong and reasonably

good arguments for claiming that we may have made progress towards the truth. (Popper, 1972, pp. 57–58.)

In making the search for verisimilitude rather than the search for truth as the aim of science (*ibid.*, p. 57), Popper has made his position almost indistinguishable from fallibilism in the strong sense. In his emphasis on the inexactness and indeterminancy of scientific knowledge, Peirce, too, comes very close to the strong form of fallibilism. He says that there are three things to which we can never hope to attain by inductive inference from sample to population, namely, »absolute certainty, absolute exactitude, absolute universality» (*CP*, 1.141). Further support for this view Peirce finds in his ontology: it is a consequence of his 'synechism' that »our knowledge is never absolute but always swims, as it were, in a continuum of uncertainty and of indeterminacy» (*CP*, 1.171).

Freeman suggest that »one of the fundamental disagreements between Peirce and Popper» is Peirce's contention that »in the long run science is *predestined* to hit on the truth»:

For Peirce seems on some occasions, but not on others (5.590–604), to adhere to the »manifest theory of truth», as Popper calls it; it is the nature of truth that it will hit us sooner or later, given enough time. Popper, on the other hand, emphatically denies this. (Freeman and Skolimowski, 1974, p. 467.)

Somewhat surprisingly, Popper regards this as »a very fair remark» (Popper, 1974b, p. 1065). By the 'manifest theory of truth' Popper means »the optimistic view that truth, if put before us naked, is always recognizable as truth» (Popper, 1963, p. 7), which is quite different from Peirce's view. Like Popper, Peirce admits that we may sometimes hit upon true theories, but again like Popper, he does not think that they can effectively be recognized as true. There is no guarantee for success after a *finite* number of steps: »we cannot be absolutely certain that our conclusions are even approximately true» (*CP*, 1.141). Peirce tried to show that success is nevertheless guaranteed in the long run, that is, as an *ideal limit* of inquiry. This view marks a difference between Peirce and Popper, but it has nothing to do with the theory of 'naked' and 'manifest' truth.

Peirce took the idea of approaching to the truth so seriously that he incorporated it into his 'definition' of truth.

The opinion which is fated to be ultimately agreed to by all who investigate, is what we mean by the truth, and the object represented in this opinion is the real. (*CP*, 5.407.) Truth is that concordance of an abstract statement with the ideal limit towards which endless investigation would tend to bring scientific belief. (*CP*, 5.565.)

Truth is objective, for »the opinion which would finally result from investigation does not depend on how anybody may actually think» (*CP*, 5.408), yet it presupposes the notion of an ideal unlimited community of investigators which is capable of advancing knowledge (*CP*, 2.654). The self-corrective nature of induction is, for Peirce, a special case of his theory of inquiry:

The validity of an inductive argument consists, then, in the fact that it pursues a method which, if duly persisted in, must, in the very nature of things, lead to a result indefinity approximating to the truth in the long run. (*CP*, 2.781.)

Already Condillac, in the eighteenth century, as well as Senebier and Prevost, in the beginning of the nineteenth century, had argued that there exists no guarantee for a new theory to be closer to the truth than its refuted predecessors (see Laudan, 1973a). Peirce's attempt to show that induction – i.e., inference from sample to population – is necessarily self-corrective in the long run is mistaken, since relative frequencies approach to the corresponding physical probabilities (or to the corresponding population characteristics) at best *with probability one.*[45] Moreover, this argument concerned only enumerative induction (i.e., statistical generalization), and Peirce's attempts to extend it to cover other kinds of »induction», like the method of hypothesis, were unsuccessful.[46] Laudan finds it hard to conceive why Peirce still insisted that all forms of scientific inference are self-corrective; he suggests that Peirce just »conveniently ignored the restricted scope» of his argument for enumerative induction and »perhaps unconsciously» slid from this argument to a general thesis (see Laudan, 1973a, p. 293). There is nothing surprising in Peirce's adherence to the general self-correcting thesis, however, because he had presented already in 'How to make our ideas clear' (1878) a general philosophical argument for this thesis, and this argument was not shaken by his failure to illustrate its content with all types of inference.

Peirce's argument for the self-correcting thesis is an application of his pragmaticist maxim:

Consider what effects, that might conceivably have practical bearings, we conceive the object of our conception to have. Then, our conception of these effects is the whole of our conception of the object. (*CP*, 5.402.)

(To separate this argument from the so called pragmatic justification of induction, we might call it the *pragmaticist argument for the progress of scientific knowledge.*) According to this maximum, the meaning of the concept 'real' wholly consists of the practical effects that the object of this concept, i.e., the reality, has. The effect which real things have upon us, Peirce says, is »to cause belief» (*CP*, 5.406); moreover, the real things »affect our senses according to regular laws» (*CP*, 5.384). In science, we let this external permanency – the reality – to control our belief formation, with the result that

different minds may set out with the most antagonistic views, but the progress of investigation carries them by a force ourside of themselves to one and the same conclusion. (*CP*, 5.407.)

This »destined centre» towards which the results of inquiry are steadily moving is the truth, and the object it represents is the real.

Peirce's theory of truth and reality has recently stimulated the supporters of the *concensus theory of truth,* especially Jürgen Habermas (see Habermas, 1970, 1973). In this theory, truth and reality are regarded as results of a consensus which is reached by means of nondistorted communication in an idealized speech community. Some of Peirce's pronouncements point towards this kind of idealistic theory (cf. »reality depends on the ultimate decision of the community», *CP*, 5.316), but nevertheless a *realistic* interpretation of his views seems better to correspond to the spirit of his philosophy. In this interpretation, truth and reality are not created by the consensus of a scientific community, but the eventual existence and possibility of such a consensus is explained by reference to the nature of the reality. By limiting and determining our scientific beliefs, the reality – the real things which are independent of our opinions and our wishes – brings the scientific community towards the truth and 'forces' it to a 'consensus'. From this perspective, Peirce appears as an remarkable figure within the modern history of critical scientific realism.[47]

Popper's theory of falsification relies upon closely similar ideas about the role of 'agreement' among the scientists. Falsification of a general hypothesis depends upon the acceptance of singular 'basic statements'; these statements are »accepted as the results of a decision or agreement; and to that extent they are conventions» (*LSD*, p. 106). These 'free decisions' may be motivated, but not justified, by observation; they are not 'arbitrary', but comparable to a deliberated verdict of a jury (see *LSD*, pp. 105–110; Popper, 1974b, pp. 1110–1111). Thus, while Peirce thinks that the scientific community will ultimately reach a consensus on the truth of scientific theories, Popper thinks that the 'scientific jury' will reach a consensus on the falsity of theories. (Popper is not suggesting a 'consensus theory of falsity', as little as Peirce is a defender of a 'consensus theory of truth'. For Popper's enthusiasm about Tarski's correspondence theory of truth, see Popper, 1972, pp. 319–335.)

Peirce's account of the relation between the reality and our knowledge about it has significant parallel in the relation between objective (physical) probabilities and subjective (epistemic) probabilities – to the extent that the former account can be regarded as a kind of generalization of the latter relation. Bruno de Finetti (1937) showed that, under fairly general conditions, two rational agents who modify their degrees of belief in accordance with Bayes's theorem and who start from different initial degrees of belief will very likely, after a sufficient number of trials, have posterior degrees of belief close to the observed relative frequency and therefore close to each other. de Finetti interprets this result as showing the *intersubjectivity* of rational learning from experience, but he refuses to assume that there exists some unknown correct value to which the observed relative frequencies, and therefore the posterior degrees of belief, would converge. The frequentist school, on the other hand, *defines* objective probabilities as limits of relative frequencies. While de Finetti's position may be characterized as 'intersubjectivity without objectivity', the frequentist position is 'objectivity *via* definition'. de Finetti errs in his denial of physical probabilities – just as idealists err in their denial of physical reality behind the appearances. The frequentists err in forgetting that relative frequencies convergence to physical probabilities (according to The Strong Law of Large Numbers) not by logical necessity but only with probability onc. They make an error when they define probabilities as limits – just as the supporters of the consensus theory of truth err when they define truth as the limit of ideal communication.

In opposition to the frequency theory, Peirce suggested that physical probabilities are dispositional properties of devices that may be used to generate sequences of events by successive trials. This 'propensity theory of probability' was later rediscovered by Popper.[48] The long run behaviour of relative frequencies in these sequences can, according to the propensity theory, be *explained* by means of the physical propensities inherent in their generating conditions – just as Peirce attempted, in his pragmaticist argument for progress, to explain the convergence of scientific opinion towards a 'predestined' limit by the effects of the real things. Moreover, de Finetti's results show that physical probabilities can be reasonably *estimated* or *approximated* by means of 'personal' probabilities[49] – just as Peirce thought that the reality can be approximated by means of the scientific knowledge which the scientific community attains.

12. Conclusion: Progress in Science

In the preceding sections, we have found a number of common elements in Whewell's, Peirce's and Popper's conceptions of science. Whewell gives an admirable formulation to the method of hypothesis in science: a scientist invents bold explanations which impose a new conceptual element on previously known facts, and he severely tests them by deriving from them predictions of new facts. Successive generalization and conceptual unification together give the characteristic integrative nature to the growth of scientific knowledge. Peirce develops an evolutionary approach to knowledge: science is an objective and co-operative enterprise which in the long run brings the scientists close to the truth. Scientific knowledge is fallible; science does not aim at ultimate explanations. Whewell's anti-inductivism, his emphasis on the relativity of facts and theories, on conceptual enrichment and on integrative growth, Peirce's fallibilism and evolutionary theory of knowledge, his emphasis on the merits of falsifiability of hypotheses and on the severity of tests – these are among those features in which Popper can be regarded as a follower of Whewell and Peirce.

Among the differences between these three philosophers, perhaps the most important is the incompatibility of Whewell's doctrine of necessary truth with Peirce's and Popper's fallibilism. On the other hand, between Peirce's and Popper's fallibilist views we found only a relatively insignificant difference in emphasis. Some differences – like Whewell's remarks on the variable role of boldness in inquiry and Peirce's account of scientific discovery – reveal insights which suggest ways of overcoming some of the weak points in Popper's philosophy of science. From the perspective of this paper, the novelty of Popper's methodology is in his idea about degrees of empirical content – a valuable insight which has duly been incorporated into the modern theory of semantic information. However, Popper's attempt to unify the nineteenth-century elements by means of the novel idea of content lays a too heavy burden on a single concept: we have seen that his attempt to reduce many of the interesting methodological concepts to the notion of content cannot give a coherent foundation to the Popperian edifice of a falsificationist methodology.

The fundamental problem in all comprehensive philosophies of science is the problem concerning the 'value' of scientific knowledge and the 'rationality' of the scientific enterprise. We may conclude this paper by comparing and contrasting Whewell's, Peirce's, and Popper's answers to this question.

All these philosophers see the rationality of science in the process rather than in the product: science is in *progress,* because scientific knowledge$_1$ is growing and moving towards a non-exhaustible limit. (For an alternative view of scientific progress, see Kuhn 1970a, p. 206.) For Whewell, this progress consists in the 'idealization of facts', i.e., in the establishment of more and more necessary truths. Through the attainment of knowledge$_3$ about nature, science has for Whewell an independent cognitive value. If the existence of factual knowledge$_3$ in science is denied, this defence of scientific rationality is not tenable any more. Peirce tried to show that the attainment of knowledge$_2$ in science is guaranteed as the ultimate limit of inquiry – but he recognized that his argument for progress embodies a *hope* (*CP*, 5.407). Even if Peirce thought that the success of inquiry is guaranteed only in ideal limit, he argued that we *may* hit upon truth, or come close to it, in a finite number of steps, and perhaps have done so in a number of cases. For Peirce, science has cognitive value in so far as we have, and ultimately will, attain or approximate truth. He also recognized that true (or approximately true) knowledge$_1$ serves as a guide to successful practice and as a rational basis of action (*CP*, 5.170, 5.199). In addition to its cognitive value, science thus has a derivative practical value.

In spite of its shortcomings, Peirce's view seems to contain an important element of truth (cf. also the preceding section). Scientists have undeniably been successful in solving theoretical and practical problems in science. Perhaps the best explanation to this success is the assumption that many theories which they employ are reasonably close to the truth. (This is not a proof, but an abductive inference!) If this assumption is valid, then it follows that the scientists will, in future, be relatively successful within certain limits. This prediction is, moreover, verified in each day in scientific practice, which gives some confirmation to the given assumption.

Popper develops the problem of rationality in the direction which is opposite to the above argument. He belittles the cognitive value of science by claiming that we usually miss the truth by a wide margin.[50] Popper admits that a high degree of corroboration is an indicator of a high degree of verisimilitude (Popper, 1972, p. 103). On the other hand, degrees of corroboration do not say anything of the future performance of a theory (*ibid.*, p. 83). But as a high degree of verisimilitude tells something about this future performance (cf. above), we cannot but conclude that for Popper most theories have a low degree of verisimilitude. As theories have this little cognitive value, Popper cannot explain their practical value either (cf. Section 6). He claims that »a *pragmatic belief in the results of science* is not irrational, because there is nothing more 'rational' than the method of critical discussion, which is the method of science» (*ibid.*, p. 27). The non-existence of more rational alternative methods does not suffice to show that the scientific methods *are* rational – this argument does not convince at all those who do not accept the presupposition that there exists even one rational method. It seems, therefore, that in his efforts to avoid the Skylla of infallibilism Popper is dangerously drifting to the Charybdis of scepticism.

NOTES

Reprinted from *Ajatus* **37**(1978) with the permission of the Philosophical Society of Finland.

1 For an extensive bibliography of the history of methodology, see Laudan (1968). An anthology of (mostly) nineteenth-century philosophy of science has been edited by Kockelmans (1968). Interesting parallels between modern issues and nineteenth century ideas are presented, for example, in Laudan (1973a, 1973b); the former article is contained in an important collection of essays on the nineteenth-century methodology: Giere and Westfall (1973).

2 The critical edition of Whewell's *Collected Works* is since 1967 being edited by Gerd Buchdahl and Laurens Laudan. Whewell's *History of the Inductive Sciences* was originally published in 1837. *The Philosophy of Inductive Sciences* was first published in 1840; the second edition (1847) has been reprinted as vols. VI and VII of the *Collected Works* in 1967. The third edition appeared as three separate works: *History of Scientific Ideas, Novum Organon Renovatum,* and *The Philosophy of Discovery*. The last mentioned work has been reprinted in 1971 from the original 1860 edition by Burt Franklin, Lenox Hill Pub. & Dist. Co., New York. A very useful collection of selections of Whewell's work has been edited by Butts (1968). For literature dealing with Whewell's philosophy, see Ducasse (1951), Butts (1970, 1973), and Laudan (1971). The first collection of Peirce's fragmentary writings was edited by Morris R. Cohen in 1923 with the title *Chance, Love and Logic.* Vols. I–VI of the *Collected Papers* of Peirce were edited by C. Hartshorne and P. Weiss in 1931–35, and vols. VII–VIII by Arthur Burks in 1958. References to the *Collected Papers* are made in the standard manner; thus, for example, (*CP*, 7.182) indicates the 182nd paragraph of the seventh volume. A handy collection of Peirce's texts is contained in J. Buchler's selection *Philosophical Writings of Peirce* (Dover, New York, 1955). A collection of Peirce's writings in German has been edited by Karl-Otto Apel: *Schriften* I–II (Suhrkamp, Frankfurt am Main, 1967 and 1970).

2a Butts fails to do justice to Ernst Mach, however. Mach's *Erkenntnis und Irrtum* (cf. Mach, 1976) contains many favourable references to Whewell and Apelt.

3 *'LSD'* refers to Popper's *The Logic of Scientific Discovery* (1959)

4 That this discovery should come as a surprise to some of Popper's readers may be partly due to Popper's tendency of seeking his forerunners among the Presocratics rather than among the nineteenth-century methodologists (see Popper, 1963, pp. 151–153, 236–237). Besides that, Popper has sometimes vigorously claimed for priority of views which – to quote von Wright's remark on the 'Popper-Hempel' theory of explanation (von Wright, 1971, p. 175) – »had been something of a philosophical commonplace ever since the days of Mill and Jevons». (Cf. note 7 to Ch. XXV in Popper, 1945.) We may add here that a detailed theory of probabilistic explanation which included deductive explanation as a limiting case was presented by Peirce in 'A Theory of Probable Inference' (1883; see *CP*, 2.694–754).

5 This is not the 'positivist' tradition of Comte and Mill, as von Wright (1971), p. 175, suggests, but rather the tradition of Whewell and Peirce. (But note that von Wright has primarily in mind problems of social science and social philosophy, which are excluded from consideration in this paper.) Similarities between Whewell's and Popper's theories of scientific method have been noticed at least by Lakatos (1970), pp. 100n, 104, Agassi (1974), pp. 184, 196, and Medawar (1974), pp. 283–286. Similarities between Peirce's and Popper's views have been suggested, among others, by Gallie and Rynin (see Popper, 1963, p. 240), Apel (1974), p. 34, Laudan (1973a), p. 306, and Giere (1975). The first systematic attempt to discuss the relation between Peirce and Popper is Freeman and Skolimowski (1974), which is dated already in 1969. Some remarks on this relation were made in my article 'Fallibilismista', *Sosiologia, Journal of the Westermarck Society* 11 (1974), pp. 274–286 (English summary on p. 316) which parts of the present paper are based on.

5a The relation between Whewell's and Popper's requirements for the testing of scientific hypotheses was noted already by Agassi (1966) who disagrees with the »Whewell-Popper formula» (see also Agassi, 1975). The same topic is discussed in Laudan (1971). For comments on Whewell, see also Butts (1977) and Laudan (1981). Hacking (1981) suggests that there is an interesting similarity between Peirce's and Lakatos's philosophies of science.

6 As logical empiricists disagreed on many important issues, tables like the one on p. 487 in Freeman and Skolimowski (1974) are more likely to be misleading than illuminating. For example, Kraft was opposed to inductivism, and Neurath did not assume a rock bottom of indubitable statements. Moreover, inductivists are, of course, very interested in the problem of the growth of knowledge.

7 See Medawar (1974), Freeman and Skolimowski (1974). The latter paper consists in three parts: the first part, 'Charles Peirce and Objectivity in Philosophy', pp. 464–482, is written by Freeman; the second part, 'Karl Popper and the Objectivity of Scientific Knowledge', pp. 483–508, is written by Skolimowski; the third part, 'Peirce and Popper Similarities and Differences', pp. 508–515, is a joint work of both authors.

8 Cf. Popper (1963), p. 240n, and Popper (1972), p. 215n. Peirce was not mentioned in the original *Logik der Forschung* (1934). In the preface to the *LSD*, Peirce and Whewell are mentioned as belonging to a group of epistemologists who have made the important discovery that »scientific knowledge can be more easily studied than common-sense knowledge». In the additions to the *LSD*, Popper also refers to an idea in Peirce's theory of probability (see *LSD*, pp. 148n, 406&n, 408). In *Conjectures and Refutations* (1963), reference is made to Peirce's idea of truth as a regulative principle (p. 226), to Quine's criticism of Peirce's idea of 'approach to the truth' (p. 231), and to Peirce's remarks on the severity of tests (p. 240). In *Objective Knowledge* (1972), all references to Peirce concern his indeterminism; here Popper says that he believes Peirce to be »one of the greatest philosophers of all time» (p. 212). As to Whewell, he is not mentioned at all in *Conjectures and Refutations* or in *Objective Knowledge*, but in the *LSD* he is mentioned (with Bacon and Mill) as an 'inductivist' (p. 419). Popper's

replies to Medawar, Freeman, and Skolimowski seem to be the only other comments that he has made on Whewell and Peirce (see Popper, 1974b, pp. 1032–1033, 1065–1066).

9 See the letter to Adam Sedgwick in Whewell (1847), p. iii.

10 Whewell thinks that his Inductive Tables serve to confirm Francis Bacon's ideas about successive floors (*tabulata*) of induction and about each science being representatible as a pyramid (see Whewell, 1860, p. 132). He also tells us that Raymond Lully employed »Genealogical Trees, each of which he dignifies with the name of the Tree of Science» (*ibid.*, p. 61), but this is hardly anything more than a coincidence of terminology.

11 It is remarkable that the Kantian doctrine of the inseparable conceptual element in human knowledge was first appropriated within the philosophy of the *physical sciences.* There is a – perhaps widespread – mistaken assumption, expressed for example by Itkonen (1974), p. 16, that the philosophers of natural science learned this principle only in this century or at least later than it was 'originated' as a part of the philosophy of the human sciences by the 'hermeneutic' philosophers of the latter part of the nineteenth century.

12 As Ideas are inseparable from Facts – as much as Facts from Ideas – it would be misleading to regard Whewell's 'necessary' truths as conventions or to regard him (without qualifications) as a representative of conventionalism. Whewell's 'necessary' truths in physics are learnt from experience, but the ground of their truth is nevertheless in our ideas. A similar view is attributed to Duns Scotus in Losee (1972), p. 39.

12a For Whewell's critical attitude towards Darwin's theory, see Hull (1975, 1977) and Ruse (1979). Fore a more detailed discussion about 'evolutionary' models of scientific change, see Chapter 4 below.

13 Cf. Popper's remarks about his Darwinian influences in Popper (1972), p. 67.

14 Physical indeterminism, which is assumed to be compatible even with the classical Newtonian mechanics, is again one of the common views of Peirce and Popper – indeed, the one most clearly recognized as such by Popper. See Popper (1972), pp. 212–215.

15 In Popper's terminology, world 1 is the physical world, world 2 is the world of thought and mental states, and world 3 is the world of objective contents of thought.

16 »If we must speak of animals as *learning* from experience, we may at least abstain from applying to them terms which imply that they learn, in the same way in which men learn astronomy from the stars, and chemistry from the effects of mixture and heat.» (Whewell, 1860, p. 243.)

17 At least in the physical sciences, the testing of a hypothesis is usually conducted in 'artificial' laboratory conditions which do not involve practical utilities and losses characteristic of the real-life conditions for action. Testing is acting in order to check a belief, not acting upon a belief.

18 Popper says that the subjective act of understanding, and thus the 'subjective knowledge' in general, should be analysed in terms of third-world objects (Popper, 1972, p. 164); among them there are the *third-world problems,* which are not »man-made» but »internally generated» (Popper, 1974b, p. 1066). In view of Peirce's theory of signs as involving a genuinely triadic relation – a sign is »something which stands to somebody for something in some respect or capacity» (*CP*, 2.228) – it may be said that Peirce would not have approved Popper's 'epistemology without a knowing subject'. The real subject of scientific knowledge is, for Peirce, an ideal scientific community. (This idea has recently impressed Apel's theory of transcendental community of communication; see Apel, 1973.)

19 For this terminology, see Levi (1967b).

19a In his present position, Levi differs both from Peirce and Popper (cf. Levi, 1980).

20 Cf. the critical remarks about Popper in Niiniluoto and Tuomela (1973), pp. 201–204.

20a For recent discussions of this issue, see Salmon (1981) and Niiniluoto (1982).
21 Thomas Reid's antihypothetical or inductivist methodology is discussed in Laudan (1969).
22 See Hintikka and Remes (1974), pp. 105–117. The analogy to geometrical analysis is one of the most interesting systematic ideas within the 'logic of discovery'. Whewell's theory of induction is obviously related to the idea of analysis as an 'upward movement', and the sagacity he claims to be involved in invention of the appropriate conceptions may be taken to correspond to the nonpredictable character of the required auxiliary constructions within analysis (cf. Hintikka and Remes, 1974, pp. 4–5, 41–48). Peirce's theory of abduction is essentially a probabilistic version of the same idea.
23 'Inductivism' is a word which is used in many different senses; some of them have been discussed in Niiniluoto and Tuomela (1973), pp. 197–200.
Popper attacks against induction, not only as a method of discovery, but also as a part of after-trial evaluation of hypotheses. Many of Popper's arguments against induction are exaggerated and mistaken; they are not discussed in this paper. (See, for example, Niiniluoto 1973, Niiniluoto and Tuomela, 1973, and the various essays on the Bayesian and Popperian methodologies in Nos. 1–2 of *Synthese* 30 (1975).)
24 Cf. Peirce's remark that almost all logicians of his time make »the fundamental mistake of confounding the logical question with the psychological question» (*CP*, 5.85).
25 It may be of some interest to note that the original German phrase for 'the logic of scientific discovery', on p. 27 of the *LSD*, is simply 'Forschungslogik'.
26 As a result, they often have failed to make any difference between inductivism and attempts to give a noninductivist account of inferential steps involved in discovery – cf. the myth of Newton as an inductivist, and the claim that Whewell is an inductivist in the *LSD*, p. 419. Peirce differs from the hypothetico-deductive conception and from Popper's deductivism most clearly in his insistence that the procedure of finding hypotheses involves genuine (probable) *inference* with its own criteria of *validity* (see *CP*, 2.715).
27 The term 'logic' is used here in a weak sense as referring to an attempt to conceptualize, in general terms, some recurring patterns of reasoning involved in scientific discovery. It does not imply any assumption of 'mechanical' or 'fool-proof' procedures for discovery. In essentially the same sense, one usually speaks of inductive 'logic' or of the 'logic' of acceptance: modern 'inductive logicians' do not assume that there are purely logical (i.e., syntactical and semantical) criteria of validity for induction, since their 'inductive probabilities' depend upon extra-logical parameters (see, for example, Niiniluoto and Tuomela, 1973, pp. 24–25, 184).
28 Cf. Popper (1963), pp. 58, 217; (1972), p. 143; and Niiniluoto (1973), pp. 420–422.
29 A similar idea is expressed by Whewell in his claim that his Inductive Tables show us how those physical sciences which have followed »the most rigorous conditions of gradual and successive generalization» possess »a solid truth, which is always ready to stand the severest tests of reasoning and experiment» (Whewell, 1860, p. 134). Cf. Popper (1957a), p. 135.
30 Schema (Q) need not be connected with the first discovery of hypothesis *A*, as explained above, but it may as well be involved in the *rediscovery* of an old hypothesis. Moreover, schemas (P) and (Q) may be involved even in the appraisal of an old and well-remembered hypothesis – cf. Copernicus' reconsideration of the Ancient hypothesis of the sun's position in the centre of the universe. Achinstein seems to be right, when he claims that there are patterns of inference to laws – like the 'inductive' and 'abductive' ones – which may be employed both in the 'context of discovery' and in the 'context of justification' (see Achinstein, 1971, pp. 110–165), even if it seems somewhat exaggerated to say that the 'logic of discovery' cannot be distinguished at

all from the 'logic of justification' (*ibid.*, p. 141). Achinstein makes also the remark that Hanson has forgotten to analyse the role of theories within the discovery of new laws (*ibid.*, p. 118). It is obvious that to support the second premise of schema (Q) one needs some general background assumptions, scientific or 'metaphysical' theories. For an attempt to analyse the role of theories as ***premises*** of inductive arguments, see Niiniluoto and Tuomela (1973).

31 Popper notes that Kepler's discoveries were inspired by metaphysical or astrological assumptions, but otherwise he attempts to tell this story as a usual application of the trial-and-error pattern – that Tycho's observation *refuted* the original circle hypothesis, and so Kepler »tried the next best solutions – the oval, and the ellipse» (Popper, 1963, p. 188). Thus, Popper does not explain the rationality of Kepler's strategy; Freeman and Skolimowski (1974), p. 511, fail to see any essential difference in Peirce's and Popper's accounts of Kepler's work.

32 In spite of some exaggeration, Whewell had a good point in his argument against Mill, that when »the question is about accounting for a large and complex series of facts» he would »propose as a challenge to any person of fertile imagination to devise» any new hypothesis to account for them, if one such hypothesis has already been invented. (Whewell, 1860, p. 271).

32a For important recent studies in scientific discovery, see Nickles (1980a, 1980b).

33 When $(h\&b) \vdash e$, then Hintikka's three measures of transmitted information give the following results for the amount of information e transmits about h relative to b:

a) $-\log P(e/b)$

b) $1 - P(e/b)$

c) $\frac{P(h/b)}{P(e/b)}(1 - P(e/b))$

(cf. measures $inf(e//h)$, $cont(e//h)$, and $cont(e///h)$ in Niiniluoto and Tuomela, 1973, p. 66).

34 This battle-cry is taken from a conversation of Popper, 'On Reason & the Open Society', in *Encounter*, vol. **38**, No. 5, May 1972, pp. 13–18.

35 The same idea is repeated in Popper (1972), p. 144.

36 In speaking about scientific knowledge, I usually assume that science seeks knowledge of the general or generic features of the world, so that 'genuine' scientific knowledge consists primarily of universal laws and theories.

37 For recent discussion about the concept of knowledge, see Hilpinen (1970) and Hintikka (1970). Hintikka notes that the principle '$Kp \supset KKp$' follows, in epistemic logic, from the rule

$$\frac{Kp \qquad Kp \supset q}{Kq}$$

38 Medawar (1974) attributes to Dugald Stewart (early nineteenth century) the view that the possibility of exceptions to scientific knowledge is not excluded. Oddly enough, Popper interprets this nice characterization of weak fallibilism as an »anti-inductivist or deductivist position» (Popper, 1974b, pp. 1032–1033). Inductivism and fallibilism in the weak sense do not exclude each other.

39 Ackerman defines fallibilism as the doctrine which claims that all knowledge$_1$ will *almost certainly* be falsified in the future (Ackerman, 1972, p. 78).

40 For example, a weak fallibilist may consistently assume that in some special areas knowledge$_2$ cannot be attained.

41 This qualification is related to Peirce's view about mathematics, and it is connected with the tradition of *maker's knowledge* in epistemology (cf. Hintikka, 1974).

42 Engels says that the dialectical philosophy destroys illusion about final and absolute truth: that which is now recognized as true will later manifest its latent false side (Engels, 1934). Lenin views truth as a process in which knowledge$_1$ approaches – rela-

tive to historical conditions – to objective truth (see Lenin, 1927, pp. 133–136, 319). In this process, knowledge_1 becomes more complete and more precise (*ibid.*, p. 99). Lenin thinks that knowledge_1 always has a false part which will ultimately be eliminated (cf. fallibilism in the strong sense). On the other hand, he assumes that there is an absolute part in knowledge which will not be eliminated in the progress of science (cf. Peirce's perfect knowledge). Against infallibilism, Lenin states that »the criterion of practice can never, in the nature of things, either confirm or refute any human idea *completely*» (*ibid.*, p. 142).

43 Wennerberg (1966), p. 155, notes the difference between successions of types (i) and (ii), which he recognizes as analogous to the difference between the following two sequences of numbers: (1) 1, 1/2, 1/3,... → 0 and (2) 1, 3, 5, 7, 0, 0, 0, ... → 0.

44 These ideas have interesting parallels in the discussion concerning the possibility of separating from our factual knowledge the contribution of our conceptual system – separating Ideas from Facts, as Whewell would say. Hintikka (1973) has shown that, in sufficiently rich languages with a noneffective notion of logical truth (like a first-order language with relation symbols), this contribution is not *effectively separable*, even if it can effectively be limited and diminished. In other words, 'purely factual' knowledge about the relationships obtaining in the reality can, perhaps, be attained, but we do not know when we have reached it (cf. fallibilism in the weak sense). There is an interesting resemblance between Peirce's theory of reality as a limit of scientific activity and Hintikka's theory of *Dinge an sich* as a limit of our conceptualizing activity (see Hintikka, 1974, pp. 193–195).

45 The whole frequency theory of probability is based upon the same mistake, as Stegmüller has recently emphasized (see Stegmüller, 1973, pp. 34–37).

46 Peirce thought that the successive approximations to truth take us, at each step, closer to the truth. He writes: »But we hope that in the progress of science its error will indefinitely diminish, just as the error of 3.14159, the value given for π, will indefinitely diminish as the calculation is carried to more and more places of decimals» (*CP*, 5.565). In a very interesting note, von Wright tells that already James Bernoulli (died 1705) had, in his correspondence with Leibniz, suggested that the approximation of the true value of π is analogous to the convergence of observed relative frequencies to the true values of probabilities (see von Wright, 1957, p. 225). von Wright notes further that Leibniz »acutely observes that the analogy is fallacious»: we do not *know* that new observations will take the relative frequencies *nearer* to the true values.

47 Peirce is interpreted precisely in this spirit by Scheffler (1967), p. 11.

48 See *CP*, 2.66 and Popper (1957b). The differences in Peirce's and Popper's theories of probability have been discussed by Settle (1974).

49 See Good (1965) and Hintikka (1971).

50 In view of the unsatisfactory character of Popper's definition of verisimilitude (cf. Miller, 1974, 1975), the ground for this claim is rather difficult to evaluate.

BIBLIOGRAPHY

Achinstein, P., *Law and Explanation,* Oxford University Press, Oxford, 1971.

Ackerman, R., *Belief and Knowledge,* Anchor Books, New York, 1972.

Agassi, J., 'The Confusion Between Science and Technology in the Standard Philosophies of Science', *Technology and Culture* **7** (1966), 348–366.

Agassi, J., 'Postscript: On the Futility of Fighting Philistines', *Philosophia* **4** (1974), 163–201.

Agassi, J., *Science in Flux,* D. Reidel, Dordrecht, 1975.

Apel, K.-O., *Transformation der Philosophie,* Bd II, Surkamp, Frankfurt a.M., 1973.

Apel, K.-O., 'From Kant to Peirce: The Semiotical Transformation of Transcendental Logic', in L.W. Beck (ed.), *Kant's Theory of Knowledge,* D. Reidel, Dordrecht-Holland, 1974, pp. 23–37.

Butts, R. (ed.), *William Whewell's Theory of Scientific Method,* University of Pittsburgh Press, Pittsburgh, 1968.

Butts, R., 'Whewell on Newton's Rules of Philosophizing', in R. Butts and J.W. Davis (eds.), *The Methodological Heritage of Newton,* University of Toronto Press, Toronto, 1970, pp. 132–149.

Butts, R., 'Whewell's Logic of Induction', in Giere and Westfall (1973), pp. 53–85.

Butts, R., 'Consilience of Inductions and the Problem of Conceptual Change in Science', in R.G. Colodny (ed.), *Logic, Laws, and Life: Some Philosophical Complications,* University of Pittsburg Press, Pittsburgh, 1977, pp. 71–88.

Chalmers, A.F., 'On Learning from Our Mistakes', *British Journal for the Philosophy of Science* **24** (1973), 164–173.

Ducasse, C.J., 'Whewell's Philosophy of Scientific Discovery', *The Philosophical Review* **60** (1951), 56–69, 213–234. Reprinted in R.M. Blake, C.J. Ducasse, and E.H. Madden, *Theories of Scientific Method: The Renaissance Through the Nineteenth Century,* University of Washington Press, Seattle, 1960, pp. 183–217.

Duhem, P., *The Aim and Structure of Physical Theory* (trans. by Ph. Wiener). Princeton University Press, Princeton, 1954. Originally published in 1906.

Engels, F., *Ludwig Feuerbach and the Outcome of Classical German Philosophy,* New York, 1934. Originally published in 1888.

Finetti, B. de, 'Foresight: Its Logical Laws, Its Subjective Sources', in H.E. Kyburg and H. Smokler (eds.), *Studies in Subjective Probability,* John Wiley and Sons, New York, 1964. Originally published in 1937.

Freeman, E. and Skolimowski, H., 'The Search for Objectivity in Peirce and Popper', in Schilpp (1974), pp. 464–519.

Giere, R., 'Popper and the Non-Bayesian Tradition: Comments on Richard Jeffrey', Synthese **30** (1975), 119–132.

Giere, R. and Westfall, R. (eds.), *Foundations of Scientific Method: the Nineteenth Century,* Indiana University Press, Bloomington, 1973.

Good, I.J., *The Estimation of Probabilities: An Essay on Modern Bayesian Methods,* MIT Press, Cambridge, Mass., 1965.

Habermas, J., 'Towards a Theory of Communicative Competence', in H.P. Dreitzel (ed.) *Recent Sociology 2. Patterns of Communicative Behavior,* London, 1970, pp. 115–148.

Habermas, J., 'Wahrheitstheorien', in H. Fahrenbach (ed.), *Wirklichkeit und Reflexion,* Pfullingen, 1973, pp. 211–265.

Hacking, I., *The Emergence of Probability,* Cambridge University Press, Cambridge, 1975.

Hacking, I., 'Lakatos's Philosophy of Science', in I. Hacking (ed.), *Scientific Revolution,* Oxford University Press, Oxford, 1981, pp. 128–143.

Hanson, N.R., 'Is There a Logic of Discovery?', in H. Feigl and G. Maxwell (eds.), *Current Issues in the Philosophy of Science,* Holt, Rinehart and Winston, New York, 1961, pp. 20–35.

Hilpinen, R., 'Knowing that One Knows and the Classical Definition of Knowledge', Synthese **21** (1970), 109–132.

Hintikka, J., '»Knowing that One Knows» Reviewed', *Synthese* **21** (1970), 141–162.

Hintikka, J., 'Unknown Probabilities, Bayesianism, and de Finetti's Representation Theorem', in R. Buck and R.S. Cohen (eds.), *Boston Studies in the Philosophy of Science,* vol. **8**, D. Reidel, Dordrecht-Holland, 1971, pp. 325–341.

Hintikka, J., *Logic, Language-Games, and Information: Kantian Themes in the Philosophy of Logic,* Oxford University Press, Oxford, 1973.

Hintikka, J., *Knowledge and the Known,* D. Reidel, Dordrecht-Holland, 1974.

Hintikka, J. and Remes, U., *The Method of Analysis: Its Geometrical Origin and Its General Significance,* D. Reidel, Dordrecht-Holland, 1974.

Hull, D., 'Darwin's Debt to Philosophy: An Examination of the Influence of the Philosophical Ideas of John F.W. Herschel and William Whewell on the Development of Charles Darwin's Theory of Evolution', *Studies in the History and Philosophy of Science* **6** (1975), 159–181.

Hull, D., 'William Whewell and the Argument from Design', *Monist* **60** (1977), 244–268.

Itkonen, E., *Linguistics and Metascience* (Studia Philosophica Turkuensis, Fasc. II), Kokemäki, 1974.

Kockelmans, J. (ed.), *Philosophy of Science: The Historical Background,* The Free Press, New York, 1968.

Kuhn, T., *The Structure of Scientific Revolutions,* 2nd ed., The University of Chicago Press, 1970. (a)

Kuhn, T., 'Logic of Discovery or Psychology of Research', in Lakatos and Musgrave (1970), pp. 1–23. (b)

Lakatos, I., 'Falsification and the Methodology of Scientific Research Programmes', in Lakatos and Musgrave (1970), pp. 91–195.

Lakatos, I. and Musgrave, A. (eds.), *Criticism and the Growth of Knowledge*, Cambridge University Press, Cambridge, 1970.

Laudan, L., 'The Clock Metaphor and Probabilism: The Impact of Descartes on English Methodological Thought, 1650–65', *Annals of Science* **22** (1966), 74–104.

Laudan, L., 'Theories of Scientific Method from Plato of Mach: A Bibliographical Review', *History of Science* **7** (1968), 1–63.

Laudan, L., 'Thomas Reid and the Newtonian Turn of British Methodological Thought', in R. Butts and J. Davis (eds.), *The Methodological Heritage of Newton,* University of Toronto Press, Toronto, (1970), pp. 103–113.

Laudan, L., 'William Whewell on the Consilience of Inductions', *The Monist* **55** (1971), 368–391.

Laudan, L., 'Peirce and the Trivialization of the Self-Correcting Thesis', in Giere and Westfall (1973), pp. 275–306. (a)

Laudan, L., 'Induction and Probability in the Nineteenth Century', in P. Suppes *et al.* (eds.), *Logic, Methodology, and Philosophy of Science IV,* North-Holland, Amsterdam, 1973, pp. 429–438. (b)

Laudan, L., 'Why Was the Logic of Discovery Abandoned?', in Nickles (1980a), pp. 173–184.

Lenin, V.I., *Materialism and Empiriocriticism,* International Press, New York, 1927. Originally published in 1909.

Levi, I., 'Information and Inference', Synthese **17** (1967), 369–391. (a)

Levi, I., *Gambling with Truth: An Essay on Induction and the Aims of Science,* Alfred A. Knopf, New York, 1967. (b)

Levi, I., 'Certainty, Probability and the Correction of Evidence', *Nous* **5** (1971), 299–312.

Levi, I., *The Enterprise of Knowledge: An Essay on Knowledge, Credal Probability, and Chance,* The MIT Press, Cambridge, Mass., 1980.

Losee, J., *A Historical Introduction to the Philosophy of Science,* Oxford University Press, Oxford, 1972.

Mach, E., *Knowledge and Error: Sketches on the Psychology of Enquiry*, D. Reidel, Dordrecht, 1976.

Medawar, P., 'Hypothesis and Imagination', in Schilpp (1974), pp. 274–291. (Reprinted from *The Art of the Soluble,* Methuen, London 1971.)

Miller, D., 'Popper's Qualitative Theory of Verisimilitude', *British Journal for the Philosophy of Science* **25** (1974), pp. 166–177.

Miller, D., 'The Accuracy of Predictions', *Synthese* **30** (1975), 159–191.

Nickles, T., (ed.), *Scientific Discovery, Logic, and Rationality,* D. Reidel, Dordrecht, 1980. (a)
Nickles, T. (ed.), *Scientific Discovery: Case Studies,* D. Reidel, Dordrecht, 1980. (b)
Niiniluoto, I., Review of Alex C. Michalos: The Popper-Carnap Controversy, *Synthese* **25** (1973), 417–436.
Niiniluoto, I. and Tuomela, R., *Theoretical Concepts and Hypothetico-Inductive Inference,* D. Reidel, Dordrecht-Holland, 1973.
Niiniluoto, I., 'On the Truthlikeness of Generalizations', in R.E. Butts and J. Hintikka (eds.), *Logic, Methodology, and Philosophy of Science V,* Part Three, D. Reidel, Dordrecht and Boston, 1977.
Niiniluoto, I., 'What Shall We Do With Verisimilitude?', *Philosophy of Science* **49** (1982), 181–197.
Peirce, C.S., *Collected Papers:* vols. I–IV ed. by C. Hartshorne and P. Weiss, 1931–35, vols. VII–VIII ed. by A. Burks, 1958. Harvard University Press, Cambridge, Mass. (Referred to as *CP.*)
Popper, K.R., *Logik der Forschung,* Julius Springer Verlag, Vienna, 1934.
Popper, K.R., *The Open Society and Its Enemies,* I–II, Routledge and Kegan Paul, London, 1945.
Popper, K.R., *The Poverty of Historicism,* Routledge and Kegan Paul, London, 1957. (a)
Popper, K.R., 'The Propensity Interpretation of the Calculus of Probability, and the Quantum Theory', in S. Körner (ed.), *Observation and Interpretation,* Butterworths Scientific Publishers, London, 1957, pp. 65–70, 83–89. (b)
Popper, K.R., *The Logic of Scientific Discovery,* Hutchinson, London, 1959. (Referred to as *LSD.*)
Popper, K.R., *Conjectures and Refutations: The Growth of Scientific Knowledge,* Routledge & Kegan Paul, London, 1963.
Popper, K.R., *Objective Knowledge: An Evolutionary Approach,* Oxford University Press, Oxford, 1973.
Popper, K.R., 'Autobiography of Karl Popper', in Schilpp (1974), pp. 1–181. (a)
Popper, K.R., 'Replies to my Critics', in Schilpp (1974), pp. 959–1197. (b)
Ruse, M., *The Darwinian Revolution: Science Red in Tooth and Claw,* The University of Chicago Press, Chicago, 1979.
Salmon, W., *The Foundations of Scientific Inference,* University of Pittsburgh Press, Pittsburgh, 1966.
Salmon, W., 'Rational Prediction', *The British Journal for the Philosophy of Science* **32** (1981), 115–125.
Scheffler, I., *Science and Subjectivity,* Bobbs-Merrill, Indianapolis 1967.
Schilpp, P. (ed.), *The Philosophy of Karl Popper,* I–II, Open Court, La Salle, Ill., 1974.
Settle, T., 'Induction and Probability Unfused', in Schilpp (1974), pp. 697–749.
Stegmüller, W., *Personnelle und Statistische Wahrscheinlichkeit. Probleme und Resultate der Wissenschaftstheorie und Analytischen Philosophie, Band IV, Teil* III, Springer-Verlag, Berlin, Heidelberg, and New York, 1973.
Wennerberg, H., *Pragmatismen,* Rabén & Sjögren, Stockholm, 1966.
Whewell, W., *The Philosophy of Inductive Sciences, Founded upon Their History,* 2nd ed., John W. Parker and Sons, London, 1847. Reprinted in the *Collected Works,* 1967.
Whewell, W., *On the Philosophy of Discovery, Chapters Historical and Critical,* John W. Parker and Sons, London, 1860. Reprinted by Lenox Hill Pub. & Dist. Co. (Burt Franklin), New York, 1971.
Wright, G.H. von, *The Logical Problem of Induction,* 2nd rev. ed., Basil Blackwell, Oxford, 1957.
Wright, G.H. von, *Explanation and Understanding,* Cornell University Press, New York, 1971.

CHAPTER 4

THE EVOLUTION OF KNOWLEDGE

»The evolutionary theory in general throws great light upon history and especially upon the history of science – both its public history and the account of its development in an individual intellect. As great a light is thrown upon the theory of evolution in general by the evolution of history, especially that of science – whether public or private.»

(Charles Peirce, 1896.)

1. DIACHRONIC THEORIES OF KNOWLEDGE

The title of this paper covers a very wide area of logical, philosophical, historical, biological, and psychological studies where the phenomenon of human knowledge is viewed from a dynamic perspective. I shall restrict my attention mainly on the question concerning scientific knowledge and its development. In particular, I shall discuss attempts by some philosophers to analyse the growth of scientific knowledge by using ideas and concepts from the biological theories of evolution.

The study of the growth of knowledge has been a central theme within the analytical philosophy of science during the latest two decades.[1] However, the emergence of this »new philosophy of science» should not lead one to the mistaken view that the »diachronic» theory of science – as contrasted to the »synchronic» approach of the logical empiricists in the thirties – was invented by Hanson, Kuhn, Toulmin, and Feyerabend in the late fifties, or possibly by Popper in 1934.[2]

In fact, descriptions of scientific change were given already by such philosophers as Bacon (1620) who realized that science is a growing enterprise which systematically aims at finding new knowledge. The general idea of progress in the eighteenth century Enlightenment (Fontenelle, Condorcet), the Romantic emphasis on historical growth and development (Herder), and the idea of evolution in nature (Buffon, Erasmus Darwin, Lamarck) had a profound influence on nineteenth century thinking, both in natural science (Hutton, Lyell, Charles Darwin), in the cultural sciences, and in philosophy.[3] As a result, some of the major epistemological trends in that age (Hegel, Marx, Engels, Spencer, pragmatists) were explicitly concerned with human knowledge as a »process», i.e., as an essentially dynamic, temporally developing phenomenon.[4] This general insight was built into several different metaphysical systems which tried to combine the evolution of knowledge with a more general vision about finalistic patterns of development of the whole reality (Hegel, Spencer, Fiske, Peirce, Bergson, Whitehead). In addition to this speculative tendency, philosophers like Whewell, Peirce, and Duhem gave detailed and sophisticated analyses of some central aspects of scientific change.[5]

The diachronic study of knowledge was thus influenced by the theory of evolution already in the nineteenth century. More recently, Darwinism has provided a source of inspiration especially for Popper's and Toulmin's views on scientific change.[6] Before considering the discussion about evolutionary epistemology by Peirce, Popper, and Toulmin (Sections 4–6), I wish to make a few more general remarks on the influence of biological models and theories to the study of science (Sections 2–3).

2. Biologism in Epistemology

One may distinguish two different kinds of applications of biological theories to epistemology: literal and analogical.

In the *literal* applications, knowledge formation is viewed as a process in which human beings acquire information about nature, i.e., as part of the activities by means of which *Homo sapiens* adapts itself to its environment. Knowledge – ranging from the most primitive inborn dispositions to the highest results of science – is a product of the general process of human evolution, and therefore it belongs to the range of the applications of the biological theory of evolution. In this literal sense, then, scientific knowledge formation *is* an evolutionary process.

There is no doubt that originally science began as a continuation to the adaptation of man to nature, and that the idea of studying science from the evolutionary perspective is worth developing. On the other hand, it is very problematic to claim that contemporary science is still a part of *biological* evolution: science is a complex social and cultural phenomenon, so that the production and cultivation of scientific knowledge is a central part of *cultural* evolution. The view that cultural evolution is reducible to biological evolution may be called *biologism.* As the controversies over Social Darwinism and over the recent sociobiology[7] show, biologism as a general doctrine is open to serious objections. Therefore, in particular, biologism in epistemology is also questionable.

An overemphasis of the evolutionary perspective in the analysis of scientific knowledge formation leads one easily to a position which is known as *instrumentalism* in philosophy of science – as is shown by the views of James, Dewey, and Mach.[8] Generally speaking, instrumentalism is the doctrine that the value of scientific knowledge is to be measured by its »instrumental» or »practical» utility for the survival of the human species. According to this view, science is nothing but a tool for human action, so that the question whether scientific theories are true (in the sense that they correspond to reality) does not arise or does not even make sense. In this respect, instrumentalism is opposed by *scientific realists* who argue that the basic aim of science is to find informative truths about reality and in this way to contribute to the formation of a scientific outlook of the world.

Another way of applying biological concepts and models to epistemology is *analogical:* it is only claimed that certain structural patterns exhibited by scientific change are similar to certain aspects of biological evolution. This view is free from biologism, and as such it is neutral in the instrumentalism-realism issue. In the next sections, we shall discuss ideas which exemplify such analogical applications.

3. Some Biological Models for the Growth of Knowledge

Perhaps the most visible feature of scientific change is its *exponential growth.*[9] This is analogous to the growth of biological populations, the only difference being that the doubling time of scientific output is between 10 and 15 years, while that of the human population on earth is »only» 35 years.

For many natural processes, exponential growth is in fact the beginning of the so-called *logistic curve.* Price (1963) illustrates this with the growth in length of a beanstalk, and suggests as his »second basic law of the analysis of science» that »all the apparently exponential laws of growth must ultimately be logistic».

The *tree of knowledge* is a metaphor which is often mentioned. A tree-like structure which grows up from a common stem into more and more branches can be used to represent the biological evolution of species and some aspects of cultural evolution (e.g. development of languages). In connection with science, this model is best applicable to the development of scientific disciplines (i.e., the emergence of special sciences or »branches» of science).

The situation changes – or turns upside down – if the development of science is considered on the level of theories and their content. As Whewell noted in 1840 – essentially relying on the Baconian idea of scientific knowledge as a pyramid – the development of physics and optics can be represented by tree-like *inductive tables* which »converge to unity» and thus grow downwards, toward a common stem.[10]

The structure of Whewell's »inductive tables» is closely similar to the *theory-nets* that Sneed and Stegmüller have used to represent Kuhnian normal science. In this account, »theory-evolution» means essentially that a theory-net grows by generating longer and longer branches or entirely new branches – which correspond to new applications of (the specializations of) the basic theoretical assumptions.[11]

4. Peirce on Evolution in Science

Peirce's philosophy – both his epistemology and »cosmology» – is full of evolutionary ideas.[12] Here we consider only the comparison of the rival theories of evolution and the development of science that he makes in his manuscript 'Lessons from the History of Science' (1896).[13]

Peirce notes first that the »main theories of the evolution of organic species are three»: 1) Darwin's theory (evolution by »successive purely fortuitous and insensible variations in reproduction»), 2) Lamarck's theory (evolution by »a succession of very minute changes» which are not fortuitous but »wholly the result of strivings of the individuals»), and 3) the theory of »cataclysmal evolution» (changes which take place chiefly in reproduction, but are neither small nor fortuitous, and are due to »sudden changes of the environment»). Peirce regards it probable that »all three of these modes of evolution have acted».

As an illustration of the »parallels» of these modes in »other departments of evolution», Peirce discusses the evolution of standards of weights and measures. If a prototype of one pound is copied from time to time so that »the measure of each copy is imperceptibly larger or smaller than its immediate prototype», and if too small copies are destroyed, the average of these standards »will slowly grow larger by Darwinian evolution». A gradual lightening of the pound by Lamarckian evolution might take place, if the owners of pounds were disposed to smooth their surface so that they become lighter. Finally, new conditions for trade may lead to a reformation of the system of measures.

Peirce then considers the development of science in the light of these three modes.

> »In the evolution of science, a Darwinian mode of evolution might, for example, consist in this, that at every recall of a judgment to the mind – say, for example, a judgment in regard to some such delicate question as the marriage of the clergy – a slight fortuitous modification of the judgment might take place; the modified judgment would cause a corresponding modification of the belief-habit, so that the next recall would be influenced by this fortuitous modification, though it would depart more or less from it by a new fortuitous modification. If, however, by such summation of modifications an opinion quite untenable were reached, it would either be violently changed or would be associationally weak and not apt to be recalled. The effect of this would be in the long run that belief would move away from such untenable positions. It is possible that such a mode of influence may affect our instinctive feelings; but there can be nothing of this sort in science, which is controlled and exact. But another sort of Darwinian evolution undoubtedly does take place. We are studying over phenomena of which we have been unable to acquire any satisfactory account. Various tentative explanations recur to our minds from time to time, and at each occurrence are modified by omission, insertion, or change in the point of view, in an almost fortuitous way. Finally, one of these takes such an aspect that we are led to dismiss it as impossible. Then, all the energy of thought which had previously gone to the consideration of that becomes distributed among the other explanations, until finally one of them becomes greatly strengthened in our minds.»[14]

Lamarckian evolution is more significant than Darwinian:

> »Lamarckian evolution might, for example, take the form of perpetually modifying our opinion in the effort to make that opinion represent the known facts as more and more observations came to be collected. This is all the time going on

in regard, for example, to our estimate of the danger of infection of phthisis. Yet, after all, it does not play a prominent part in the evolution of science. The physical journals – say, for example, Poggendorff's [*Annalen der Physik*] and *Beiblätter* – publish each month a great number of new researches. Each of these is a distinct contribution to science. It represents some good, solid, well-trained labor of observation and inference. But as modifying what is already known, the average effect of the ordinary research may be said to be insignificant. Nevertheless, as these modifications are not fortuitous but are for the most part movements toward the truth – could they be rightly understood, all of them would be so – there is no doubt that from decade to decade, even without any splendid discoveries or great studies, science would advance very perceptibly. We see that it is so in branches of physics which remain for a long time without any decisive conquests. It was so, for example, in regard to the classification of the chemical elements in the lapse of time from Berzelius to Mendeléeff, as the valuable history of Venable shows. This is an evolution of the Lamarckian type.»

However, the »usual process of scientific evolution» is cataclysmal, »not by insensible steps»:

»But this is not the way in which science mainly progresses. It advances by leaps; and the impulse for each leap is either some new observational resource, or some novel way of reasoning about the observations. Such novel way of reasoning might, perhaps, be considered as a new observational means, since it draws attention to relations between facts which would previously have been passed by unperceived.»

Peirce's discussion is interesting especially for the reason that he makes an explicit distinction between genuinely *evolutionary* (Darwin, Lamarck) and *revolutionary* (cataclysmal) changes in science. It is related to more recent discussions: Lamarckian evolution, as described by Peirce, might constitute a part of Kuhn's normal science, while the cataclysmal evolution, with its emphasis on new ways of perceiving old facts, correspond to Kuhn's »scientific revolutions». But it also reflects the controversy between *uniformitarians* (Hutton, Lyell, Darwin) and *catastrophists* (Cuvier, Agassiz) in natural history, including geology and biology. In the field of intellectual history, Peirce is mainly on the side of the catastrophists against the uniformitarian evolutionists.

Toulmin argues that these two approaches to natural history gradually converged towards each other: the uniformitarians had to assume ever more »catastrophic» natural causes, and the catastrophists had to introduce more and more small-scale »micro-catastrophes». The same development, Toulmin (1972) suggests, has taken place in the accounts of scientific change, with the result that it has become difficult to distinguish normal from revolutionary change. Toulmin himself is a consistent neo-Darwinian and evolutionist in his theory of science, while Popper tries to combine Darwinism with a revolutionist view of scientific change.

5. Popper on the Darwinian Selection of Theories

The idea that competing scientific theories have to struggle for survival in a selection process is an old one. The Darwinist T.H. Huxley expressed it in 1880 a follows:

»The struggle for existence holds as much in the intellectual as in the physical world. A theory is a species of thinking, and its right to exist is coextensive with its power of resisting extinction by its rivals.»[15]

This idea is in harmony with the hypothetico-deductive view of science which gradually gained popularity in the nineteenth century, especially through Whewell's masterful formulation in 1840. According to this view, science proceeds by a trial-and-error method: a hypothesis is proposed to account for known facts, this hypothesis is then tested by deducing consequences from it, and those hypotheses which do not bear the test will be rejected.[16] The idea that the method of science is structurally similar to Darwinian selection – either natural or artificial – is thus very plausible for those who favour the hypothetico-deductive view of science.[17]

Popper, who defends the method of hypothesis against all forms of inductivism in *Logik der Forschung,* also notes this analogy:[18]

»It may now be possible for us to answer the question: How and why do we accept one theory in preference to others?

The preference is certainly not due to anything like an experiential justification of the statements composing the theory; it is not due to a logical reduction of the theory to experience. We choose the theory which best holds its own in competition with other theories; the one which, by natural selection, proves itself the fittest to survive. This will be the one which not only has hitherto stood up to the severest tests, but the one which is also testable in the most rigorous way. A theory is a tool which we test by applying it, and which we judge as to its fitness by the results of its applications.»

The same theme is elaborated in *Conjectures and Refutations* (1963), *Objective Knowledge: An Evolutionary Approach* (1972), 'Autobiography' (1974), and *The Self and Its Brain* (1977, with John Eccles). The following is a characteristic statement of his view:[19]

»The method of trial and error is not, of course, simply identical with the scientific or critical approach – with the method of conjecture and refutation. The method of trial and error is applied not only by Einstein but, in a more dogmatic fashion, by the amoeba also. The difference lies not so much in the trials as in a critical and constructive attitude towards errors; errors which the scientist consciously and cautiously tries to uncover in order to refute experimental tests which his theories and his ingenuity permit him to design.

The critical attitude may be described as the conscious attempt to make our theories, our conjectures, suffer in our stead in the struggle for the survival of the fittest. It gives us a chance to survive the elimination of an inadequate hypothesis – when a more dogmatic attitude would eliminate it by eliminating us. (There is a touching story of an Indian community which disappeared because of its belief in the holiness of life, including that of tigers.) We thus obtain the fittest theory within our reach by the elimination of those which are less fit. (By 'fitness' I do not mean merely 'usefulness' but truth; see chapters 3 and 10, below.) I do not think that this procedure is irrational or in need of any further rational justification.»

Perhaps the clearest summary of Popper's analysis of scientific progress in terms of the Darwinian theory of natural selection is contained in his Herbert Spencer Lecture of 1973.[20] Popper distinguishes three levels of adaptation: 1) genetic adaptation, 2) adaptive behavioural learning, and 3) scientific discovery. The fundamental similarity of all three levels is summarized as follows:

»On all the three levels which I am considering, the genetic, the behavioural, and the scientific levels, we are operating with inherited structures which are passed on by instruction; either through the genetic code or through tradition. On all the three levels, new structures and new instructions arise by trial changes from *within the structure:* by tentative trials which are subject to natural selection or the elimination of error.»

The main difference between these levels is in the »trial changes» or »variations» of instructions. On the genetic level, mutations are blind in two senses: they are not goal directed, and »the survival of a mutation cannot influence the further mutations». On the behavioural level, new types of behavioural patterns are goal directed, and animals can learn to avoid patterns which are not successful. On the scientific level, discoveries of new tentative theories are revolutionary and creative, and these theories can be formulated linguistically which makes them open to interpersonal criticism – and therefore they can even »die in our stead».

One important question for Popper's account is the following: Is adaptation on the scientific level really *Darwinian?* Popper's positive answer explicitly associates this issue with his anti-inductivism:

»Thus *there is no induction:* we never argue from facts to theories, unless by way of refutation or »falsification». This view of science may be described as selective, as Darwinian. By contrast, theories of method which assert that we proceed by induction or which stress *verification* (rather than *falsification*) are typically Lamarckian: they stress *instruction* by the environment rather than *selection* by the environment.»[21]

It is clear that Popper's position here has to exclude as Lamarckian *any* attempt to develop a rational »logic of discovery», not only the inductivist one (which claims that induction is a method of discovery). Popper has, indeed, claimed that there is always an irrational element in scientific discoveries,[22] but a general trend among philosophers today is to take seriously the idea that there are some rational principles for the heuristic search of scientific hypotheses. Popper himself has admitted that scientific discoveries are not random in the same sense as mutations – they are »blind» and somehow guided background knowledge.[23] In other words, there seems after all to be an important Lamarckian element in the method of science.

Another problem in Popper's account is his purely negative attitude to the testing of theories: »selection» means for him »falsification» and »elimination», but he does not have much positive to say of those theories which so far have survived in their struggle for existence. In *Objective*

Knowledge Popper suggests that such theories can be claimed to have a higher degree of verisimilitude (truthlikeness) than their rivals, but this concept is laden with many problems.[24] Occasionally, Popper suggests that »the always changing, always inconclusive results of the critical discussion constitute what may be called 'the science of the day'»,[25] but here his Darwinian analogy leads to problematic consequences. It certainly is an exceptional situation in science that only one of the rival theories remains unfalsified, just as it would be highly unusual that only one organic species survives in some ecological environment. But if »the science of the day» is allowed to contain rival – and thus contradictory – theories, we seem to end up with a position which is very close to Feyerabendian anarchism.

A general problem for the analogy between natural selection and the testing of hypothetical theories can be formulated as follows: Do theories correspond to individual organisms or to species? In the former case, the falsification of a theory in tests would correspond to the perishing of an organism; in the latter case, to the extinction of a species. Similarly, in the former case, the discovery of a new theory would correspond to the birth of an individual; in the latter case, to speciation or the emergence of a new species. As natural selection operates on the level of individuals rather than groups,[26] the proposed analogy would seem to require the acceptance of the former alternative. But if theories correspond to individual organisms, what on earth could correspond to populations or to species? And alternatively if theories are correlated with species, and if different linguistic formulations of the same theory are correlated with organisms belonging to the same species, we are faced with the problem that the »killing» of one formulation of the theory is sufficient to refute at the same time all of its other formulations.

We shall see in the next section that similar problems arise with Toulmin's evolutionary model, even though he tries to state it in more detail than Popper does.

6. Toulmin on the Evolution of Conceptual Populations

Toulmin's book *Human Understanding* (1962) is a remarkable attempt to describe the evolution of human knowledge, and to explain its rationality dynamically, by making use of the neo-Darwinian population genetics.[27] Toulmin is a sophisticated instrumentalist, but he explicitly rejects biologism. In his evolutionism, he is anti-revolutionist, and he seriously tries to avoid cultural relativism. Lakatos has characterized his project as a »Darwinian synthesis of Hegel and Wittgenstein».

In Toulmin's model, a collection of »concepts», »methods», and »fundamental aims» corresponds to a population of genotypes which constitutes a gene pool. Due to the relation between the elements of such collections, one may distinguish various »intellectual disciplines» which have their own characteristic concepts, methods, and aims. In Toulmin's schema, these disciplines are historical entities which correspond to biological species. The basic task of the model is to give an evolutionary account of the historical development of such disciplines through conceptual change. New variants of concepts, methods, and aims are generated within the discipline (»conceptual variation»), and they may win a place in science in an evaluation process which involves both intellectual factors (»intellectual selection») and professional demands (»intellectual ecology»). This duality between »the disciplinary and human aspects of intellectual change» – or between »reasons» and »causes», between »internal» and »external» history – is a counterpart of the duality between »genealogical» and »ecological» accounts of organic speciation.

One problem with Toulmin's model is the vagueness of some of his key terms, such as »concept» and »discipline». This is, in fact, an intended feature of his presentation, because he thinks that the »intellectual content of an entire science» can be represented in a logical system »only in quite exceptional circumstances».[28] It would be an interesting task to reformulate Toulmin's model by replacing his »disciplines» with historical entities which have more structure – e.g. by Lakatos's »research programmes». In other words, we could correlate organic species with research programmes, so that individual organisms would correspond to the variants of the basic theory within the programme which have been accepted by the members of the scientific community. Evolution – as a change in the constitution of the gene pool – would be correlated with the changes in the status of research programmes within the scientific community. In particular, speciation would correspond to the birth of a new programme, and extinction of species to the situation where a programme loses all its supporters.

Suppe's basic criticism against Toulmin's model is that it seems to be compatible with *any* pattern of conceptual change. Therefore, it fails to explain any particular patterns; it is »incapable of accounting for the epistemic success of science»; and it says »precious little about *rationality* in the growth of scientific *knowledge*». Similarly, Lakatos argues that in Toulmin's view all change within the scientific community is progress – which is another way of saying that there is no way of distinguishing progressive and degenerative intellectual change in Toulmin's framework.

This criticism is related to the fact that Toulmin does not have much to say about the criteria of »fitness» for conceptual variants and for scientific disciplines. In the model proposed above, fitness would mean the *acceptance* of a research programme by scientists. It is thus essentially a model for belief formation within scientific communities. But this is not the only relevant question about scientific change. At least if we are scientific realists, we are also interested in the question whether the beliefs of scientific communities are true. Thus, it would seem more natural to analyse fitness in terms of *rational acceptability*[29] or *truth* than actual acceptance. In the last section of this paper, brief comments are made on the possibility of combining evolutionary models of science with a realist view of science.[30]

7. Evolution and Truth

Kuhn has used analogies from the theory of evolution on two occasions. At the end of his *Structure of Scientific Revolutions* (pp. 170–172), Kuhn suggests that we should learn »to substitute evolution-from-what-we-do-know for evolution-toward-what-we-wish-to-know». The development of science is not evolution *towards* anything. Pre-Darwinian evolutionary theories (Lamarck, Spencer) took evolution to be a goal-directed process, but Darwin's *Origin of Species* »recognized no goal set either by God or nature». Similarly, the evolution of science should not be viewed as a progress towards »some one full, objective, true account of nature».

In his 'Postscript-1969', Kuhn says that »scientific development is, like biological, a unidirectional and irreversible process». This means that, in »an evolutionary tree» representing the development of scientific specialties (cf. Section 3), one can distinguish the earlier theories from the later ones by some objective criteria.[31]

Toulmin protests that biological evolution is not unidirectional in Kuhn's sense – this idea, he claims, belongs the pre-Darwinian »providentialist view of evolution».[32] He seems to exaggerate this point, however. There are unidirectional developmental processes which are not teleological or goaldirected, as the following, often mentioned example shows: if a person becomes better and better as a piano player, there is no reason to suppose that he or she approaches some ideal or perfect piano player as a goal.

There is a sense in which one can say that biological evolution is after all goal-directed: species adapt themselves to their environments. However, Hull argues that this indicates a difference between the growth of science and the evolution of species: while scientific theories may approximate with ever-increasing accuracy the »eternal and immutable regularities» in nature, species have to adapt themselves to »ever-changing environments», and »no sooner do they become adapted reasonably well to a particular environment, than it moves out from under them».[33] From this perspective, one might claim that the evolution of species is similar to the progress of those historical sciences which have a changing social reality as their object.

In spite of such doubts as Kuhn's, it is possible to defend the idea that scientific progress means the same as »convergence towards the truth», i.e., that science makes cognitive progress when new theories are more truthlike representations of reality than the old ones.[34] Whether science in fact makes progress in this sense becomes then a factual question. Some philosophers would claim that this is so as a matter of necessity – thus, for example, Lenin says that »the fact that we are approaching nearer to» the objective, absolute truth is »unconditional».[35] Peirce says that science is »destinied» with probability one to make progress at least in the long run: men have a natural propensity of proposing hypotheses which with high probability contain the true one, and further scientific tests should be designed so that a false hypothesis will be rejected with a high probability.[36]

Peirce's assumptions would give a probabilistic explanation to the fact that the applications of the scientific method have been cognitively successful, and the cognitive success of science would in turn be the most natural explanation of the practical success of science in the level of its applications.

Notes

Reprinted from *Evolution – Different Aspects, Common Problems* (ed. by Y. Haila *et al.*), Helsinki, 1982, with the permission of Tutkijaliitto.

1 See, for example, Popper (1959, 1963, 1972), Kuhn (1962, 1977), Lakatos (1970, 1976), Lakatos and Musgrave (1970), Toulmin (1972), Feyerabend (1975), Diederich (1974), Harré (1975), Howson (1976), Stegmüller (1976), Laudan (1977), Krajewski (1977), Rescher (1978), Niiniluoto and Tuomela (1979), and the reviews in Suppe (1977) and Böhme (1977).

2 Popper's *Logik der Forschung* appeared in 1934. Its English translation is contained in Popper (1959).

3 See, for example, Randall (1940), Toulmin and Goodfield (1965) and Toulmin (1972), Ch. 5.

4 Due to its Hegelian background, this dynamic perspective has been characteristic to dialectical materialism. Lenin formulates the idea as follows: »In the theory of knowledge, as in every other branch of science, we must think dialectically, that is, we must not regard our knowledge as ready-made and unalterable, but must determine how *knowledge* emerges from *ignorance,* how incomplete, inexact knowledge becomes more complete and more exact» (Lenin, 1927, p. 99). For attempts to analyse scientific change in the terms of dialectics, see Kedrov (1977, 1978) and Bogomolov (1977). Cf. also Krajewski (1977).

5 Cf. Chapter 3 above for a discussion on Whewell and Peirce. See also Skagestad (1979).

6 A very useful review of evolutionary epistemology, with a large bibliography, is Campbell (1974). For criticism, see Thagard (1980).

7 Sociobiology is an attempt to study the biological foundations of animal social behaviour. See Ruse (1979).

8 See Mach (1905). Skagestad (1979) argues plausibly that Peirce, the founder of American pragmatism, had some reservations for Darwin's biological theory just because he favoured the extension of the evolutionary account to cover cultural development and science but did not share the instrumentalist outlook of other pragmatists. Peirce argued that the Darwinian type of evolution does not play a very significant role in the development of culture (cf. Section 4).

9 See Price (1963) and Rescher (1978).

10 Cf. Popper's article 'Evolution and the Tree of Knowledge', in Popper (1972), and Chapter 3 above.

11 See Stegmüller (1976) and Moulines (1979).

12 For the influence of the theory of evolution to pragmatism, see Wiener (1949).

13 See Peirce (1931–35), vol. 1, 103–109.

[14] Popper compares new ideas and genetic mutations in Popper and Eccles (1977), p. 540.

[15] Quoted by Toulmin (1972), p. 319.

[16] Peirce also supported the method of hypothesis, but he supplemented it with an account of »abduction», i.e., discovery of hypotheses. In his famous papers in 1877–78, Peirce described human belief formation as a trial-and-error procedure.

[17] For references to late nineteenth century and early twentieth century discussions of this idea, see Campbell (1974).

[18] Popper (1959), p. 108; see also p. 42.

[19] Popper (1963), p. 52.

[20] See Popper (1975).

[21] Popper (1974), p. 68; see also *ibid.*, p. 134, and (1975), p. 78. Cf. the critical remarks in O'Hear (1980), pp. 171–176. Cohen (1980) points out difficulties in reconciling Popper's »third-world» and »Darwinian» epistemologies.

[22] See Popper (1959), p. 32.

[23] See Popper (1974), pp. 1061–1062.

[24] See Niiniluoto (1978) and Chapter 3 above.

[25] See Popper (1974), p. 68; cf. (1963), p. 51.

[26] Cf. Ruse (1979), pp. 13–16.

[27] See also Toulmin (1967a, 1967b). For critical evaluations of Toulmin's ideas, see Suppe (1977) and Lakatos (1978).

[28] See Toulmin (1972), p. 128. Toulmin suggests (*ibid.*, p. 127) that the title of Kuhn's book, *The Structure of Scientific Revolutions*, should have been *Revolutions in Scientific Structure*. The latter title captures nicely the essential thrust of the »structuralist» approach to theory change by Sneed and Stegmüller. In this respect, Toulmin's position is anti-structuralist.

[29] Cf. Suppe (1977), p. 680, note 196.

[30] Popper (1972) and Campbell (1974) defend this possibility.

[31] See Kuhn (1970), pp. 205–206.

[32] See Toulmin (1972), p. 323.

[33] See Hull (1979), p. 460.

[34] See Niiniluoto (1979, 1980).

[35] See Lenin (1927), p. 135.

[36] Cf. Levi (1980).

Bibliography

Bogomolov, A.S., 'Dialectical Contradiction and its Solution. Concerning the Central Problem of Dialectical Logic', in *Philosophy in the USSR: Problems of Dialectical Materialism*, Progress, Moscow, 1977, pp. 134–156.

Butts, R.E. and Hintikka, J. (eds.), *Historical and Philosophical Dimensions of Logic, Methodology and Philosophy of Science*, Reidel, Dordrecht and Boston, 1977.

Böhme, G., 'Models for the Development of Science', in I. Spiegel-Rösing and D. de Solla Price (eds.), *Science, Technology and Society*, Sage Publications, London, 1977, pp. 319–351.

Campbell, D., 'Evolutionary Epistemology', in Schilpp (1974), pp. 413–463.

Cohen, L.J., 'Some Comments on Third World Epistemology', *The British Journal for the Philosophy of Science* **31** (1980), 175–180.

Diederich, W. (ed.), *Theorien der Wissenschaftsgeschichte: Beiträge zur diachronen Wissenschaftstheorien*, Suhrkamp, Frankfurt am Main, 1974.

Feyerabend, P., *Against Method: Outline of an Anarchistic Theory of Knowledge,*

NLB, London, 1975.
Harré, R. (ed.), *Problems of Scientific Revolutions: Progress and Obstacles to Progress in the Sciences,* Clarendon Press, Oxford, 1975.
Howson, C. (ed.), *Method and Appraisal in the Physical Sciences: The Critical Background to Modern Science, 1800–1905,* Cambridge University Press, Cambridge, 1976.
Hull, D., 'Laudan's Progress and Its Problems', *Philosophy of the Social Science* **9** (1979), 457–465.
Kedrov, B.M., 'Evolution of the Concept of Matter in Science and Philosophy', in Butts and Hintikka (1977), pp. 187–208.
Kedrov, B.M., 'Lenin on the Dialectical Treatment of the History of the Natural Sciences', in M.E. Omelyanovsky (ed.), *Lenin and Modern Natural Science,* Progress, Moscow, 1978, pp. 97–133.
Krajewski, W., *Correspondence Principle and the Growth of Knowledge,* Reidel, Dordrecht, 1977.
Kuhn, T.S., *The Structure of Scientific Revolutions,* University of Chicago Press, Chicago, 1962. 2nd ed. 1970.
Kuhn, T.S., *The Essential Tension: Selected Studies in Scientific Tradition and Change,* The University of Chicago Press, Chicago, 1977.
Lakatos, I., 'Falsification and the Methodology of Scientific Research Programmes', in Lakatos and Musgrave (1970), pp. 91–196.
Lakatos, I., 'History of Science and Its Rational Reconstructions', in Howson (1976), pp. 1–39.
Lakatos, I., 'Understanding Toulmin', in I. Lakatos, *Philosophical Papers,* vol. 2, Cambridge University Press, Cambridge, 1978, pp. 224–243.
Lakatos, I. and Musgrave, A. (eds.), *Criticism and the Growth of Knowledge,* Cambridge University Press, Cambridge, 1970.
Laudan, L., *Progress and Its Problems: Towards a Theory of Scientific Growth,* Routledge and Kegan Paul, London, 1977.
Lenin, V.I., *Materialism and Empirio-Criticism,* International Publishers, New York, 1927.
Levi, I., 'Incognizables', *Synthese* **45** (1980), 413–426.
Mach, E., *Knowledge and Error: Sketches in the Psychology of Enquiry,* Reidel, Dordrecht, 1976. (Originally published in 1905.)
Moulines, C.U., 'Theory-Nets and the Evolution of Theories: The Example of Newtonian Mechanics', *Synthese* **41** (1979), 417–440.
Niiniluoto, I., 'Truthlikeness: Comments on Recent Discussion', *Synthese* **38** (1978), 281–330.
Niiniluoto, I. and Tuomela, R. (eds.), *The Logic and Epistemology of Scientific Change,* Acta Philosophica Fennica 30, North-Holland, Amsterdam, 1979.
O'Hear, A., *Karl Popper,* Routledge and Kegan Paul, London, 1980.
Peirce, C.S., *Collected Papers* (ed. by C. Hartshorne and P. Weiss), vols. 1–5, Harvard University Press, Harvard, 1931–35.
Popper, K.R., *The Poverty of Historicism,* Routledge and Kegan Paul, London, 1957.
Popper, K.R., *Conjectures and Refutations,* Routledge and Kegan Paul, London, 1963.
Popper, K.R., 'Autobiography', in Schilpp (1974), pp. 1–181.
Popper, K.R., 'Replies to My Critics', in Schilpp (1974), pp. 961–1200.
Popper, K.R., 'The Rationality of Scientific Revolutions', in Harré (1975), pp. 72–101.
Popper, K.R., *Objective Knowledge: An Evolutionary Approach,* Oxford University Press, Oxford, 1972. Revised ed. 1979.
Popper, K.R. and Eccles, J.C., *The Self and Its Brain,* Springer International, Berlin, 1977.
Price, D.J. de Solla, *Little Science, Big Science,* Columbia University Press, New York, 1963.

Randall, J.H., *The Making of the Modern Mind,* Houghton Mifflin, Cambridge, Mass., 1940.

Rescher, N., *Scientific Progress: A Philosophical Essay on the Economics of Research in Natural Science,* Blackwell, Oxford, 1978.

Ruse, M., *Sociobiology: Sense or Nonsense,* D. Reidel, Dordrecht, 1979.

Schilpp, P. (ed.), *The Philosophy of Karl Popper,* I, Open Court, LaSalle, 1974.

Skagestad, P., 'C.S. Peirce on Biological Evolution and Scientific Progress', *Synthese* **41** (1979), 85–114.

Stegmüller, W., *The Structure and Dynamics of Theories,* Springer, Berlin, 1976.

Suppe, F., 'Afterword 1977', in F. Suppe (ed.), *The Structure of Scientific Theories,* 2nd edition, The University of Illinois Press, Urbana, 1977, pp. 617–730.

Thagard, P., 'Against Evolutionary Epistemology', in P.D. Asquith and R.N. Giere (eds.), *PSA 1980,* vol. 1, Philosophy of Science Association, East Lansing, 1980, pp. 187–196.

Toulmin, S., 'Conceptual Revolutions in Science', in R.S. Cohen and M. Wartofsky (eds.), *Boston Studies in the Philosophy of Science,* vol. VII, Reidel, Dordrecht, 1967, pp. 331–355. (a)

Toulmin, S., 'The Evolutionary Development of Natural Science', *American Scientist* **55** (1967), 456–471. (b)

Toulmin, S., *Human Understanding,* vol. 1, Clarendon Press, Oxford, 1972.

Toulmin, S. and Goodfield, J., *The Discovery of Time,* Hutchinson, London, 1965.

Wiener, P.P., *Evolution and the Founders of Pragmatism,* Harvard University Press, Cambridge, Mass., 1949.

CHAPTER 5

SCIENTIFIC PROGRESS*

1. INTRODUCTION

Science is often distinguished from other human enterprises by its progressive nature. Science, it is said, has "a progressive and public character"[1], and an essential feature of scientific knowledge is its "continued growth"[2]. Just like Kant who complained that metaphysics has not yet found "the sure path of science"[3], so many philosophers have thought that science can be effectively demarcated from religion, art, philosophy and politics by reference to its method; this method is thought to be sufficient to guarantee continuous progress for science or at least to give clear criteria for recognizing 'progressive' developments within science.[4]

This view has recently been challenged by philosophers who argue that there is no deep difference between science and many fields of non-science.[5] But along with other philosophers, most of them would agree that there is some sense in which science can be said to make 'progress'. Even Feyerabend, perhaps the most radical critic of the distinction between science and non-science, goes so far as to claim that his "anarchism helps to achieve progress in any one of the senses one cares to choose".[6]

In many discussions of the development of science, the notion of 'progress' is either left unanalysed or characterized only in vague terms. Semantical, epistemological, and factual questions concerning progress are also often confused with each other. The definitions of progress which in fact have been proposed so far, while disagreeing greatly with each other, have widely different implications for our understanding of the nature of science. The study of this concept seems therefore to be a very important task for philosophers of science – and it is no wonder that increasing attention has been paid to it in recent years.

'Progress' is a normative or goal-relative – rather than purely descriptive – term: saying that the step from one theory *A* to another theory *B* constitutes progress means that *B* is an improvement of *A* in some respect. In this sense, 'progress' can be contrasted with such neutral terms as 'development', and a philosophical analysis of scientific progress is tantamount to a specification of the aims of science.

One can raise three different questions concerning scientific progress. First, what is meant by the notion of 'progress'? In what different senses can one say that the development of science is 'progressive'? Secondly, how can we identify 'progressive' developments within science? What are the most reliable indicators of scientific progress? Thirdly, has science made and will it make progress? The third question is a factual one, given some answer to the first conceptual question and to the second methodological question.

In this paper, I shall discuss and evaluate the most important types of answers which have been given to these questions. In particular, my aim is to show that it is possible to give a systematic defense of the 'realist' theory progress which claims that science makes progress so far as it succeeds in gaining true or highly truthlike information about reality. This view of progress has been the target of forceful attacks by Kuhn, Feyerabend, Stegmüller, Laudan, and many others. A recent commentator has even argued that "the rate of improvement" of "the various formal analyses of science in terms of truth, approximation to the truth, et cetera" is "diminishing rapidly".[7] This paper is an attempt to support the opposite conclusion.

2. PROGRESS AS CUMULATION

The idea of progress is of a relatively recent origin. In his classical study of this idea, J. Bury argued that the conception of progress in the historical development of mankind was, in spite of its anticipation by some medieval and renaissance thinkers, established only in the 17th and 18th century.[8] An essential ingredient of this conception was the view that scientific knowledge grows by cumulation. Associated with the methodological optimism of classical empiricists (Francis Bacon) and rationalists (Descartes) there was the view that the Scientific Method, if properly used, guarantees that science grows by accumulating reliably established truths.[9] Scientific progress, in this view, means that new truths are added to the body of accepted results of scientific inquiry.

The *accumulation-of-truths view* of progress was effectively criticized by many philosophers and scientists from the 18th century onwards,[10] and it is nowadays widely regarded as giving a naive and oversimplified picture of the development of science.[11] First, it makes unreasonable assumptions of the epistemological status of scientific knowledge: the results of factual sciences are not absolutely certain or conclusively established, but rather they are always to some degree conjectural and corrigible in principle. Secondly, it also makes unreasonable assumptions of the linear or continuous character of scientific growth: the earlier results of science are not only preserved and reorganized in the future development of science, but they may also be reinterpreted, re-evaluated and rejected by new theories and conceptual frameworks. Thirdly, it makes unreasonable assumptions concerning the truth-value of scientific statements: even the best theories in science are in some respects idealized and their claims about the reality may strictly speaking be false.

The traditional view of progress is 'realist' in the sense that it takes truth (or informative truth) to be the aim of science and claims that science makes progress in so far as it succeeds in realizing this aim. Against this sort of a *realist view*, some critics have claimed that the notion of truth need not or should not be used in the characterization of progress. In the sequel, our main interest will be in the contrast between the various non-cumulative versions of the realist view and the non-realist accounts of scientific progress.[12]

3. PROGRESS AS APPROACH TO THE TRUTH

If scientific knowledge is always corrigible, subject to further revision, and possibly false, does it still make sense to speak of scientific progress in the realist sense? An answer to this question can be based on the suggestion of the 17th century physicists Robert Boyle and Robert Hooke who compared the method of science to the rule called *regula falsi* in arithmetic.[13] This rule is a method of solving a linear equation by starting from a conjectural false guess of the solution. Boyle referred to Bacon's remark that "truth does more easily emerge out of error than confusion", and accordingly suggested that false assumptions may be useful in the search for truth in science. In the 18th century, this analogy between the methods of arithmetic and science was extended to cover the so-called iterative methods of approaching the solutions of equations by means of successive approximation. David Hartley, Joseph Priestley, and Georges LeSage argued, on the basis of this analogy, that the method of science is essentially self-corrective and brings scientific theories gradually to closer approximations to the truth.[14]

The *approach-to-the-truth view* of progress received support also from a quite different source, viz. from rationalist and idealist metaphysics. Already in the 15th century, Nicholas of Cusa argued in his doctrine of "learned ignorance" that the absolute truth about the infinite universe (or God) is beyond the reach of our finite intellect: God is "infinitely greater than anything that words can express", but

> "it is by the process of elimination and the use of negative propositions that we come nearer the truth about Him".

For example,

> "it is truer to assert that God is intelligence and life than to assert that He is earth, stone or anything material".[15]

The ideas of the infinity of the world (Spinoza[16]), of the objective reality of truth as divine thought (Leibniz[17]), and of the self-determined 'dialectical' development of Objective Spirit towards its perfection (Hegel), had an important influence on many 19th century philosophers. The British Hegelian Bradley developed, on the basis of a coherence theory, a doctrine of absolute and relative truths and of "degrees of truth"[18], while Engels and, following him, Lenin attempted to combine in their "dialectical materialism" the views that the nature is infinite and "inexhaustible", and that

> "the *limits* of approximation of our knowledge to the objective, absolute truth are historically conditional, but the existence of such truth is *unconditional*, and the fact that we are approaching nearer to it is also unconditional".[19]

Similar ideas were expressed by many French philosophers in the 19th century. Laplace contrasts the human mind (which is capable of making progress in its search for truth) with a vast intelligence "from which it will always remain infinitely removed".[20] Cournot says that, even though "we do not have the ability to attain absolute reality", we may be able to gradually move from one stage to another "in the series whose last term would be absolute reality".[21] Claude Bernard argues that scientific theories are "only partial and provisional truths" and "literally speaking" false, but nevertheless we are "constantly nearing" the "absolute truth" about "complex and infinitely varied phenomena".[22] Duhem suggests that

"to the extent that physical theory makes progress, it becomes more and more similar to a natural classification which is its ideal end".[23]

Perhaps the most important defender of the self-corrective nature of science in the 19th century was Charles Peirce who thought that truth can be defined as the *limit of inquiry*. From a historical perspective, Peirce seems to be a point where the methodological and the metaphysical traditions in the approach-to-the-truth view meet each other. This is illustrated by the fact that in his article 'Methods for Attaining Truth' (1898) on self-corrective methods Peirce refers both to iterative methods in mathematics and to Hegel (CP 5.574-604). (This is also an instance of the two different tendencies in Peirce's thinking that Goudge has called Peirce's 'naturalism' and 'transcendentalism'.) The Peircean account of scientific progress will be discussed in detail in the next section.[24]

4. TRUTH AS THE LIMIT OF INQUIRY

As we mentioned in Section 4, the first defenders of the self-corrective nature of science compared the method of hypothesis in science to the mathematical methods for solving equations.[25] The latter methods typically satisfy the following principles:

(1) *Monotone convergence*: Each step in the iteration brings us closer to the true solution.

(2) *Fixed point*: True solution is a fixed point of the iteration, i.e., if the true solution is once found, it will not be lost later.

(Here (2) follows from (1).) Moreover, they usually satisfy one of the following conditions:

(3) *Finite success*: The true solution is found in a finite number of steps.

(4) *Success in limit*: The true solution, even though it may in some cases be found in a finite number of steps, is the limit of the successive approximations obtained in the different steps in the iteration.

Assume now that analogous principles could be formulated for the scientific method – let us call them (S1)–(S4). Is there any evidence for the validity of these principles? Peirce once claimed the following[26]:

"But we hope that in the progress of science its error will indefinitely diminish, just as the error of 3.14159, the value given for π, will indefinitely diminish as the calculation is carried to more and more places of decimals." (5.565)

If this analogy were successful, scientific method would satisfy principles (S1) and (S4). However, several philosophers already in the 18th and 19th century pointed out that in fact there is no guarantee for monotone convergence (S1) in science.[27] Peirce, too, argued on many occasions against principles (S1), (S2), and (S3). He realized that convergence to the truth is "irregular" even in the case of enumerative induction (2.775). Even though the opinion of a scientist is "constantly gravitating" to a final conclusion, "the individual may not live to reach the truth" (8.12, cf. 2.654). According to Peirce, induction leads to approximate truth only "in the long run" (1.67, cf. 2.781 and 5.380), so that success is guaranteed only as an "ideal limit" of "endless investigation" (5.565). Moreover, Peirce rejects the idea that the truth, once found, could be effectively recognized as the truth (cf. (S2)).[28]

It is important to observe that the validity of (S1) is not necessary for the validity of (S4). In mathematics, converging sequences need not be monotone – it is sufficient that the upper bound of the absolute error decreases at each step and eventually tends to zero. For this reason, arguments against (S1) and (S2) do not refute (S4). In spite of his rejection of (S1)–(S3), Peirce could thus consistently remain a defender of the limit principle (S4) for scientific knowledge.

In fact, Peirce took the limit principle so seriously that he defined truth as the limit of inquiry. On Peirce's pragmatist theory of meaning, the "absolutely incognizable" reality is "absolutely inconceivable" or merely "a meaningless word" (5.310). Therefore, the reality must have the capacity of influencing our opinion. According to Peirce, "real things" cause our beliefs in such a way that, in scientific research,

"different minds may set out with most antagonist views, but the process of investigation carries them by force outside of themselves to one and the same conclusion ... The opinion which is fated to be ultimately agreed to by all who investigate, is what we mean by the truth, and the object represented in this opinion is the real." (5.407)[29]

Even if Peirce himself was convinced that "that the rule of induction will hold good in the long run may be deduced from the principle that reality is only the object of the final opinion to which sufficient investigation would lead" (2.693), he tried to substantiate his general argument for scientific progress by giving a direct proof for the self-corrective nature of induction.

Enumerative induction, as an inference from sample to a statistical generalization about a population, involves as an essential element the problem of the estimation of physical probabilities. Already James Bernoulli suggested that the convergence of observed relative frequencies to the true values of probabilities is analogous to the approximation of the true value of π (cf. above), but Leibniz was quick to observe the fallacy of this analogy: the convergence of relative frequencies does not satisfy the monotonicity principle (1).[30] But does this convergence satisfy the limit principle (4), as Peirce repeatedly claims? The answer to this question depends on the underlying theory of probability.

In the frequency theory of Venn, Peirce, von Mises, and Reichenbach, probabilities are defined as limits of relative frequencies, so that the limit principle (4) becomes valid *per definitionem* for those cases where limits in fact exist. Reichenbach, who thought that all scientific methods are ultimately reducible to enumerative induction, used this fact as the basis of his pragmatic justification of induction:

> "The rule of induction is justified as an instrument of positing because it is a method of which we know that if it is possible to make statements about the future we shall find them by means of this method."

The critics of Reichenbach have acutely observed, however, that "his justification of the principle of induction reduces to the bare tautology that if there is a limit, there is a limit".[31]

According to the propensity theory, which Peirce himself developed in the first decade of this century, probabilities are numerical tendencies which exist objectively as dispositional physical properties of chance set-ups (2.664). In this theory, the limit principle (4) is not valid any more. The strongest result which can be proved is Borel's Theorem (1909):

(5) The relative frequency of event A in a sequence of independent trials of an experiment *converges almost surely* (i.e., with probability one) to the probability of A.[32]

Another way of writing (5) is the following:

(5′) *Probable success in limit*: With probability one, the true solution is the limit of successive approximations obtained in the different steps in the iteration.

Even if this result makes success in the limit extremely probable, it is logically weaker than the limit principle (4). Thus, Peirce's paradigm example for principle (4), viz. enumerative induction, after all fails to satisfy it.[33]

It is remarkable that Peirce seems to have been aware of this difficulty:

> "I take it that anything may fairly be said to be *destined* which is sure to come about although there is no necessitating reason for it. Thus, a pair of dice, thrown often enough, will be sure to turn up sixes some time, although there is no necessity that they should. The probability that they will is 1; that is all." (4.547n)

If we assume that, instead of the problematic (S4), Peirce was attempting to defend a principle (S5) which claims that science is "destined" to converge to the truth, i.e., that *very probably* science is successful in the limit, his programme is justifiable at least in some important cases.[34] Moreover, this programme cannot be said to "trivialize" the thesis of the self-corrective nature of science – instead this charge (due to Laudan) should be directed against attempts, such as the recent *consensus theory of truth*[35], which *define* truth in terms of a limiting consensus of an ideal scientific community. The analysis of scientific progress by means of such a consensus theory would be as circular as Reichenbach's pragmatic justification of induction.

5. CAN SCIENCE GROW FOREVER?

One important aspect of the approach-to-the-truth view is the assumption that, in some sense, the world is infinite but human knowledge is finite. Therefore, it is argued, there will be no end for science: as creations of scientists, theories can never exhaust the whole universe in its all variety. A clear expression of this view was given by John Herschel in 1831.[36] Herschel reminds the reader of Newton's famous remark of himself as a boy in the shore of a wide and unexplored ocean. According to Herschel, the world is an "inexhaustible store which only awaits our continued endeavors".

Science is

> "essentially incomplete, and incapable of being fully embodied in any system, or embraced by any single mind."

Therefore,

> "in whatever state of knowledge we may conceive man to be placed, his progress towards a yet higher state need never fear a check, but most continue till the last existence of society."

More recently, similar arguments against the possibility of a "final state of knowledge" on the basis of the "infinite complexity of the world" have been presented by Popper, Bohm, and Vigier.[37]

On the other hand, there are philosophers and scientists – among them George Gore in 1878 – who have claimed that scientific inquiry will ultimately come to a terminal point, either because all scientific problems become solved or because the resources for investigation will be exhausted.[38]

William Kneale has argued that the principle of perpetual revolution in science cannot be justified by appeal to the infinite complexity of nature.[39] In other words, the infinity of nature does not preclude the possibility that science comes to an end through the acceptance of a wholly true single theory. Kneale distinguishes between three different versions of the infinity doctrine:

(i) the world contains an infinite multiplicity of particulars
(ii) the natural phenomena have an infinite variety
(iii) there is in the world an infinite number of distinguishable layers of fine structure.

Of these possibilities, (i) is irrelevant since "theories are not concerned with particulars as such", and (ii) does not rule out a final theory since e.g. Newton's law of gravitation covers "infinitely many different mechanical phenomena". Further, (iii) would in fact mean that there is no single or unitary theory about the whole world but at best an infinite conjunction of explanatory theories. In this case, Kneale argues, there would be no point in speaking about "approximations to truth" (as Bohm and Vigier want to do).[40]

Kneale's argument do not seem quite conclusive. In the first place, it is not clear that those who think that science will continue and grow forever have to assume a strong principle of the unity of science. It is an important feature of scientific progress that advances towards generality have at the same time been "steps towards simplification", as Herschel said.[41] But this tendency towards unification – "consilience of inductions", as Whewell called it[42] – may have limits which are due to the existence of different irreducible levels or 'layers' of the world. Then, instead of being interested in one single overall theory which is an "organic unity",[43] we would consider separately theories for the different 'layers' of reality – and there might be at least one 'layer' where scientific inquiry may continue forever without reaching a final state.

In the second place, it is not quite clear what logical requirements should be imposed on a theory. In the sixties and the seventies, logicians have developed different sorts of infinitary languages, such as (a) languages with arbitrary large infinite vocabularies, (b) languages which infinite conjunctions and disjunctions and with quantification over infinite sequences of variables, and (c) languages with formulas which have infinitely long nested subformulas.[44] How complex the world ever is, it seems that there is a suitable infinitary language in which we can express 'theories' about the whole world – and if no restrictions are given to the notion of a theory, these 'theories' may also be infinitely complex. But, on the other hand, as soon as restrictions are given, the possibility arises that there is no single theory of the admissible kind which would completely describe the reality.

For example, as Gödel showed in 1931, there cannot be any *recursive* axiomatization of the class of all truths about natural numbers. Nevertheless, we may say that our knowledge in arithmetic (if it has to be expressed as a recursive theory) can grow indefinitely and make progress towards the so-called complete arithmetic without ever reaching it.[45]

Kneale's other argument for the possibility of the terminal state of science is the following: Bohm and others can show at most that no theory is *known* to be final. In other words, we may find the final theory without knowing that it is final. Let us distinguish two kinds of successions of theories:

(6) $T_1, T_2, \ldots, T_{n-1}, T_n, T_n, T_n, \ldots$

where T_n is true, and

(7) $T_1, T_2, \ldots, T_{n-1}, T_n, T_{n+1}, \ldots$

where T_n is false and $T_n \neq T_{n+1}$ for all n. Then Kneale seems to be right: the 'fallibilist' view of the corrigibility of scientific knowledge is not incompatible with the possibility of theory successions of type (6). But, on the other hand, there are philosophers – I have called them "strong fallibilists" elsewhere[46] – who would claim that, as all theories are strictly speaking false, only sequences of type (7) are possible.

Sequences (6) and (7) may represent successive solutions to a single scientific problem. In the case of (6), the correct solution is found in a finite time, while in (7) the solution is at best the limit of endless investigation. For example, (7) might represent a sequence of successive and increasingly accurate point estimates of some unknown physical constant, or it might correspond to successive steps in a problem involving enumerative induction (cf. Section 4). Now, in order that science would come to an end in a finite time, it should be the case that all sequences of problem solutions are of type (6). If there is even *one* sequence of type (7), then science will continue for ever – and there is no way of ruling out this possibility. Thus one need not accept 'strong fallibilism' to defend the view that science will not reach a terminal state: it is enough that there exist problems which may require an indefinite time for their solution.

One further problem with Kneale's view is the fact that, *if* we find the 'final' true theory without knowing it, it seems that we may also lose it later. To find this theory is not the same thing as to accept it permanently.[47]

Robert Almeder has argued that

> "if the process of scientific inquiry were to continue forever, then science would ultimately terminate in the acceptance of a single theoretical framework better than all conceivable other, and ... there is some evidence in favor of the view that science will continue unto eternity but no evidence in favor of the contrary view."[48]

Almeder thinks that it is implausible to suppose that we can approach the final theory as a limit without actually accepting it already after some finite time. He says that there is "no real difference between endorsing a final theory and approaching it by a margin of error infinitely small". Almeder seems thus to think that principle (S4) or (S5), i.e., success in limit, entails (S3), i.e., finite success. Indeed, he explicitly supports this view by claiming that (5) entails (3). This argument is mistaken, since there *is* a difference – as the mathematical notion of limit shows us – between reaching a limit in a finite time and approaching a limit indefinitely, i.e., between sequences of types (6) and (7).[49]

Even if Almeder were right in thinking that a final theory will ultimately be found, his argument does not show that this theory could not be an incomplete true theory. Indeed, to claim that science, pursued indefinitely, *will* terminate in a final and irreversible answer to any question that may be asked and admits of an answer sounds like an instance of the notorious Principle of Plenitude (i.e., the thesis that all genuine possibilities will eventually be realized).

We may thus conclude that there are no strong arguments for denying the possibility that science will grow forever – or at least "till the end of society", as Herschel put it.

6. ARE THERE IDEAL CONCEPTUAL STRUCTURES AND THEORIES?

What does it mean to say that truth is the ideal *limit* of endless investigation? In the preceding section, we have spoken of theses (S1)–(S5) as if their content would be more or less unproblematic. It is by no means clear, however, what 'nearness to the truth' means in connection with scientific theories. Indeed, Quine has argued that this idea involves a false numerical analogy.[50] Moreover, Quine adds, even if we could speak of 'limits' of theories, what would guarantee the existence and the uniqueness of such limits?

If the accumulation-of-truths view were acceptable, Quine's problems could be easily solved. In speaking of 'the truth' as the aim of inquiry we would mean 'the truth, the whole truth, and nothing but the truth', i.e., the class T of all true (general) statements about the world. A true theory h (as a deductively closed set of statements) would then simply be a subset of the class T; a true theory h' would be closer to T than another true theory h if and only if h is contained in h',(See Fig. 1) or h' is more informative than h. Moreover, to say that T is the limit of the true theories $h_1, h_2, \ldots$ would mean that T is the set-theoretical union of $h_1, h_2, \ldots$.

For false theories the situation is much more problematic, however, since they always have a non-empty intersection with both T and $-T$ (the complement of T). In 1865, Bernard defined progress as follows:

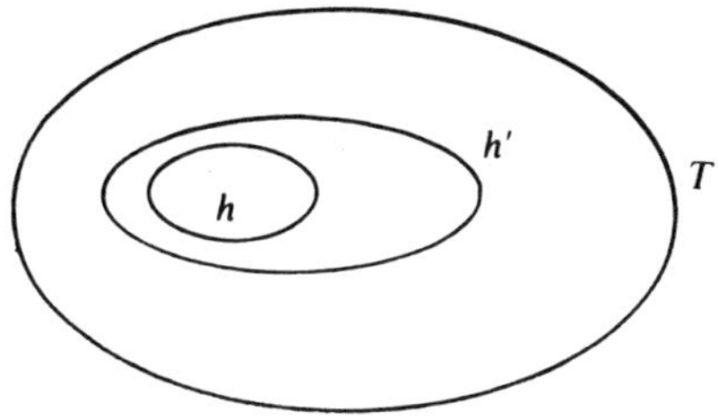

Fig. 1.

"the sum of truths grows larger in proportion as the sum of error grows less".[51] A more explicit version of essentially the same idea was given by Karl Popper in 1960: h' is *more truthlike* than h if and only if

(a) $h \cap T \subseteq h' \cap T$
(b) $h' \cap -T \subseteq h \cap -T$

where one of the inclusions (a) and (b) is strict.[52] (See Fig. 2.) However, it turned out that this definition does not work in the intended way: h' is more truthlike than h, in this sense, *only if* h' is true.[53]

Another approach to defining the notion of 'more true' is due to Wilfrid Sellars – who views this enterprise as an attempt to "revitalize central themes in nineteenth-century Idealism".[54] Sellars says that a proposition is true if it is correctly S-assertible, i.e., assertible in accordance with the semantical rules describing a conceptual structure. More precisely, a proposition p in a conceptual structure CS_i is *true in* CS_i if p is S-assertible in CS_i. Further, p in CS_i is *true quoad*

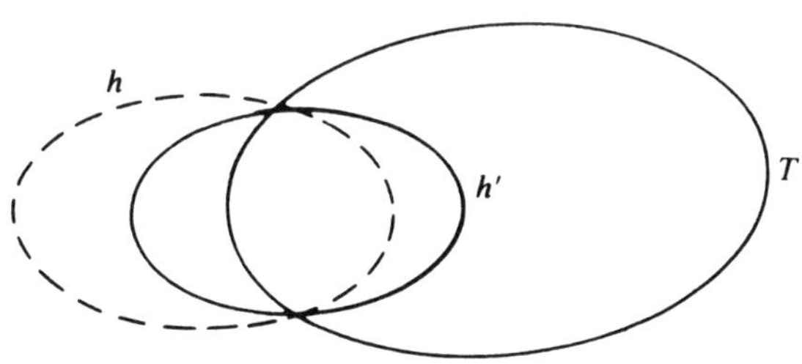

Fig. 2.

CS_j if the 'counterpart' proposition of p in CS_j is true in CS_j. For Sellars, truth as such does not admit of degrees, but still

> "one conceptual framework can be more 'adequate' than another, and this fact can be used to define a sense in which one proposition can be said to be 'more true' than another."

The adequacy of a conceptual system CS_i depends on the adequacy of the 'pictures' of genuinely extra-linguistic objects that the atomic statements of CS_i can give.

A language CSP which enables its users to form "*ideally* adequate pictures of objects" is called "Peirceish" by Sellars. Then a proposition p in CS_i (e.g., in *our* conceptual system CSO) is true *quoad* CSP if its counterpart in CSP is S-assertible by the users of CSP. This notion of ultimate truth is, for Sellars, a "regulative ideal": the ideal "Peirceish community" need not exist.

Assuming that the notion of a 'linguistic picture' is viable,[56] one may ask: When is one picture 'more adequate' than another? What is an 'ideally adequate' picture of objects? Is it possible, and on what conditions, to say that of two propositions in CSO which are false *quoad* CSP one is 'more true' than another? If Sellars does not have answers to these questions – as it seems to me – then the whole talk about the Peirceish conceptual structure does not give us much help in analysing the idea of scientific progress.

Rosenberg has tried to supplement Sellars' view by pointing out that there may be successions of theories which generate converging (in Cauchy's sense) sequences of numbers.[57] The ideal Peirceish theory could then be identified with the limit of such succession. However, it is not clear that all interesting theories are of the required kind: qualitative theories seem to be excluded in any case. Moreover, assume that a succession of theories satisfies up to the present time Cauchy's criterion for all the sequences of numbers that it generates. What guarantees that it will satisfy this condition in the future? What excludes the possibility that this sequence has two or more alternative converging continuations? And even if the existence and the uniqueness of the limit theory were guaranteed, this theory might concern only some restricted aspects of the world – unless some strong principle for the unity of science is presupposed.

Michael Hooker has also suggested that the notion of an ideal Peirceish theory is contradictory in the sense that this theory should both *be* a member of the series of theories which converges towards it (otherwise it would not be a candidate for replacing these theories) and *not be* a member of this series (otherwise it could be replaced by a better theory).[58]

To avoid the difficulties with the notion of "Peirceish", we might suggest the following two theses:

(6) There is no unique, ideally adequate conceptual framework for describing the world.

(7) The comparative notion of 'more true' can be defined relative to each choice of a conceptual framework for describing the world.

To deny thesis (6) would amount to the claim that there is an ideal language which has a privileged status in the description of the reality. This seems to be at least part of the view that Putnam calls "metaphysical realism" as opposed to "internal realism".[59] However, one can argue against such metaphysical realism that the world is not carved up into 'pieces' and 'facts' without human conceptualizing activity. THE WORLD *exists* independently of any conceptualization (to claim otherwise would lead us to linguistic idealism), but it can be approached and described only *via* different conceptual frameworks.

THE WORLD conceptualized by language L will be called the *L-world.* More precisely, let L be a language with vocabulary λ, and let $\mathcal{M}_L$ be the class of L-structures (in the ordinary model-theoretical sense). Thus, $\mathcal{M}_L$ is the class of those set-theoretical structures which contain interpretations for the terms in λ. (Hence, one may speak of the truth of sentences of L in members of $\mathcal{M}_L$ in Tarski's sense.) If L is a semantically determinate language which does not contain vague or 'fuzzy' terms, then there is a *unique* member W_L of $\mathcal{M}_L$ which represents THE WORLD as far it can be represented in L.

Each L-world W_L is, in an obvious sense, a fragment of the reality and can be the object of scientific investigation. Moreover, the L-worlds and THE WORLD constitute a partially ordered structure where L_j-world is higher than L_i-world if language L_j is an extension of language L_i (see Fig. 3.).

The theses (6) and (7) together suggest that one can make pictures only of the L-worlds but not of THE WORLD. To think that THE WORLD looks like something, and that one can make pictures of it, already involves a presupposed way of composing this world into facts. On the other hand, more or less adequate pictures can be made

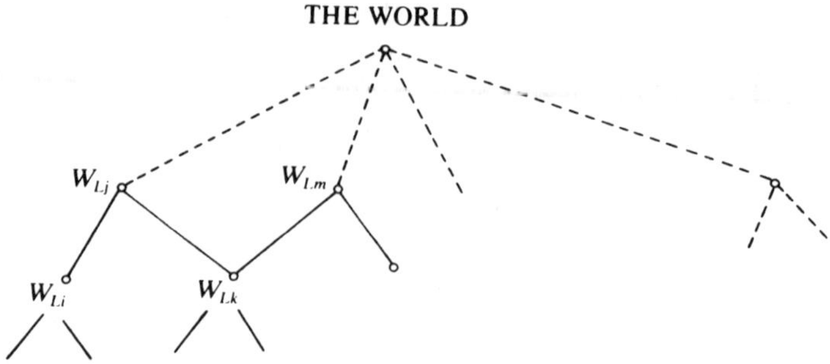

Fig. 3.

of the L-worlds. But as the L-worlds are fragments of THE WORLD, pictures of the former can, in a derivative sense, be said to be pictures of the latter.

It is worth pointing out that, independently of the validity of (6), our position seems to be better off than the Sellarsian view. Namely, if (6) is valid, then Sellars' treatment is inadequate; and if (6) is not valid, i.e., if the Peircean framework CSP exists, then we may utilize our solution of (7) and define the notion of truthlikeness *relative to CSP*.

It is important to add that thesis (6) is compatible with the view that, for inquiries which have more restricted or 'local' aims than the description of the whole truth,[60] there may exist languages which include all (or almost all) relevant factors:

(8) For each scientific problem or problem area, there exists an ideal or practically ideal language.

The aim of a whole field of science (e.g., genetics) may be to investigate certain objects (e.g., animals) from a certain viewpoint. In this case, the ideal conceptual structure L for this field should include all those features of the objects under study which are relevant for this general aim.[61] The task of this field would not, then, be the description of THE WORLD but rather of the L-world W_L. In practice, we do not have absolute criteria for choosing L, so that the study of the L-world W_L starts with some language L' in which we attempt to include at least what we expect to be the most relevant features of the objects. This method of *idealization* is then complemented by a process of 'concretization' which means that new factors are added to the language L' – with the aim of reaching eventually the most adequate language L and the most truthlike theory in L.[62]

In the next section, we shall outline an approach to 'truthlikeness' which takes the theses (6), (7), and (8) as its starting point.

The following mathematical analogy may clarify the underlying idea (cf. the different branches of Fig. 3). Let us consider the sequence of natural numbers

$$1, 2, 3, \ldots, n, \ldots$$

which has the infinite number ω as its 'ideal limit'. Here ω is not a natural number any more. If we ask how close to ω two numbers m and n are, we cannot answer this by measuring the distance of m and n from ω – for the simple reason that $\omega - m = \omega - n = \omega$. But, in another sense, we can say that m is closer to ω than n if and only if $k - m \geqslant k - n$ for some (and hence all) natural number $k \geqslant m, k \geqslant n$.

7. DEGREES OF TRUTHLIKENESS

Let L be a first-order language with a finite vocabulary λ. Here λ may consist of several families of predicates (in Carnap's sense) where some of these families may be based upon quantitative concepts. Then each generalization h in L has a distributive normal form, i.e., h is logically equivalent to a finite disjunction of mutually exclusive constituents.[63] Assuming that one can define a distance $d(C_i, C_j)$ between two constituents, the distance $d(h, C_j)$ of the generalization h from C_j can be defined as a function of the distances $d(C_i, C_j)$, where C_i is in the normal form of L.[64]

For each language L, there is one and only one constituent which is (actually) true. Let us denote it by C_*. As C_* expresses the whole truth about the L-world W_L (cf. Section 6), it seems natural to say that the *degree of L-truthlikeness* of h is measured by

$$(9) \qquad M(h, C_*) = 1 - d(h, C_*).$$

(Cf. Fig. 4.) Given a suitable choice of d, this measure M has intuitively satisfactory properties. For example, C_* itself has the maximum degree of L-truthlikeness. If h' is logically stronger than h, and both h' and h are true, then $M(h, C_*) \leqslant M(h', C_*)$.

If C_* is unknown (as it in most cases is), we cannot apply formula (9). However, given some evidence e which can be used to evaluate

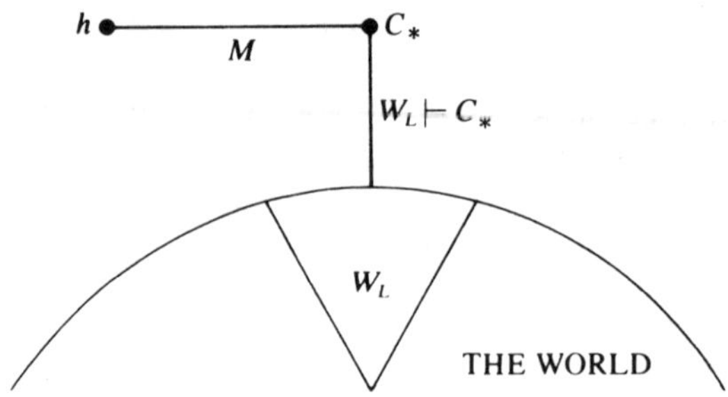

Fig. 4.

the inductive probabilities $P(C_i/e)$ of the constituents C_i, we can propose to estimate the value of $M(h, C_*)$ by the following:

(10) $$\text{ver}(h/e) = \sum_i P(C_i/e)M(h, C_i)$$

where i ranges over the indices of all constituents in L.[65] Thus, $\text{ver}(h/e)$ is the *estimated degree of L-truthlikeness* of h on the basis of evidence e.

In Hintikka's system of inductive logic, when the size n of evidence e (in L) grows without limit, there is a unique constituent C_c such that $P(C_c/e) \to 1$. This C_c is the constituent which says that the L-world is of the same kind as sample e, i.e., that those and only those Q-predicates are instantiated in the L-world which are exemplified in e. Here c indicates the number of different *kinds* of individuals in sample e. Hence,

(11) $\text{ver}(h/e) \to M(h, C_c)$, when $n \to \infty$.

Here $C_c = C_*$ if e is both true (no empty Q-predicates are claimed to be non-empty be e) and fully informative (all non-empty Q-predicates are exemplified in e). In particular,

(12) $\text{ver}(C_c/e) \to 1$, when $n \to \infty$ and c remains fixed.

On the basis of measures M and ver, we can say that a theory h' in L is *closer to the truth* than another theory h in L if $M(h', C_*) > M(h, C_*)$. Similarly, h' *seems on e to be closer to the truth* than h if $\text{ver}(h'/e) > \text{ver}(h/e)$.[66]

A sequence of theories $h_1, h_2, \ldots$ in L can be said to *converge* *towards the truth* if $M(h_i, C_*) \to 1$ when $i = 1, 2, \ldots$. Similarly, this sequence *seems to converge* towards the truth if $\text{ver}(h_i/e) \to 1$ when $i = 1, 2, \ldots$.

The notion of L-truthlikeness is relative to language in the sense that there are cases where h has greater L-truthlikeness than h' but h' has greater L'-truthlikeness than h, when L' is an extension of L.[67] We might say that h' in L has, in an absolute sense, greater truthlikeness than h in L if h' has greater L'-truthlikeness than h in all extensions L' of L.

8. IN DEFENCE OF THE REALIST THEORY OF PROGRESS

The theory of truthlikeness, which was outlined in Section 7, gives us a systematic tool for defending a realist theory of scientific progress. Let h and h' be rival theories which are concerned with the same problem area of science. Let L be the ideal language for this problem area (cf. (9)). Then h' is closer to the truth than h if and only if h' has a greater degree of L-truthlikeness than h; in this case, the step from h to h' is *progressive*.[68] In practice, L may be unknown to us, and the evaluation of h and h' has to be made by appeal to evidence e in our conceptual structure, say L'. Then we can say that, relative to L', the step from h to h' *seems progressive* on evidence e if and only if $\text{ver}(h'/e)$ is greater than $\text{ver}(h/e)$ in L'. The rational appraisal of theories is, in this sense, historical – bound to the best conceptual systems that we so far have been able to find.

One difficulty – which is related to the problem of 'incommensurability' – should be mentioned. If a language consists of a vocabulary and a set of meaning postulates, then two languages L and L' with partly overlapping vocabularies may contain contradictory sets of meaning postulates (MP and MP′, say), so that there cannot be any common extension of L and L' which preserves the meaning postulates of both languages. In this case, the following can be suggested: let h be a theory in L and h' in L', and let L'' be a common extension of L and L'. Then L'' is chosen to contain only the meaning postulates, if any, in $\text{MP} \cap \text{MP}'$, and h' is said to be more truthlike than h relative to L'' if and only if h' *together with* postulates $\text{MP}' - \text{MP}$ has a greater degree of L''-truthlikeness (relative to $\text{MP} \cap \text{MP}'$) than h *together with* postulates $\text{MP} - \text{MP}'$. In other words, we treat the specific meaning postulates of L (resp. L') as a part of the theory h (resp. h').

This account of progress seems to give an answer to many well-known arguments against the approach-to-the-truth view. For example, we have answered Laudan's challenge that

"no one has been able even to say what it would mean to be 'closer to the truth', let alone to offer criteria for determining how we could assess such proximity."[69]

Laudan adds that

"we apparently do not have any way of knowing for sure (or even with some confidence) that science is true, or probable, or that it is getting closer to the truth. Such aims are *utopian*, in the literal sense that we can never know whether they are being achieved."[70]

But these aims do not seem *more* problematic than the task of finding a correct estimate of some unknown quantity. In particular, we can never know for sure whether the task of simple enumerative induction has been achieved. Such 'utopian' tasks belong to the everyday practice of science.

Kuhn argues that we need not assume that the evolution of science is directed toward some goal:

"Does it really help to imagine that there is some one full, objective, true account of nature and that the proper measure of scientific achievement is the extent to which it brings us closer to that ultimate goal? If we can learn to substitute evolution-from-what-we-do-know for evolution-toward-what-we-wish-to-know, a number of vexing problems may vanish in the process. Somewhere in this maze, for example, must lie the problem of induction."[71]

In our account, there is no *unique* goal for science, since, for *any* L-world, knowing the whole truth about it is a legitimate goal for science. However, science is a goal-directed process: there indeed are things which we wish to know but do not yet know. But, of course, the estimates of truthlikeness, i.e., values of ver, are based on evidence which we already know. Kuhn argues further that

"there is, I think, no theory-independent way to reconstruct phrases like 'really there'; the notion of a match between the ontology of a theory and its 'real' counterpart in nature now seems to me illusive in principle."[72]

If the word 'theory-independent' is replaced by 'language-independent', this quotation seems to agree with our treatment of the difference between the L-worlds and THE WORLD (cf. (7)). But, as we have seen, the truth of (7) does not preclude the possibility, denied by Kuhn, of seeing (from our conceptual system) a "coherent direction of ontological development" in actual historical cases of theory succession. We need not assume, though, that such successions are uniformly or monotonely convergent.

Are later results of science worthier of acceptance than earlier ones because they are 'truer'? Or are later results 'truer' because they are better grounded and thus worthier of acceptance? According to Rescher's "methodological pragmatism", the latter alternative is the correct one – but 'truer' is a methodological notion equivalent to 'presumably truer'.[73] This argument leads Rescher to conclude that the progress of science should be analyzed only on the "pragmatic" level, where progress means "the increasing success of applications in problem solving and control".[74]

In view of our results – in particular, our distinction between 'true' and 'estimated' verisimilitude, i.e., between functions M and ver – Rescher's considerations do not constitute an argument against the realist view of progress. The pragmatic success of theories increases their degree of corroboration which can be shown to covary with their estimated verisimilitude or their degree of 'presumed truth'.[75] There are evidential situations e and hypotheses h such that ver(h/e) is high. In such cases, it is rational for us to claim that the unknown degree of truthlikeness $M(h, C_*)$ is also high, but this estimate may of course be wrong (and corrigible by further evidence). Thus, in addition to the methodological level of estimated progress, it is meaningful to speak of 'real' progress on the semantical level of the content of theories.

Most philosophers would agree that science makes progress in the scope of its practical applications. But why is science pragmatically successful? This problem has puzzled both Kuhn:

> "how a value-based enterprise of the sort I have described can develop as a science does, repeatedly producing powerful new techniques for prediction and control. To that question, unfortunately, I have no answer at all, but that is only another way of saying that I make no claim to have solved the problem of induction."[76]

and Laudan:

> "Until and unless we can show why science can be an effective instrument for the solution of problems, then its past success at problem solving can always be viewed as an accidental piece of good fortune which may, at any time, simply dry up."[77]

The realist theory of progress is able to provide an answer to this query: pragmatically successful theories have a high degree of estimated truthlikeness, and their continued success can be *explained* by the *hypothesis* that they in fact are close to the truth at least in the relevant respects.[78]

Isaac Levi has recently argued, in his 'Truth, Fallibility and the Growth of Knowledge', that

> "it is fallibilism together with the doctrine that getting closer to the truth is the ultimate aim of inquiry which renders knowledge irrelevant to practice and truth irrelevant to the revision of knowledge".

Against Peirce and Popper (and his own earlier views), Levi endorses the doctrine of *infallibilism* which says that what we know or believe is, from our point of view, infallibly and certainly true. He argues further that infallibilism is consistent with *corrigibilism* (i.e., the view that our knowledge is always open to revision) if we abandon the idea that all specific inquiries should promote the ultimate aim of getting closer to the truth. Thus, replacement of a theory by another which contradicts it may be rational for an infallibilist – to whom this move seems to be a deliberate replacement of a certain truth by a falsity – if he extols "the virtues of myopia" and holds that avoidance of error is an invariant feature of the *proximate* (rather than ultimate) goals of specific inquiries.

Levi's argument shows, I think, successfully that the combination

infallibilism & corrigibilism & myopic realism

is a coherent alternative to

fallibilism & corrigibilism & long-sighted realism.

But, on the other hand, I do not find his reasons for rejecting the latter alternative compelling – at least when this alternative is presented in the form that has been defended in this paper.

First, the notion of estimated verisimilitude gives us a unified way of appraising theoretical revisions of knowledge. It gives emphasis to both information *and* truth, so that "avoidance of error" is not the only relevant consideration. Indeed, according to the measures mentioned in Section 7, there are false theories which have a greater degree of truthlikeness than some true theories. Moreover, it is even possible that a step from one theory h to another h', where h and h' are incompatible, may seem progressive relative to an evidential situation which contains h. Second, Levi himself admits that the problems concerning the rationality of the contraction of knowledge systems do not arise for a fallibilist. Thirdly, I take it that decision theory – or the theory of rational decision making under *uncertainty* – is sufficient to show how fallible knowledge is relevant to the guidance of practical action. In contrast with Levi's present position, I thus admit that there is "a gulf between theory and practice", but this gulf is not wider – and not narrower – than the difference between epistemic and practical utilities.

9. SCIENTIFIC PROGRESS AND PROBLEM-SOLVING

In arguing against the approach-to-the-truth view, Kuhn characterizes his own position as follows:

> "Later scientific theories are better than earlier ones for solving puzzles in the often quite different environments to which they are applied. This is not a relativist's position, and it displays the sense in which I am a convinced believer in scientific progress."[79]

The idea that scientific progress can be defined by means of the *problem-solving ability* of theories has been further developed by Laudan in his *Progress and its Problems.* For Laudan, science is "a problem-solving rather than a truth-seeking progress"[80], and he assumes that scientific progress can be analyzed without the notions of 'truth' and 'confirmation'.[81] Laudan summarizes his definition of progress as follows:

> "progress can occur if and only if the succession of scientific theories in any domain shows an increasing degree of problem solving effectiveness",

where

> "the overall problem-solving effectiveness of a theory is determined by assessing the number and importance of the empirical problems which the theory solves and deducting therefrom the number and importance of the anomalies and conceptual problems which the theory generates."[82]

In the preceding section, we have found reasons to think that the problem-solving ability of theories is not unconnected with the truthlikeness of these theories: a problem about a domain gets a good solution (which guarantees continued problem-solving ability) only if the theory gives a sufficiently truthlike description of this domain. For this reason, we might suggest that the *value* of the solution that a theory h gives relative to a domain z is measured either by $M(h, C_z)$, where C_z is the true constituent about domain z in an appropriate language L, or by $\text{ver}(h/e)$, evaluated in L relative to our evidence e. Let t be the degree of truthlikeness of a tautology,[83] and let J be the set of all problems (domains) which h is supposed to solve. Then the problem-solving effectiveness of h relative J could be defined either by

$$(10) \qquad \sum_{z\in J} a_z(M(h, C_z) - t)$$

or by

$$(11) \qquad \sum_{z\in J} a_z(\text{ver}(h/e) - t),$$

where $a_z \geqslant 0$ is the *weight* given to problem z. (The penalties for the conceptual anomalies of h are ignored in these formulas.)[84]

This notion of problem-solving capacity can be applied also to Sneed's and Stegmüller's set-theoretical reconstruction of "Kuhn-theories" as nets of systematically related 'solutions' to theoretical problems.[85] A "theory-element" is a pair $\langle K, J\rangle$ consisting of a class of domains J and a mathematical core K; the claim associated with $\langle K, J\rangle$ is that the members of J satisfy the fundamental law of K and the "constraints" of K binding them together.[86] In this sense, the core K is supposed to solve several problems at the same and its effectiveness can be measured by (10) or (11). As a special case where J contains only one element z, the evaluation of (10) and (11) reduces to the problem of truthlikeness with respect to one domain.

NOTES

* The first version of this paper was read at the symposium on 'Aspects of Scientific Realism', organized by the Philosophical Society of Finland in Helsinki, May 17, 1979. In writing this paper I have benefited from the critical remarks of many colleagues who do not share my views. Among them I should like to mention especially Larry Laudan and Isaac Levi.

[1] See Randall and Buchler (1942), p. 57.

[2] See Popper (1963), p. 125.

[3] See the Introduction to the 2nd edition of *Kritik der reinen Vernunft* (1787).

[4] An authoritative representative of this view is George Sarton who wrote in 1936 in *The Study of the History of Science* as follows:

> "*Definition.* Science is systematized positive knowledge, or what has been taken as such at different ages and in different places.

> *Theorem.* The acquisition and systematization of positive knowledge are the only human activities which are truly cumulative and progressive.
> *Corollary.* The history of science is the only history which can illustrate the progress of mankind. In fact, progress has no definite and unquestionable meaning in other fields than the field of science." (Sarton, 1957, p. 5.)

In his Herbert Spencer Lecture of 1973, Karl Popper argued that "progress in science can be assessed rationally" and that "science seems to be the only field of human endeavour of which this can be said" (Popper, 1975, p. 83). For an attempt to defend progress as a demarcation criterion for science, see also Quay (1974). (The term 'science' is used in a wide meaning in this paper, so that it covers not only natural science but also systematic social and human sciences.)

[5] Cf. Kuhn (1970), Ch. XIII, Feyerabend (1975), p. 306, and Rorty (1978).

[6] Feyerabend (1975), p. 27.

[7] Hull (1979), p. 465. I have earlier discussed the problems of scientific progress in Niiniluoto (1978a), (1978c). Section 4 of this paper is based upon Niiniluoto (1978b).

[8] See Bury (1932). Cp. Zilsel (1945), who claims that the idea of scientific progress was created by the "artisans" of the 16th century. See also Crombie (1975). One of the early expressions of the idea that science is a cumulative enterprise was the image, originally due to Bernard of Chartes in the 12th century, of the modern scholars as dwarfs on the shoulders of giants (see Molland, 1978).

[9] Bacon compares his method to a "machine" which produces results when the "stuff" for it is "gathered from the facts of nature" (Bacon, 1960, p. 6). In the same spirit, Comte describes the cumulative ideal of science (or "philosophy" in its "positive stage") as follows: "Its character will be henceforth unchangeable, and it will then have only to develop itself indefinitely, by incorporating the constantly increasing knowledge that inevitably results from new observations or more profound meditations" (Comte, 1970, p. 13).

[10] See, for example, Laudan (1973).

[11] See, for example, Kuhn (1970) and Feyerabend (1975). In spite of these criticisms, the cumulative view may be valid in some fields (such as some branches of mathematics) or in some periods of science (such as Kuhnian normal science). Perhaps it should be noted that, in speaking about cumulation in science, we mean the *content* of the results of science (i.e., science as a cognitive enterprise), rather than simply the amount of scientific output (books, articles, journals, etc.). The latter amount has increased exponentially since the latter part of the 18th century. In his interesting discussion of the quantitative growth of science, Rescher (1978a) suggests that the number of "first-rate" results has increased linearly rather than exponentially. However, no cognitive criteria are given by him for defining what results are "first-rate". (Cf. Section 8.)

[12] It should be noted that this use of the term 'realism' in connection with scientific progress is not intended to be the same as in the realism-instrumentalism issue concerning scientific theories. (This is not sufficiently emphasized in Niiniluoto, 1978b.) A person may support a realist interpretation of theories without accepting a realist theory of progress (cf. Kuhn, Feyerabend, and also Laudan, 1980); and a person may support an instrumentalist interpretation of theories even if he supports some sort of a realist view of scientific progress (cf. Duhem). What I want to suggest with my choice

of the terminology, however, is that a consistent philosopher who defends the realist view of theories should also subscribe to a realist view of progress in science, and *vice versa*.

[13] Boyle's discussion is in *Certain Physiological Essays* of 1661 (see Boas Hall, 1966, pp. 123–124). See also Laudan (1966). Cf. Bernard (1957), p. 40.

[14] A useful account of this tradition has been given by Laudan (1973). However, there is not much evidence to show that Peirce knew Hartley's and LeSage's work on progress. It is perhaps interesting to add that Mach, in his *Erkenntnis und Irrtum* of 1905, refers to Priestley and LeSage, and further explains the idea of coming nearer to the truth by means of an example of a procedure for approximating to the roots of an equation (see Mach, 1976, pp. 176, 183–184).

[15] See the chapter on Cusanus in de Santillana (1956), pp. 47–63. For Cusanus, the truth itself is something indivisible, not capable of more or less, whereas our understanding of the truth can always be further developed. See Duhem (1969), pp. 57–58.

[16] Spinoza's doctrine on the infinite structure of God turns out to involve non-trivial set-theoretical assumptions (see Friedman, 1976).

[17] See, for example, Martin (1967), Ch. VII.

[18] See Bradley (1893). For a revival of the coherence theory of truth as a criterion of truth, see Rescher (1973).

[19] See Lenin (1927), Ch. 2.5 and 5.2.

[20] See Laplace (1951), p. 4.

[21] See Cournot (1956), pp. 15–16.

[22] See Bernard (1957), pp. 12, 31, 35–36, 40–42, 54.

[23] See Duhem (1954), p. 298. Cf. also pp. 26–28, 177, 297, 334–335.

[24] Cf. Laudan (1973), Niiniluoto (1978a), (1978b), Rescher (1977), (1978b), Skagestad (1979), and Levi (1980).

[25] For iterative methods, see Henrici (1964).

[26] 5.565 refers to paragraph 565 in the fifth volume of Peirce's *Collected Papers* (1931–35).

[27] See Laudan (1973).

[28] "There is nothing, then, to prevent our knowing outward things as they really are, and it is most likely that we do thus know them in numberless cases, although we can never be absolutely certain of doing so in any special case." (5.311)

[29] For an earlier but similar argument, to the effect that external influence on our opinions forces us to a consensus, see Cournot (1956), p. 103.

[30] See von Wright (1957), p. 225.

[31] See Reichenbach (1949), p. 475, and Lenz (1966), p. 435–440.

[32] This result says that the set of infinite sequences with inappropriate limiting behaviour has the measure zero. Borel's theorem is a special case of the Strong Law of Large Numbers. Its weaker counterpart is Bernoulli's Theorem: for any $\epsilon > 0$, $P(|rf_n(A) - P(A)| > \epsilon) \to 0$, when $n \to \infty$, where $rf_n(A)$ is the relative frequency of A in n independent trials.

[33] Isaac Levi (1980) has argued convincingly that, at least in the articles 'The Probability of Induction' and 'The Theory of Probable Inference', Peirce regarded induction as interval estimation essentially in the style of Neyman and Pearson. According to Levi, induction is self-corrective for Peirce in the following sense: either the conclusion reached via an inductive rule is correct or, if wrong, the revised estimate

emerging from a new attempt at estimation based on a different sample will with a preassigned probability k be correct. This guarantees that, if the scientist would repeat drawing samples from the same population *ad infinitum*, the relative frequency of correct inferences would converge on k in the long run; but it does not show that "an inductive rule will, in the messianic long run, reveal the true value".

If Levi is right, then Peirce did not support the limit thesis (4). But this, it seems to me, is contradicted by the evidence that Peirce repeatedly spoke about convergence *towards the truth* rather than towards some limiting *truth-frequency* among inferences – and this happens both in his general theory of reality and truth and in his theory of induction. For example, in 1902, he says:

> "The validity of induction of an inductive argument consists, then, in the fact that it pursues a method which, if duly persisted in, must, in the very nature of things, lead to a result indefinitely approximating to the truth in the long run." (2.781)

Moreover, convergence towards the truth was, for Peirce, related to the solution of specific problems or to the evaluation of single hypotheses (cf. 8.12, 5.565), and not only to the 'messianic' idea of approaching to the final and complete true theory of the world. For these reasons, Levi's observations seem to indicate that Peirce had difficulties in trying to reconcile his new theory of inductive inference with his commitment to the thesis that truth is the limit of endless scientific investigation – but they do not show that Peirce would have abandoned the thesis (4) (or its weaker variant (5)).

Michael Friedman (1979) has recently argued that, instead of long-run justifications, one should try to show that scientific method tends to produce true hypotheses "in the class of actual and physically possible inferences". This seems to be (even though Friedman does not notice it) very close to Peirce's idea that the 'probability of induction' should be defined as the proportion in which a particular method of inference produces true conclusions from true premises. According to Peirce,

> "all human certainty consists merely in our knowing that the processes by which our knowledge has been derived are such as must generally have led to true conclusions." (2.692)

Essentially the same idea is incorporated in the Neyman–Pearson theory of statistical testing: the probability of rejecting the true hypothesis is chosen to be low, and the probability of rejecting the false hypothesis should be made as high as possible.

Friedman's more specific suggestion is the following: it should be possible to derive from confirmation theory law-like "reliability statements" of the form

(A) The probability that S is true, given that S is accepted (rejected) by method M, is r,

where r is an objective physical probability (distinct from mere actual relative frequency). In order that this is possible, he further claims, the theory of truth should be supplemented with something like Putnam's causal theory of reference. This argument is not convincing, however. There is no analysis available which would allow us to assign physical probabilities to scientific hypotheses – and if r in (A) were interpreted as an epistemic probability, then all inductive rules of acceptance M which take high probability as a necessary condition for acceptance would automatically satisfy the condition: if S is accepted by M, then the probability that S is true is high.

Friedman seems to confuse (A) with a perfectly legitimate physical probability statement

(B) The probability that method M accepts (rejects) S, given that S is true, is r.

Such statements constitute the essential ingredients of the 'orthodox' theory of statistical tests and, as Levi argues, of Peirce's mature theory of induction – and these theories have been created and practiced for a long time before 'causal theories of reference' were invented.

[34] Bruno de Finetti proved in the 1930's a theorem which erroneously may seem to establish Peirce's limit thesis in its strong form. de Finetti showed that, assuming fairly general conditions of 'exchangeability', the subjective probabilities of agents starting from different non-dogmatic prior beliefs will, on the basis of common evidence, tend towards observed relative frequencies and thus also towards each other. (See de Finetti, 1964.) de Finetti interprets this result only as showing the intersubjective nature of rational beliefs, but some other theorists have suggested that his results show precisely how subjective probabilities can be used to estimate physical probabilities (see Good, 1965).

An interesting variant of de Finetti's result has been given by Savage (1954), pp. 46–50. Let $e_1, e_2, \ldots$ be the successive independent results of a random experiment, and let h be a hypothesis such that $P(e_n/h) \neq P(e_n/{\sim}h)$. Then the following holds:

(A) If $P(h) > 0$, then $P\left[\frac{P(h/e_1 \ldots e_n)}{P({\sim}h/e_1 \ldots e_n)} \xrightarrow[n\to\infty]{} \infty/h\right] = 1.$

If we define $h^* = h$ if h is true and $h^* = {\sim}h$ if h is false, then (A) entails

(B) If $P(h^*) > 0$, then $P(P(h^*/e_1 \ldots e_n) \approx 1) \xrightarrow[n\to\infty]{} 1.$

These results, which concern the approach of beliefs towards certainty, are confused with results concerning approach to the truth by Aulin-Ahmavaara (1977), pp. 210–211. He seems to think that, instead of (B), the following result is provable:

(C) If $P(h^*) > 0$, then $P(h^*/e_1 \ldots e_n) \xrightarrow[n\to\infty]{} 1.$

There is a crucial difference between (B) and (C), however.

[35] In the consensus theory, truth is defined as the limit of ideal communication (see Habermas, 1973). This definition seems to be problematic at least in the following sense: to say whether the conditions for communication really were 'ideal', we should know whether the limit result is really true (in some realist sense). Cf. also Niiniluoto (1980).

[36] See Herschel (1831), p. 360.

[37] See Popper (1963), p. 125, Bohm (1957) and Vigier (1957).

[38] Ravetz (1975) argues that, for Bacon, the significance of scientific progress lies in the fact that it helps to bring about the Millennium – progress is good evidence that the 'last days' are coming closer. For an interesting discussion of the future of scientific progress, with many bibliographical references, see Rescher (1978a), Chs. I–III. Rescher's own well-argued position is that, even though the advance of science will

decelerate, there are "unlimited horizons" for further progress in science. See also Rescher (1979).

[39] See Kneale (1967).

[40] Kneale also suggests that Bohm and Vigier confuse (i) and (ii) – or move without justification from (i) to (ii).

[41] See Herschel (1831), p. 360.

[42] See Whewell (1840). Cp. Niiniluoto (1978a).

[43] See Kneale (1967), p. 36.

[44] For infinitary languages of type (a) and (b), see Monk (1976), and of type (c), see Hintikka and Rantala (1976).

[45] Complete arithmetic is just the class of true statements about natural numbers. As it is not recursively axiomatizable, one may say – as Sellars (1968), p. 135, does – that "in arithmetic there is no end to the series of 'more adequate' axiomatic systems". But it might be equally natural to say that complete arithmetic is the ideal theory which axiomatic systems of arithmetic approximate from within (cf. Fig. 1 in Section 6).

[46] See Niiniluoto (1978a). Many philosophers (like Peirce and Popper) have combined both 'weak' and 'strong' fallibilistic elements in their epistemology.

[47] Kneale suggests that the final theory would be "irrefutable", but it is not clear why this should be the case.

[48] See Almeder (1973). See also Almeder (1975).

[49] A more charitable interpretation of Almeder's discussion of enumerative induction would be the assumption that he is speaking of more and more precise interval estimates rather than point estimates. In other words, the thesis about finite success would be the following: given $\epsilon > 0$, we shall find in a finite number of steps an interval of length less than 2ϵ which covers the unknown true value. However, while Borel's Theorem entails that this kind of success is overwhelmingly probable, it does not guarantee that it also will be achieved in all cases.

[50] See Quine (1960), p. 23.

[51] See Bernard (1957), p. 42.

[52] See Popper (1962), (1972).

[53] See Miller (1974) and Tichy (1974).

[54] See Sellars (1968), Ch. 3. It is interesting to note that both Popper's theory of verisimilitude and the approach to truthlikeness in Section 7 below are related to the coherence theory of truth (even if they both presuppose the correspondence theory in the Tarskian form): for example, Popper's definition, given Fig. 2, essentially measures the degree of 'compatibility' or 'coherence' between two sets of *sentences* (h and T). Cf. also Fig. 4 below.

[55] Sellars thinks that in the evolution of theories one can find 'counterparts' of propositions in earlier and later theories, but it is not quite clear how this notion is to be defined. Cf. Burian (1979), pp. 203–205.

[56] For a lucid analysis of this question, see Rosenberg (1974).

[57] See Rosenberg (1974), pp. 54–55, and the critical remarks in Hooker (1978).

[58] See Hooker (1978).

[59] See Putnam (1978). Putnam's internal realism combines three theses: (i) there is an epistemically ideal theory, (ii) the epistemically ideal theory coincides with the true theory, and (iii) the truth or approximate truth of theories in mature science explains the success of science.

Michael Bradie (1979), who rejects the task (iii) as unnecessary, has claimed that Putnam's position is indistinguishable from a "sophisticated pragmatism" which replaces 'truth' with 'warranted assertibility (in the long run)'. However, there is a crucial difference between pragmatists who accept (ii) as a *definition* of truth and (internal) realists who accept (ii) as a principle which gives a *de facto* valid *methodological characterization* of truth. Among the latter we may include Peirce whose fundamental point is that the ideal theory is destined to correspond with the reality just because it is in effect produced or caused by the reality. (Thus, to define 'Peircean realism' simply by thesis (i), as Putnam does, is clearly too weak to do justice to Peirce's realism.)

Moreover, Bradie's reasons for rejecting (iii) are not convincing. He claims that only the success of particular theories has to be explained, and that this is to be done by other scientific theories. But, to explain the success of a theory T_1 by another theory T_2, a *potential* explanation is not enough: T_2 has to be true or at least more truthlike than T_1 – so that Bradie's suggestion gives no escape from a realist notion of truth.

According to the view defended in Sections 4 and 5 above, thesis (ii) is justifiable only in some special cases. No non-circular epistemic characterization of the 'ideal theory' is sufficient to guarantee its truth (cf. also Koethe, 1979). For this reason, I think, both pragmatists, consensus theorists, and Putnam are wrong. In defending the view that truth is "radically non-epistemic", I thus support part of the view that Putnam calls 'metaphysical realism'.

To classify Sellarsian realism in these terms is difficult. On the one hand, 'truth' is an 'internal' notion for Sellars, since it is relative to a conceptual framework. On the other hand, Sellars has a notion of an ideal conceptual framework which transcends the internal viewpoint, since its ideality is defined in terms of correct picturing of non-linguistic objects. Thus, truth relative to this ideal framework is not characterized in purely epistemic terms.

[60] For a discussion of 'inquiries', 'problems', and 'questions' in science, see Niiniluoto (1976). I am not suggesting that the problem of finding the true constituent of some language L is the only legitimate way of defining a scientific inquiry. For other ways of setting up 'ultimate partitions' of hypotheses, see Levi (1976).

[61] I am not suggesting that we can *know* for sure that some particular language L is 'ideal' in this restricted sense. In this paper, I cannot say much of the important problem of how a conceptual framework should be chosen. What we would need is a definition for the *theoretical effectiveness* of a language. It would tell, following Whewell's principle, how simply and how effectively one can express true general sentences in the language (cf. Niiniluoto, 1978c). In this sense, the criteria for the choice of a language are partly empirical or *a posteriori*. Moreover, Goodman's problem of the simplicity of conceptual systems seems to be a part of this more general question.

[62] For an interesting analysis of the method of idealization, see Nowak (1976) and Krajewski (1977). This analysis is used for a defense of a realist view scientific progress in Nowak (1975).

[63] The theory of distributive normal forms is due to Hintikka. If L is a polyadic language, then the normal form is relative to a given 'depth' which indicates the number of layers of connected quantifiers in a statement. Cf. Niiniluoto (1978d).

[64] For details, see Niiniluoto (1977), (1978d), (1979). It seems that there are many

alternatives in defining the distance measure d, and it is not certain that the best ones have been found so far. For the argument of this paper, the most important thing is that there are at least *some* reasonable alternatives available. For example, it seems reasonable to define

$$d(h, C_*) = \gamma d_{\min}(h, C_*) + (1-\gamma) d_{\max}(h, C_*)$$

where $0 < \gamma < 1$,

$$d_{\min}(h, C_*) = \min_{i \in I_h} d(C_i, C_*)$$

$$d_{\max}(h, C_*) = \max_{i \in I_h} d(C_i, C_*)$$

and

$$\vdash h \equiv \bigvee_{i \in I_h} C_i.$$

[65] For detailed studies of (10), see Niiniluoto (1977) and (1978c). Lakatos has suggested that some weak, conjectural "metaphysical principle of induction" has to be assumed in order to connect "scientific standards with verisimilitude" (Lakatos, 1974, pp. 260–262). In our approach, such an assumption is provided by the premise that there are rational systems of inductive probabilities $P(C_i/e)$ for constituents. For a review of this problem, see Niiniluoto (1981).

[66] This notion has interesting connections with Lakatos' notion of progress (see Niiniluoto, 1978c). Cf. the papers by John Watkins, John Worrall, and Elie Zahar in Radnitzky and Andersson (1978). It can also be shown that Hintikka's measure of corroboration, viz.

$$\text{corr}(h/e) = \min\{P(C_i/e) \mid i \in I_h\}$$

(cf. note 64), covaries with ver(h/e), if e and h are compatible. Hence, degrees of corroboration may serve as an indicator of scientific progress.

[67] Cf. Niiniluoto (1977), pp. 127–129; (1978d), pp. 308–313. Note that truth itself is not relative to language in this sense: for example, if C_* is the true constituent of L, and if L' is an extension of L, then C_* is entailed by the true constituent C'_* of L'. For comparative purposes, degrees of truthlikeness may be taken to be relative to agreed meaning postulates (cf. Niiniluoto, 1977, p. 145; Niiniluoto, 1978). However, if we have to compare sentences in different conceptual frameworks, the rival meaning postulates have to be taken into account in the comparison (see the next Section).

[68] If h and h' belong to different languages, the ideal language L is one of the common extensions of these languages (cf. Fig. 3). Therefore, h and h' (or their 'counterparts') are expressible in L, too.

[69] See Laudan (1977), pp. 125–126.

[70] *Ibid.*, p. 127.

[71] See Kuhn (1970), p. 171.

[72] *Ibid.*, pp. 206–207.

[73] See Rescher (1977), p. 179.

[74] *Ibid.*, p. 185.

[75] See Niiniluoto (1977), p. 140. Cf. note 66.

[76] See Kuhn (1978), pp. 332–333.

[77] See Laudan (1977), p. 224.

[78] This argument has been given by Shimony (1976), Niiniluoto (1978a), p. 323, and Putnam (1978), pp. 18–21. As true theories have only true deductive consequences, it is clear that true theories are also pragmatically successful for predictive purposes. It is not equally clear why this should be so for highly truthlike but false theories. To outline an argument to this effect, assume that h is the disjunction of constituents $C_1, \ldots, C_m$. If $M(h, C_*)$ is high, say larger than $1 - \epsilon$, then

$$M(h, C_*) = 1 - d(h, C_*) \geq 1 - \epsilon$$

or

$$d(h, C_*) = \tfrac{1}{2} d_{\min}(h, C_*) + \tfrac{1}{2} d_{\max}(h, C_*) \leq \epsilon$$

(cf. note 64). This entails that

$$d(C_i, C_*) \leq 2\epsilon \text{ for all } i = 1, \ldots, m.$$

In other words, *no* constituent in the normal form of h can be at a great distance from the true constituent C_* – and the larger this distance is, the greater disappointments we may encounter in dealing with the world.

[79] See Kuhn (1970), p. 206.

[80] See Laudan (1977), pp. 7, 70. For discussions about Laudan's views, see Musgrave (1979) and the Laudan Symposia in *PSA 1978*, vol. 2, and *Philosophy of the Social Sciences* 9 (1979). For a definition of 'progress' in mathematics in terms of problem-solving ability, see Hallett (1979).

[81] Laudan also argues that empirical problem-solving differs from explanation in important respects (*ibid.*, pp. 22–26), but here he seems to forget about the notions of 'approximate' and 'potential' explanations.

[82] *Ibid.*, p. 68. Lugg (1979) argues that Laudan's view is ambiguous between 'total problem-solving effectiveness' and 'problem-solving capacity relative to a specified set of problems'. However, Laudan (1981) makes it clear that he wishes to appraise theories relative to the finite number of problems that they have solved up to a certain time. In fact, we have here (as in Section 8) the possibility of distinguishing 'real' and 'apparent' progress, i.e., progress defined by the unknown potential ability and progress defined by the known performance so far. But here it would be problematic whether the latter notion may be regarded as an 'estimate' of the former (cf. below). The same distinction can be made, if the notion of empirical problem-solving is replaced by explanation (cf. note 81). This would bring us close to the notion that the Finnish philosopher Eino Kaila (1939) called 'relative simplicity': a theory has high relative simplicity if it is conceptually simple (cf. Laudan's 'conceptual problems') and has high explanatory power. As (actual) explanatory power and degree of confirmation may also be taken to be proportional to each other, this concept of relative simplicity is also related to Sober's (1975) criterion 'confirmation + simplicity' for theory appraisal.

[83] For some natural choices of measure M, we have $t = \frac{1}{2}$.

[84] Essentially this proposal has been given in Niiniluoto (1981).

[85] Stegmüller (1976) thinks that to speak of verisimilitude is questionable "teleological metaphysics", but I try to argue in Niiniluoto (1981) that the problem concerning progress for Sneedian Kuhn-theories contains as a special case the ordinary problem of truthlikeness.

[86] I assume here that J is a class of "potential", rather than "partial potential", models in Sneed's sense. Cf. Niiniluoto (1981) for the reasons of this stipulation.

BIBLIOGRAPHY

Almeder, R., 'Science and Idealism', *Philosophy of Science* **40** (1973), 242–254.

Almeder, R., 'Fallibilism and the Ultimate Irreversible Opinion', in N. Rescher (ed.), *Studies in Epistemology* (APQ Monograph Series No. 9), Blackwell, Oxford, 1975, pp. 33–54.

Aulin-Ahmavaara, Y., 'A General Theory of Acts, with Application to the Distinction between Rational and Irrational "Social Cognition"', *Zeitschrift für allgemeine Wissenschaftstheorie* **8** (1977), 195–220.

Bacon, F., *The New Organon*, Bobbs-Merrill, Indianapolis, 1960.

Bernard, C., *An Introduction to the Study of Experimental Medicine*, Dover, New York, 1957.

Bieri, P., Horstmann, R.-P., and Krüger, L. (eds.), *Transcendental Arguments and Science*, D. Reidel, Dordrecht, Boston, and London, 1979.

Boas Hall, M., *Robert Boyle on Natural Philosophy*, Indiana University Press, Bloomington, 1966.

Bohm, D., *Causality and Chance in Modern Physics*, Routledge and Kegan Paul, London, 1957.

Bradie, M., 'Pragmatism and Internal Realism', *Analysis* **39** (1979), 4–10.

Bradley, F. H., *Appearance and Reality*, Clarendon Press, Oxford, 1893.

Burian, R., 'Sellarsian Realism and Conceptual Change in Science', in Bieri *et al.* (1979), pp. 197–225.

Bury, J. B., *The Idea of Progress*, Macmillan, New York, 1932. (Dover, New York, 1955).

Comte, A., *Introduction to Positive Philosophy*, Bobbs-Merrill, Indianapolis, 1970.

Cournot, A. A., *An Essay on the Foundations of our Knowledge*, The Liberal Arts Press, New York, 1956.

Crombie, A. C., 'Some Attitudes to Scientific Progress, Ancient, Medieval and Early Modern', *History of Science* **13** (1975) 213–230.

Duhem, P., *The Aim and Structure of Physical Theory*, Princeton University Press, Princeton, 1954.

Duhem, P., *To Save the Phenomena*, The University of Chicago Press, Chicago, 1969.

Feyerabend, P., *Against Method*, NLB, London, 1975.

de Finetti, B., 'Foresight: Its Logical Laws, Its Subjective Sources', in H. E. Kyburg and H. E. Smokler (eds.), *Studies in Subjective Probability*, Wiley, New York, 1964, pp. 93–158.

Friedman, J., 'The Universal Class has a Spinozistic Partitioning', *Synthese* **32** (1976), 403–418.

Friedman, M., 'Truth and Confirmation', *The Journal of Philosophy* **76** (1979), 361–382.

Good, I. J., *Estimation of Probabilities*, The M.I.T. Press, Cambridge, Mass., 1965.

Habermas, J., 'Wahrheitstheorien', in H. Fahrenbach (ed.), *Wirklichkeit und Reflexion*, Pfullingen, 1973, pp. 211–265.

Hallett, M., 'Towards a Theory of Mathematical Research Programmes', *The British Journal for the Philosophy of Science* **30** (1979), 1–25.

Harré, R. (ed.), *Problems of Scientific Revolution: Progress and Obstacles to Progress in the Sciences*, Clarendon Press, Oxford, 1975.

Henrici, P., *Elements of Numerical Analysis*, Wiley, New York, 1964.

Herschel, J., *A Preliminary Discourse on the Study of Natural Philosophy*, Longman, Rees, Orme, Brown & Green and Taylor, London, 1831.

Hintikka, J. and Rantala, V., 'A New Approach to Infinitary Languages', *Annals of Mathematical Logic* **10** (1976), 95–115.

Hooker, M., 'Peirce's Conception of Truth', in J. C. Pitt (ed.), *The Philosophy of Wilfrid Sellars: Queries and Extensions*, Reidel, Dordrecht, Boston, London, 1978, pp. 129–133.

Hull, D., 'Laudan's Progress and Its Problems', *Philosophy of the Social Sciences* **9** (1979), 457–465.

Kaila, E., *Inhimillinen tieto*, Otava, Helsinki, 1939.

Kneale, W., 'Scientific Revolutions for Ever?', *The British Journal for the Philosophy of Science* **19** (1967), 27–42.

Koethe, J., 'Putnam's Argument Against Realism', *The Philosophical Review* **88** (1979), 92–99.

Krajewski, W., *Correspondence Principle and the Growth of Knowledge*, Reidel, Dordrecht and Boston, 1977.

Kuhn, T. S., *The Structure of Scientific Revolutions*, 2nd ed., The University of Chicago Press, Chicago, 1970.

Kuhn, T. S., *The Essential Tension*, The University of Chicago Press, Chicago, 1978.

Lakatos, I., 'Popper on Demarcation and Induction', in P. A. Schilpp (ed.), *The Philosophy of Karl Popper*, Open Court, LaSalle, 1974, pp. 241–273.

Laplace, P. S., *A Philosophical Essay on Probabilities*, Dover, New York, 1951.

Laudan, L., 'The Clock Metaphor and Probabilism: The Impact of Descartes on English Methodological Thought, 1650–65', *Annals of Science* **22** (1966), 73–104.

Laudan, L., 'Peirce and the Trivialization of the Self-Correcting Thesis', in R. N. Giere and R. S. Westfall (eds.), *Foundations of Scientific Method: The Nineteenth Century*, Indiana University Press, Bloomington, 1973, pp. 275–306.

Laudan, L., *Progress and Its Problems*, Routledge and Kegan Paul, London, 1977.

Laudan, L., 'The Philosophy of *Progress* ...', in P. D. Asquith and I. Hacking (eds.), *PSA 1978*, vol. 2. Philosophy of Science Association, East Lansing, 1981, pp. 530–547.

Lenin, V. I., *Materialism and Empirio-Criticism*, International Publishers, New York, 1927.

Lenz, J., 'Reichenbach's Defense of Induction', in M. H. Foster and M. L. Martin (eds.), *Probability, Confirmation, and Simplicity*, The Odyssey Press, New York, 1966, pp. 435–440.

Levi, I., 'Acceptance Revisited', in R. J. Bogdan (ed.), *Local Induction*, Reidel, Dordrecht and Boston, 1976, pp. 1–71.

Levi, I., 'Truth, Fallibility, and the Growth of Knowledge', forthcoming.

Levi, I., 'Induction as Self Correcting According to Peirce', in D. H. Mellor (ed.), *Science, Belief, and Behaviour*, Cambridge, 1980, pp. 127–140.

Lugg, A., 'Laudan and the Problem-Solving Approach to Scientific Progress and Rationality', *Philosophy of the Social Sciences* **9** (1979), 466–474.

Mach, E., *Knowledge and Error*, Reidel, Dordrecht and Boston, 1976.

Martin, G., *Leibniz: Logic and Metaphysics*, Barnes & Noble, New York, 1967.

Miller, D., 'Popper's Qualitative Theory of Verisimilitude', *The Br:tith Journal for the Philosophy of Science* **25** (1974), 166–177.

Molland, A. G., 'Medieval Ideas of Scientific Progress', *Journal of the History of Ideas* **39** (1978), 561–578.

Monk, J. D., *Mathematical Logic*, Springer-Verlag, New York, Berlin, Heidelberg, 1976.

Musgrave, A., 'Problems with Progress', *Synthese* **42** (1979), 443–464.

Niiniluoto, I., 'Inquiries, Problems, and Questions: Remarks on Local Induction', in R. J. Bogdan (ed.), *Local Induction*, Reidel, Dordrecht and Boston, 1976, pp. 263–296.

Niiniluoto, I., 'On the Truthlikeness of Generalizations', in R. E. Butts and J. Hintikka (eds.), *Basic Problems in Methodology and Linguistics*, Reidel, Dordrecht and Boston, 1977, pp. 121–147.

Niiniluoto, I., 'Notes on Popper as Follower of Whewell and Peirce', *Ajatus* **37** (1978), 272–327. [1978a.]

Niiniluoto, I., 'On the Realist Theory of Scientific Progress', *Section Papers, 16th World Congress of Philosophy*, Düsseldorf, 1978, pp. 463–466. [1978b.]

Niiniluoto, I., 'Verisimilitude, Theory-Change, and Scientific Progress', in I. Niiniluoto and R. Tuomela (eds.), *The Logic and Epistemology of Scientific Change* (Acta Philosophica Fennica 30, 1978), North-Holland, Amsterdam, 1979, pp. 243–264. [1978c.]

Niiniluoto, I., 'Truthlikeness: Comments on Recent Discussion', *Synthese* **38** (1978), 281–330. [1978d.]

Niiniluoto, I., 'Degrees of Truthlikeness: From Singular Sentences to Generalizations', *The British Journal for the Philosophy of Science* **30** (1979), 371–376.

Niiniluoto, I., 'The Growth of Theories: Comments of the Structuralist Approach', in *Proceedings of Second International Congress for History and Philosophy of Science, Pisa, 1978*, D. Reidel, Dordrecht, 1981. (Also Chapter 6 below.)

Niiniluoto, I., 'On Truth and Argumentation in Legal Dogmatics', *Rechtstheorie, Beiheft* 2, 1981, pp. 53–76.

Niiniluoto, I., 'Inductive Logic as a Methodological Research Programme', in *Scientia: Logic in the 20th Century,* 1983, pp. 77–100.

Nowak, L., 'Relative Truth, the Correspondence Principle, and Absolute Truth', *Philosophy of Science* **42** (1975), 187–202.

Nowak, L., 'Essence-Idealization-Praxis', *Poznan Studies in the Philosophy of the Sciences and the Humanities* **2** (1976), 1–28.

Peirce, C. S., *Collected Papers* (ed. by C. Hartshorne and P. Weiss), Harvard University Press, Cambridge, Mass., 1931–35.

Popper, K. R., 'Some Comments on Truth and the Growth of Knowledge', in E. Nagel, P. Suppes, and A. Tarski (eds.), *Logic, Methodology, and Philosophy of Science*, Stanford University Press, Stanford, 1962, pp. 285–292.

Popper, K. R., *Conjectures and Refutations*, Routledge and Kegan Paul, London, 1963.

Popper, K. R., *Objective Knowledge*, Oxford University Press, Oxford, 1972.

Popper, K. R., 'The Rationality of Scientific Revolutions', in Harré (1975), pp. 72–101.

Putnam, H., *Meaning and the Moral Sciences*, Routledge and Kegan Paul, London, 1978.

Quay, P. M., 'Progress as a Demarcation Criterion for the Sciences', *Philosophy of Science* **41** (1974), 154–170.

Quine, W. V. O., *Word and Object*, The M.I.T. Press, Cambridge, Mass., 1960.

Radnitzky, G. and Andersson, G., *Progress and Rationality in Science*, Reidel, Dordrecht and Boston, 1978.

Radnitzky, G. and Andersson, G. (eds.), *The Structure and Development of Science*, D. Reidel, Dordrecht, 1979.

Randall, J. H. and Buchler, J., *Philosophy: An Introduction*, Barnes & Noble, New York, 1942.

Ravetz, J., '*. . . et augebitur scientia*', in Harré (1975), pp. 42–57.

Reichenbach, H., *Theory of Probability*, University of California Press, Berkeley, 1949.

Rescher, N., *The Coherence Theory of Truth*, Oxford University Press, Oxford, 1973.

Rescher, N., *Methodological Pragmatism*, Blackwell, Oxford, 1977.

Rescher, N., *Scientific Progress*, Blackwell, Oxford, 1978. [1978a.]

Rescher, N., *Peirce's Philosophy of Science*, University of Notre Dame Press, Notre Dame, 1978. [1978b.]

Rescher, N., 'Some Issues Regarding the Completeness of Science and the Limits of Scientific Knowledge', in Radnitzky and Andersson (1979), pp. 19–40.

Rorty, R., 'From Epistemology to Hermeneutics', in I. Niiniluoto and R. Tuomela (eds.), *The Logic and Epistemology of Scientific Change* (Acta Philosophica Fennica **30**, 1978), North-Holland, Amsterdam, 1979.

Rosenberg, J., *Linguistic Representation*, Reidel, Dordrecht and Boston, 1974.

de Santillana, G., *The Age of Adventure: The Renaissance Philosophers*, Mentor Books, The New American Library, New York, 1956.

Sarton, G., *The Study of the History of Science*, Harvard University Press, Harvard, 1936. (Dover, New York, 1957.)

Savage, L. J., *The Foundations of Statistics*, Wiley, New York, 1954. (Dover, New York, 1972.)

Sellars, W., *Science and Metaphysics*, Routledge and Kegan Paul, London, 1968.

Shimony, A., 'Comments on Two Epistemological Theses of Thomas Kuhn', in R. S. Cohen *et al.* (eds.), *Essays in Memory of Imre Lakatos*, Reidel, Dordrecht and Boston, 1976, pp. 569–588.

Skagestad, P., 'C. S. Peirce on Biological Evolution and Scientific Progress', *Synthese* **41** (1979), 85–114.

Sober, E., *Simplicity*, Oxford University Press, Oxford, 1975.

Stegmüller, W., *The Structure and Dynamics of Theories*, Springer-Verlag, Berlin, New York and Heidelberg, 1976.

Tichý, P., 'On Popper's Definition of Verisimilitude', *The British Journal for the Philosophy of Science* **25** (1974), 155–160.

Whewell, W., *Philosophy of the Inductive Sciences*, Parker and Sons, London, 1840. (Reprinted 1967.)

Vigier, J.-P., 'The Concept of Probability in the Frame of the Probabilistic and the Causal Interpretation of Quantum Mechanics', in S. Körner (ed.), *Observation and Interpretation*, Butterworth, London, 1957, pp. 71–77.

von Wright, G. H., *The Logical Problem of Induction*, 2nd ed., Blackwell, Oxford, 1957.

Zilsel, E., 'The Genesis of the Concept of Scientific Progress', *The Journal of the History of Ideas* **6** (1945), 325–349.

CHAPTER 6

THE GROWTH OF THEORIES: COMMENTS ON THE STRUCTURALIST APPROACH

I. INTRODUCTION: THE BACKGROUND AND THE RISE OF THE SNEED-STEGMÜLLER-PROGRAMME

The nature and function of scientific theories is perhaps the most central problem within the philosophy of science. To study the basic features of theories and their role in scientific inquiry, a philosopher should be able to construe them as entities which he can handle and investigate. It is therefore no wonder that questions concerning the most appropriate ways of reconstructing scientific theories have a crucial importance in discussions about the proper method which philosophers of science should follow.

In this century, perhaps most philosophers of science have employed the *logistic method* for the analysis of the structure of scientific theories. This programme was initiated in Gottlob Frege's *Begriffschrift* (1879), where Frege defined a "formula language" by means of which he attempted to "provide a more detailed analysis of the concepts of arithmetic and a deeper foundation for its theorems". Frege was also confident that his ideography could be successfully applied to the study of the foundations of the differential and integral calculus, to geometry, and eventually "to the pure theory of motion and then to mechanics and physics" (see Frege, 1967, pp. 7–8). The new symbolic logic of Frege, Peano, and Russell provided philosophers with a tool for formalizing the informal axiomatic method which had been successfully applied in mathematics and physics. David Hilbert coined the term 'metamathematics' for the proof-theoretical study of mathematical theories, where theories are construed as *sets of sentences in a formal* (usually first-order) *language.* In the 1920's, the same approach was extended from mathematical to empirical theories by the logical positivists who treated theories as consisting of a set of axioms in an uninterpreted theoretical language (pure theory) together with a set of sentences which, perhaps partially, define or interpret theoretical terms by means of observational terms (correspondence rules). Emended with some insights and concepts from Alfred Tarski's logical semantics, this 'partial interpretation view' was further developed and modified by Rudolf Carnap, Carl G. Hempel, Ernest Nagel, and others; in its established form it is usually referred to as the 'Received View' or the 'Standard Conception' of scientific theories.[1]

The Received View was heavily attacked from many angles during the 1960's. One source of its troubles is the questionable assumption that theories contain, in their object language, a class of syntactic entities (correspondence rules) which serve special semantic *and* methodological functions (interpretation of theoretical terms). Another is the theoretical-observational dichotomy which has convincingly been shown to be untenable – at least as a general semantic division. Thirdly, it has been argued that the 'semantical empiricism' underlying the partial interpretation view does not do justice to the antecedent meaning of terms occurring in scientific theories. Fourthly, in the logistic reconstruction theories are treated as 'Finished Research Reports' (to use N.R. Hanson's words), not as historically developing and changing creations of scientific communities. The latter conception of theories was, however, strongly suggested by the work of such historians and philosophers as Thomas Kuhn, Stephen Toulmin, Paul Feyerabend, and Imre Lakatos. Finally, some philosophers argued that the first-order formalizations of physical theories are either too clumsy to be useful or else restricted to relatively uninteresting simple situations, while some others entirely denied the value of the formalization and the axiomatization of theories.

At the end of the 1960's, it was quite generally agreed that there is something wrong with the Received View. Several different attitudes towards the treatment of theories prevailed among philosophers of science at that time. Let us mention some of the most important ones.

(1) Many philosophers continued to work within the Carnap–Reichenbach–Hempel–Scheffler–Braithwaite–Nagel tradition, but only cautiously and with growing scepticism. Wolfgang Stegmüller's survey *Theorie und Erfahrung* (1970) is an excellent document of this attitude. Two serious blows to this approach were given by Hempel's rejection of the Received View in 1969 (see Hempel, 1970) and by Carnap's death in 1970.

(2) Some philosophers suggested that, apart from a few elementary results, the available logical techniques have played "a surprisingly

small role" in the methodological literature dealing with scientific theories (Hintikka and Tuomela, 1970, p. 298). Among these available but unused logical tools one can mention the logical theory of definition (cf. Rantala, 1977) and Tarski's model theory. Since the 1950's, the latter has been applied with very impressive results to the study of mathematical theories, especially of algebraic theories and of set theory. Model-theoretic methods have later been used also in the philosophy of science, most notably by Polish philosophers – in particular, by Marian Przełecki in *The Logic of Empirical Theories* (1969).[2] Raimo Tuomela's *Theoretical Concepts* (1973) applies these strong logical tools to the syntactic and semantic analysis of first-order theories; in his 'realist' view of theories Tuomela also tries to take into account the criticism which has been directed against the Received View, thus separating the logistic approach from the typical philosophical views which were combined with it by the logical empiricists.

(3) In his article 'Deterministic Theories' (1961), Richard Montague gave an explicit first-order axiomatization to Newton's mechanics. As he had to build all the required mathematics into the postulates, his axiomatization looked formidably complex. Some philosophers have suggested that, in axiomatizing specific physical theories, one should not imitate the metamathematicians (i.e., the logistic approach) but rather the mathematicians or the physicists themselves. Inspired by von Neumann's treatment of quantum mechanics, Evert Beth outlined a 'semantic' approach to physical theories (see Beth, 1961) which has been further developed by Bas van Fraassen (1970); related proposals have been made, independently, by Erhard Scheibe (1973) and Frederick Suppe (1972, 1974). Another approach, which follows the practice of mathematicians to define theories directly by means of their models (e.g., "a triple $\langle S, \mathcal{A}, P\rangle$ is a probability space if and only if..."), was started with the axiomatization of Classical Particle Mechanics by J. McKinsey, A. Sugar, and P. Suppes in 1953 and with the axiomatization of the Rigid Body Mechanics by Ernest Adams in 1955. This approach is characterized by the slogan: *To axiomatize a theory is to define a set-theoretical predicate*; it has been defended and applied by Patrick Suppes in a number of works and articles.[3]

(4) Several philosophers have searched for conceptual units which are more comprehensive and more flexible than the notion of an axiomatized theory. Of these attempts, which are mainly inspired by a

desire to be able to discuss scientific change, one can mention W.V. Quine's 'holistic' network model for theories (cf. Hesse, 1974), Kuhn's (1962) 'paradigms', Toulmin's (1972) 'conceptual systems', Lakatos's (1970) 'research programmes', Dudley Shapere's 'scientific domains' (Shapere, 1974), and Larry Laudan's 'research traditions' (Laudan, 1977). Most of these concepts, especially Kuhn's, contain pragmatic notions which refer to the community of investigators who support a scientific theory; they also emphasize the role of fundamental theories as deep commitments to comprehensive world views.

(5) Finally there is the view (sometimes combined with (4)) which is sceptical over the possibility of analysing the axiomatic structure of theories. Thus, Peter Achinstein suggests that theories can be characterized only by first defining the pragmatic concept '*A* has a theory *T*' and then saying that theories are, roughly speaking, those entities held by some people at some time (see Achinstein, 1968, Ch. 4).

The set-theoretic approach of Suppes and Adams has made rapid progress in the 70's. Joseph Sneed studied its foundations during 1966–70 and published his results in *The Logical Structure of Mathematical Physics* (1971). In this technical work, Sneed proposed a new theory-relative notion of theoretical functions; he defined theories of mathematical physics as pairs consisting of a mathematical structure ('core') and its putative applications.[4] He further introduced the notion of 'constraint' (which essentially binds together some of the applications of a theory) and proposed an 'emended Ramsey view' for expressing the empirical content of theories. In the final chapter, 'The Dynamics of Theories', Sneed discussed some aspects of the development of theories – how people come to have them and how they cease to have them.

The first reactions to Sneed's work concerned his notion of Ramsey-eliminability.[5] At the end of 1972, his book was discussed and compared to the model-theoretic approach by the Polish philosophers (cf. Przełecki, 1974, and Wojcicki, 1974b). At the same time, Stegmüller read Sneed's book – hoping to find in it something interesting about the Ramsey view of theories.[5a] What he discovered there was quite different: first, a new conception of scientific theories, viz. a *non-statement* or *structuralist view* (i.e., theory is a mathematical structure together with a set of applications) as opposed to the traditional *statement view* (i.e., theory is a set of statements), and, secondly, a sketch of the dynamics of theories which, he thought, for the first time made possible the reconstruction of Kuhn's conception of the development of science.[6] In 1973, Stegmüller published a sequel *Theorienstruktur und Theoriendynamik* to his *Theorie und Erfahrung*; it presented Sneed's formalism by using a simplified and more perspicious notation together with a more detailed discussion of the development of theories by using this formalism. Stegmüller also made a number of general comments concerning Kuhn's, Popper's, Lakatos's, and Feyerabend's views on theory-change. He has later clarified, extended, and defended in several papers this new programme for studying the structure and the growth of scientific theories.[7]

Stegmüller's enthusiasm brought Sneed to Munich in 1974–75 and to work on physical theories again. Together with Wolfgang Balzer, Sneed started to revise, generalize and simplify his formalism. The results have been published in Sneed (1976) and Balzer & Sneed (1977).[8] Other collaborators of this programme have done technical work on the Sneedian approach in thermodynamics (Moulines, 1975b), on the approximative application of theories (Moulines, 1976), and on the concept of reduction (Mayr, 1976).[8a]

One of the most impressive achievements of the set-theoretical approach has been Kuhn's very favorable response to it. Stegmüller sent in early 1974 his book on theory dynamics to Kuhn. On September 1975, Sneed, Stegmüller, and Kuhn presented papers in a Symposium on Theory Change in the London, Ontario, Congress on Logic, Methodology, and Philosophy of Science.[9] Even though Kuhn made the reservation that the Sneed formalism "currently does virtually nothing to clarify the nature of revolutionary change", he added that "to a far greater extent and also far more naturally than any previous mode of formalization, Sneed's lends itself to the reconstruction of theory dynamics" (Kuhn, 1976, p. 184). He also said that "Stegmüller, approaching my work through Sneed's, has understood it better than any other philosopher who has made more than passing reference to it" (*ibid.*, p. 179).

Even though Sneed is not the first who has proposed formal models for scientific change,[10] his set-theoretic treatment of theories is the first one which has gained any support from a leading exponent of the historical approach. Against those who are opposed to all kinds of formalization, Kuhn's comments suggest that a sufficiently rich formal description of the structure of theories which at the same time includes a reference to a number of pragmatic factors may be a valuable tool in the analysis of theory dynamics. The admission that such formal models may be relevant from the viewpoint of the historians gives a serious challenge to logically minded philosophers of science to work in this direction. (Note that the more structure one puts in the description of a theory, the more possibilities one has for distinguishing between different types of theory change.) It also indicates that the diverse approaches within the philosophy of science can not only complement but also co-operate with each other.

How adequately does Sneed's set-theoretical formalism represent the structure of scientific theories and the patterns of their growth? To what direction can it be further developed? Is there an essential difference between the set-theoretic and the model-theoretic formalizations? Can the Sneedian approach be extended from mathematical physics to other fields of science? To what extent can one reconstruct actual examples from the history of science by means of it? What new insights does this formalism give about theory dynamics? To what extent does it allow for the formalization of Kuhn's conception of scientific change? Does it favor, in some systematic sense, Kuhnian ideas in contrast with such rival accounts as Popper's or Lakatos's? These are examples of questions which are raised by the work of Sneed, Stegmüller, and their collaborators. One cannot attempt to answer in detail to all these questions in a single paper; therefore this paper contains only a preliminary survey of them.

2. THE STRUCTURALIST CONCEPTION OF THEORIES

In this section, I shall outline the elements of the structuralist notion of a scientific theory. At the same time, I shall make some technical and philosophical comments which seem to me to give good reasons for modifying and complementing some of the basic definitions.

(a) *Core*

A non-empty class of $m + k$-tuples of the form $\langle n_1, \ldots, n_m, t_1, \ldots, t_k \rangle$, $m > 0$, $k \geq 0$, is called a $m + k$-matrix, if $n_1, \ldots, n_m, t_1, \ldots, t_k$ are sets, relations, or functions. Here $n_1, \ldots, n_m$ are the non-theoretical components and $t_1, \ldots, t_k$ are the theoretical components of the matrix. (For this distinction, see below.) Let M_p be a $m+k$-matrix. Let M_{pp} be the corresponding $m+0$-matrix which is obtained from M_p by the restriction function, i.e. by $r: M_p \to M_{pp}$ such that $r(\langle n_1, \ldots, n_m, t_1, \ldots, t_k\rangle) = \langle n_1, \ldots, n_m\rangle$. If $M \subseteq M_p$, then the quadruple $F = \langle M_p, M_{pp}, r, M\rangle$ is a *frame*. Here M_p is the class of the *potential models* of F, M_{pp} is the class of the *partial potential models* of F, and M is the class of the *proper models* of F.

A *core* $K = \langle F, C\rangle$ is obtained from a frame $F = \langle M_p, M_{pp}, r, M\rangle$ by adding to it a *constraint* C for M_p, i.e., a class $C \subseteq \text{Pot}(M_p)$ such that (i) $\emptyset \notin C$, (ii) if $x \in M_p$, then $\{x\} \in C$, and (iii) if X, $Y \in \text{Pot}(M)$, $X, Y \neq \emptyset$, $X \in C$, and $Y \subseteq X$, then $Y \in C$.[12]

Comments: The intuitive idea is that M_p is "the set of all possible models for the *full* conceptual apparatus of a theory" including theoretical components, M_{pp} is the set of the corresponding non-theoretical models, M is the set of those potential models which are not excluded by the fundamental law of the frame, and finally C is a set of restrictions which rule out certain combinations of components in different potential models (Balzer and Sneed, 1977, p. 198). However, these ideas are only partly formalized by the given definitions – which thus seem to be too general. In the first place, the components of an element of a matrix are allowed to be any sets, relations or functions whatsoever.[13] For example, one element of a matrix may contain only binary relations as components, while another element of the same matrix may contain only 5-place relations. Secondly, the definition allows M_p to be any $m+k$-matrix of any non-empty size. For example, M_p might contain only one element. These unintended cases can be eliminated by requiring that an $m+k$-matrix is a collection of many-sorted[14] structures of a given *type*, where a type is essentially a finite set of functions which indicate what sort of set-theoretical entities can be chosen as the different components, and by assuming that M_p is the class of *all* structures of the given type (cf. Rantala, 1978a). With these stipulations (which we shall always assume below), the representation of the 'fundamental law' by means of a set M of models in M_p becomes identical to the standard 'semantic' treatment of propositions as classes of models,[15] which has the virtue that one can speak of propositions independently of their various linguistic formulations.

Let us denote the types of M_p and M_{pp} by τ_p and τ_{pp}, respectively. Then τ_p and τ_{pp} are finite set-theoretical entities, while M_p and M_{pp} are proper classes. From the set-theoretical viewpoint, it therefore would be much more convenient to operate with types than with the corresponding proper classes. Moreover, when we later say that a person holds a theory, it might seen more realistic to assume that, in using a core, this person is operating with the finite types rather than with the proper classes. We could thus define cores as quadruples $\langle \tau_{pp}, \tau_p, M, C \rangle$, where τ_p is a $m+k$-type, τ_{pp} is a $m+0$-type which is a restriction of τ_p, M is a class of structures of type τ_p and C is a constraint for structures of type τ_p. However, here M and C are still quite complicated set-theoretical entities.[16] This is one reason – but, as we shall see later, not the only reason – for replacing M and C with their finite formulations.

Let us say that a pair of first-order languages $\langle L_0, L \rangle$, where L_0 is of type τ_{pp}, L is of type τ_p, and L_0 is a sublanguage of L, is a *language-pair for core* $K = \langle M_{pp}, M_p, M, C \rangle$.[16a] Here L_0 is a non-theoretical language and L is the full theoretical language; the terms belonging to L but not to L_0 constitute the *theoretical vocabulary* of L. All the ordinary model-theoretical notions can now be defined in the usual way. If a sentence σ of L is true in a structure $z \in M_p$, we shall write $z \models \sigma$ or $z \in \mathrm{Mod}(\sigma)$. If Σ is a set of sentences in L, then $\mathrm{Mod}(\Sigma) = \bigcap_{\sigma \in \Sigma} \mathrm{Mod}(\sigma)$. If there is a set of sentences Σ in L such that $M = \mathrm{Mod}(\Sigma)$, then we shall say that Σ is a *formulation of the fundamental law M.*[17]

The formulation of the constraint C is not possible in a language L for core $K = \langle M_{pp}, M_p, M, C \rangle$ (cf. Przełecki 1974). To do this, we need a language L^* which is interpreted on the domain M_p and which contains the vocabulary of set theory (cf. Harris, 1978). As L^* speaks of the models of the language L, it is essentially a semantic metalanguage for L. If f is a function symbol in L and if z is a variable in L^* ranging over the elements of M_p, then f^z is a function symbol in L^* which denotes the interpretation of f in z. One can now *formulate constraints in* L^*: for example,

$$\mathrm{Dom}(z_1) = \mathrm{Dom}(z_2)$$

says that models z_1 and z_2 have the same domain, and

$$\forall z_1 \forall z_2 (f^{z_1} \restriction \mathrm{Dom}(z_1) \cap \mathrm{Dom}(z_2) = f^{z_2} \restriction \mathrm{Dom}(z_1) \cap \mathrm{Dom}(z_2))$$

says that function f has the same value for the same object in all models where that object appears.[18] In this way, the constraint C corresponds to a set of sentences in language L^*.

A Sneedian core $K = \langle M_{pp}, M_p, M, C\rangle$ can thus be replaced, if one wishes, by a seven-tuple $\langle \tau_{pp}, \tau_p, L_0, L, L^*, \Sigma, \Gamma\rangle$, where τ_{pp} and τ_p are types, τ_{pp} is a restriction of τ_p, L_0 is a first-order language of type τ_{pp}, L is a (first-order) language of type τ_p, L^* is a set-theoretical metalanguage for L, Σ is a formulation of M in L, and Γ is a formulation of C in L^*.

(*b*) *K-theoreticity*

The notion of a core presupposes a distinction between non-theoretical and theoretical components. For Sneed, this distinction is relative to a theory and to the existing expositions of their applications. There are two difficulties in giving a precise definition of the notion of T-theoreticity.[19] First, in the pure 'non-statement view' of theories which tries to avoid all linguistic or syntactic entities it becomes difficult to say *what* is T-theoretical, for the various models of a physical theory contain only specific functions. Moreover, it would obviously be impossible to use Sneed's criterion for showing separately of all the theoretical components in all the elements of the proper class M_p that they are T-theoretical. Therefore, we should like to say that there is some single entity which is T-theoretical and which is interpreted in the different models by the specific functions. The most natural way of doing this is to introduce a language for a core (see above), but this step already goes beyond the pure non-statement view of theories (cf. Stegmüller's difficulties on this matter in Stegmüller, 1976a, pp. 41–45.) Secondly, Sneed and Stegmüller define the notion of T-theoreticity before they have given a general definition of theories, and it turns out that their later definition of a theory (or theory-element) does not coincide with those entities T which are presupposed in the definition of T-theoreticity (cf. Harris, 1978) – the notion of an 'application' of a theory is different in these cases. The following reconstruction of the criterion tries to avoid these difficulties.

Let $E_t(K)$ be the set of the *existing theoretical applications* of a core K at time t. This means that (i) $E_t(K) \subseteq M_p$, (ii) for each $z \in E_t(K)$ there is an exposition of z, (iii) each existing exposition for

z gives such a description of z that it can be rationally believed that $z \in M$, (iv) $E_t(K) \in C$. In other words, $E_t(K)$ contains those structures in M_p which have successfully been shown (at time t) to be models of the core K, and $E_t(K)$ itself satisfies the constraints of K.

Let P be a term which belongs to the theoretical vocabulary of a language L for core K. Let P^z be the interpretation of P in model $z \in M_p$; here P^z is either a n-place relation or a n-place (real valued) function. Let K' be a core which may be identical or non-identical with K. Then P is *measured in a K'-dependent way in z at time t* iff there are objects $a_1, \ldots, a_n$ in the domain of P^z such that in each exposition of z the methods of establishing the truth or falsity of $\langle a_1, \ldots, a_n \rangle \in P^z$ or $P^z(a_1, \ldots, a_n) = x$ $(x \in \boldsymbol{R})$ presuppose that $E_t(K') \neq \emptyset$. Further, P is *K'-theoretical at time t* iff P is measured in a K'-dependent way at time t in each theoretical application $z \in E_t(K)$. In particular, if K' is the same as K, P is K-theoretical at time t.

One way of modifying these definitions is to replace the talk of the 'expositions' (i.e., textbooks, lecture notes, etc.) by the talk of the *actions* (measurement, determination of truth-values) of typical scientists within certain scientific communities (see Tuomela, 1973, 1978b), Instead of saying that there is an exposition for $z \in M_p$, we could say that the scientists at time t are able to 'handle' the structure z by means of their shared knowledge and abilities. In this case, we may relativize the set $E_t(K)$ to a certain *scientific community SC*, and denote it by $E_{SC,t}(K)$. In fact, this modification would bring the structuralist conception closer to the Kuhnian way of viewing science, since Kuhn has repeatedly emphasized the importance of the research activity of the scientists as compared to the expositions which record the scientific achievements (see, for example, Kuhn, 1962, p. 1).

Stegmüller has argued that the notion of presupposition used above has to be analysed in the sense that 'A presupposes B' means the same as 'B is a logical consequence of A' (see Stegmüller, 1976a, pp. 45, 55). This interpretation is extremely strong, and one may doubt whether there exist, in this sense, any K'-theoretical terms in scientific theories. To see this, let us mention some typical examples of theoretical terms in science. Radio astronomy studies radio sources (supernovas, quasars, pulsars) by means of radio telescopes. In using a radio telescope, the astronomers assume that it is functioning in accordance with the theory of electromagnetic radiation. Thus, the term 'radio source' is theoretical relative to the theory of electromagnetism. Similarly, many terms in astronomy are theoretical relative to optics. For example, in using telescopes to detect double stars we assume that optics can be succesfully applied to these instruments.[20] However, it is sufficient to presuppose (in Stegmüller's strong sense) that *something like our optics* holds of the telescope – in spite of some changes in our optics we might still be able to argue that we detect double stars (rather than something created by the instrument). The same remark applies also to the case where the scientists measure theoretical functions: an application of classical particle mechanics CPM to a projectible problem may involve the belief that CPM itself applies *at least approximately* (see Sneed, 1971, p. 32) to the systems which are used to measure the mass function. Again, we can only infer that some theory which is approximately equal to *CPM* is applicable.

For these reasons, I shall assume that the notion of 'presupposition' is interpreted in a pragmatic sense, so that it refers to the activities which the scientists in a community *SC* actually are capable of performing in a given situation. I shall leave it as an open problem whether Stegmüller's strong interpretation of 'presupposition' could be maintained by using in the above definition of K-theoreticity some suitable notion of existing theoretical *approximative* applications (cf. Moulines, 1976).

Let P be a term which occurs in a language for core K, and assume that P is K-theoretical. According to Sneed, mass and force in *CPM* are examples of such theoretical terms. Let us say that P is *strongly K-theoretical* iff for each theoretical application $z \in E_t(K)$ the measurement of P in z at time t presupposes not only that $E_t(K) \neq \emptyset$ but also that $z \in E_t(K)$. Do scientific theories contain any strongly theoretical terms? The answer seems to be positive: for example, if a medical theory is only able to describe the symptoms of a desease, then the only method of diagnosing that a person has this disease is to argue from the symptoms by assuming the truth of the theory in this situation. More generally, many dispositional terms in scientific theories seem to be strongly theoretical – at least so long as the scientists have not yet been able to discover an independent way of determining whether the base of the disposition is actualized in some situation or not.[20a]

(c) *Possible applications*

A core K can be applied to a non-theoretical structure $z \in M_{pp}$ if z can be expanded to a structure in M_p which satisfies both the fundamental laws of K and the constraint C for M_p. This idea is made precise in the following definitions. Let $\bar{r}: \text{Pot}(M_p) \to \text{Pot}(M_{pp})$ be the function defined by

$$\bar{r}(X) = \{r(z) \mid z \in X\}, \quad \text{for } X \subseteq M_p.$$

Then the class of the *sets of the possible non-theoretical applications* of core K is defined by

$$(1) \qquad A(K) = \{\bar{r}(Z) \mid Z \in \text{Pot}(M) \cap C\}.$$

Thus, $A(K) \subseteq \text{Pot}(M_{pp}) - \{\emptyset\}$ (cf. Fig. 1).

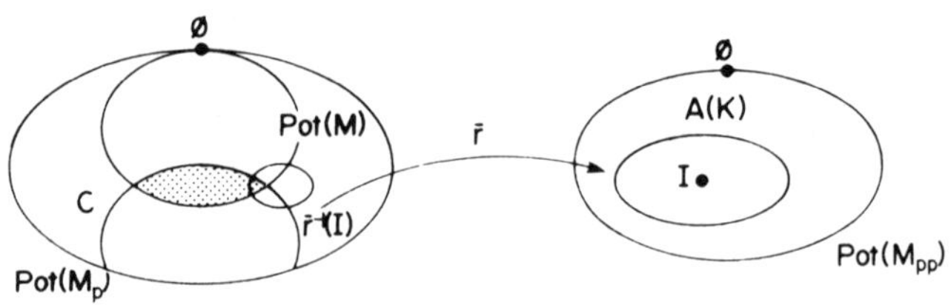

Fig. 1.

In addition to the non-theoretical applications, we can define the class of the *sets of the possible theoretical applications* of core K simply by

$$(2) \qquad AT(K) = \text{Pot}(M) \cap C.$$

Then, by (1), $A(K) = \bar{\bar{r}}(AT(K))$.

Classes $A(K)$ and $AT(K)$ are objectively determined as soon as core K is given. However, it should be remembered that the non-theoretical/theoretical distinction within K is relative to the set $E_{SC,t}(K)$ of the existing theoretical applications of K.[21]

(d) *Intended applications and theory-elements*

Sneed (1976) defines 'theory-elements' as pairs $\langle K, I\rangle$, where K is a core and $I \subseteq M_{pp}$ is a non-empty set of the 'intended applications' of K. Following Moulines, a pragmatic characterization of theory-elements can be given as follows.

Let SC be a scientific community and t a point (or an interval) of time. Let $I_{SC,t}(K)$ be the set of the non-theoretical structures in M_p to which SC intends to apply K at time t. Then $I_{SC,t}(K)$ is the set of the *intended non-theoretical applications* of K relative to SC and t. (If no confusion is to be expected, we shall sometimes refer to this set simply by 'I'.) The pair $\langle K, I_{SC,t}(K)\rangle$ is a *theory-element* relative to SC and t.

Some of the applications of K in $I_{SC,t}(K)$ have been sufficiently confirmed by the community SC before time t. If $C_{SC,t}(K)$ is the set of these *confirmed* (non-theoretical) *applications* or K at time t, then $C_{SC,t}(K) \subseteq I_{SC,t}(K)$ and the community SC believes that $C_{SC,t}(K) \in A(K)$. The community claims something more, however, namely that all the intended applications of K are possible applications of K. In general, the *claim* of a theory-element $\langle K, I\rangle$ is simply

(3) $\qquad I \in A(K)$.

Note that the claim (3) is vacuous if $M = M_p$ and $C = \text{Pot}(M) - \{\emptyset\}$, i.e., if core K does not contain any non-trivial laws or constraints.

Formula (3) implicitly contains existential quantified variables ranging over theoretical components. Therefore, it may be regarded as a modification of the ordinary Ramsey-sentence for (finitely axiomatizable) first-order theories (cf. Sneed, 1971). This suggests that we might regard (3) as an expression of the *non-K-theoretical content* of theory-element $\langle K, I\rangle$. It should be remembered, however, that the non-K-theoretical terms may be K'-theoretical for some other cores K' at time t.[21a]

Does (3) express the whole claim that the supporters of K in SC wish to make? Sneed and Stegmüller seem to think so, since for them the intended applications of a core K are always members of the class M_{pp} of partial potential models. But why could we not have 'intended applications' in class M_p as well? A scientific realist might indeed argue that it is precisely the K-theoretical terms which are the most interesting elements connected with a core K: theories are attempts to obtain 'deep' theoretical knowledge of the reality, and one should not assume that the theoretical terms refer only to fictitious entities in the various applications of a theory. A putative description of a model which includes the theoretical relations and functions is more informative than, and hence preferable to, a purely non-theoretical description. Moreover, we defined above the notion of K-theoreticity by referring to the set of the existing theoretical applications of K at time t, i.e., to set $E_{SC,t}(K) \subseteq M_p$. The definition of this set explicitly involves the assumption that the community SC believes that $E_{SC,t}(K) \in AT(K)$ (cf. (2)). (Another natural requirement is the following: $\bar{r}(E_{SC,t}(K)) \subseteq C_{SC,t}(K)$.)

These remarks suggest that it is natural to define a set of the *intended theoretical applications* of core K relative to SC and t. Let us denote this set by $J_{SC,t}(K)$ or J. The triple $\langle K, I_{SC,t}(K), J_{SC,t}(K)\rangle$ is a *full theory-element* relative to SC and t, and the *theoretical claim* of this theory-element is

$$(4) \qquad J_{SC,t}(K) \in AT(K).$$

It is required that the set $J_{SC,t}(K)$ satisfies the conditions

$$(5) \qquad E_{SC,t}(K) \subseteq J_{SC,t}(K)$$

and

$$(6) \qquad \bar{r}(J_{SC,t}(K)) = I_{SC,t}(K).$$

In other words, J is an element in the inverse image $\bar{r}^{-1}(I)$ of I. The claim (3) is equivalent to $\bar{r}^{-1}(I) \cap \text{Pot}(M) \cap C \neq \emptyset$, while (4) says that J which by (6) is chosen as a member of $\bar{r}^{-1}(I)$ is also a member of $\text{Pot}(M) \cap C$. It thus follows immediately from (6) that the theoretical claim (4) entails the claim (3), i.e., the full theoretical content of a full theory-element is at least strong as its non-theoretical content. On the other hand, it need not be the case that (3) entails (4).

Against this suggestion it might be objected that the claim (4) could not possibly have more 'empirical' or 'testable' content than (3). The reason for this objection is the peculiar character of the K-theoretical terms, i.e., the difficulty which Sneed calls 'the problem of theoretical terms' (see Sneed, 1971, p. 38; Stegmüller, 1976a, Chapters 3 and 4). To make the statement (4) we have to pick up a set J in $\bar{r}^{-1}(I)$. But how can we ever know that any such set J belongs to $AT(K)$? No matter whether J is defined extensionally (i.e., by enumerating its members) or intensionally (i.e., by the means of a set-theoretical property), it seems that, to settle this question, one should be able to determine the values of the theoretical components of the members of J. As these components are K-theoretical, this determination is possible only on the condition that K has some successful theoretical applications. This condition presupposes again that K has successful applications, and so on. It therefore seems that we have ended up either with a vicious circle or with an infinite regress (cf. Stegmüller, 1976a, p. 58). Note also that if the theoretical components are strongly K-theoretical, then the choice of their values in an application $z \in M_p$ guarantees that z is a model of K. But if these values are chosen in this way, then the claim that $z \in M$ seems to become true *a priori*. For these reasons, Stegmüller has argued that the Ramsey–Sneed solution is "absolutely compulsory" (Stegmüller, 1978a, p. 44), i.e., that it is the only known way of solving the problem of theoretical terms.[21b]

As an answer to these objections, I shall present six different arguments to defend the theoretical claims and their testability. Let us first note that in making the distinction between non-K-theoretical and K-theoretical terms it was already assumed that K has or at least may have some successful theoretical applications, viz. the members of $E_{SC,t}(K)$. Now there are two possibilities: either the claim that these applications are successful is based on a viciously circular argument or it is not. In the former case, it seems that the whole idea of separating K-theoretical and non-K-theoretical terms was after all groundless. In the latter case, it seems that one can go on to extend the set $E(K)$ of the existing theoretical applications of K and thus justify claims of the form (4) without bothering too much about the 'problem of theoretical terms'.[22]

Secondly, it should be noted that already the problem concerning the testability of the claim $I \in A(K)$, which involves only non-K-theoretical terms, may be problematic. To see this, assume that the values of all the non-K'-theoretical functions (or relations) in a core K' are known. Then, by the definition of K'-theoreticity, all the methods for measuring the K'-theoretical functions presuppose that K' has successful applications. This does not imply – even if there were expositions of K' – that there in fact are any methods for measuring these functions. Moreover, in those cases where K' can be used for making such measurements, there need not be any guarantee that this can be done in a unique or exact way: the K'-theoretical functions may happen to be definable or identifiable in K' by the non-K'-theoretical functions only in some weak sense.[23] Let K now be another core which contains terms which are non-K-theoretical but K'-theoretical. In this case, which is not untypical,[24] there are special problems already in ascertaining the values of these non-K-theoretical functions – contrary to what Stegmüller claims (Stegmüller, 1976a, p. 57). To call partial possible models 'observable facts' (cf. Stegmüller, 1976a, p. 59; Balzer and Sneed, 1977, p. 199) is therefore very misleading – it may give the wrong impression that the distinction between K-theoretical and non-K-theoretical functions is a revived version of the old theoretical – observational dichotomy.

Thirdly, it was seen above that the measurement of the K-theoretical components in an application $z \in M_p$ of K normally presupposes only that something like K has a successful application z'. If we used K itself to fill in the theoretical components of z', an infinite regress or a vicious circle would be started. However, it may be sufficient to apply some $K' \neq K$ to z'; and I can see no reason to exclude the possibility that P is not measured in a K'-dependent way in z'. In this case, the vicious circle resulting from the K-theoreticity of the theoretical components of z is blocked. (We may of course have troubles which are similar to those mentioned in the second point above, but they do not show that there is some difference in the testability of theoretical and non-theoretical claims.) Another way of putting this argument is the following: we may have good reasons (perhaps from other areas of theoretical science) to claim that *some* theory in a larger class $\mathcal{K}$ of theories (where $K \in \mathcal{K}$) can be applied to z'. If now all the members of $\mathcal{K}$ give essentially the same description of z', then we have good – and non-circular – grounds for deciding whether K can be successfully applied to z or not.[24a]

Fourthly, it seems that there are examples involving strongly theoretical terms in which the testing (i.e., confirmation or disconfirmation) of a theoretical claim is possible. Assume that a psychologist proposes a theory that certain persons who behave in certain ways in certain situations suffer from a mental disorder, which he calls P-fobia, and that this disorder has been caused by some injury in the earlier history of the person. Write Px for 'x has P-fobia', O_1x for 'x has been injured earlier', and $O_2x, \ldots, O_nx$ for various predicates expressing different behavioral patterns of x. Then a formulation of the theory of P-fobia can be given by the conjunction σ of the following two sentences

$$(7) \qquad (x)(Px \supset O_1x)$$
$$(x)(O_2x \supset (Px \supset O_3x \& \ldots \& O_nx)).$$

Let us call these sentences σ_1 and σ_2. The potential models are now structures of the form $z = \langle D_z, O_1^z, \ldots, O_n^z, P^z \rangle$, where D_z is a domain of individuals, the partial potential models are of the form $r(z) = \langle D_z, O_1^z, \ldots, O_n^z \rangle$, and the class M of proper models is simply $\text{Mod}(\sigma)$. As the constraint C we may assume a principle to the effect that an individual b has P-fobia in one of the models of the theory if and only if it has P-fobia in all such models. This defines the core K of the P-fobia theory. Here P is strongly K-theoretical and $O_1, \ldots, O_n$ are non-K-theoretical predicates.

What are the intended applications of K? In the Sneedian approach, it is assumed that "what the theory is about" is a class I of *partial* potential structures (cf. Balzer and Sneed, 1977, p. 199). In this case, there does not seem to be any natural choice for such a class I. Moreover, it is more natural (at least from the realist viewpoint) to say that the theory of P-fobia is a theory *about P-fobia* rather than about observable behavior. In other words, this theory should be applicable to all cases which are instances of P-fobia – and these applications are theoretical rather than non-theoretical. Thus, we can say that the set $J(K)$ of the intended theoretical applications of K is the class of all actual and potential victims of *P-fobia*, i.e., the set of all individuals who now suffer, or have suffered, or will suffer, or would suffer, from this disease. Set-theoretically, $J(K)$ can be defined as the set of all structures of the following form

$$z = \langle \{a_z\}, O_1^z, \ldots, O_n^z, P^z \rangle$$

where a_z is a human being, O_i^z is $\{a_z\}$ or $\emptyset$ depending on whether a_z has or does not have the property expressed by $O_i (i = 1, \ldots, n)$, and $a_z \in P^z$ (that is, the individual a_z has P-fobia in z). It follows immediately that $J(K)$ satisfies the constraint C.

How can we now test the claim that $J(K) \in AT(K)$, or equivalently $J \subseteq M = \text{Mod}(\sigma)$? Note first that if z is a fully specified model in M_p then the claim $z \in M$ is a priori true or false – it is just an exercise in logic to determine which one of these alternatives is valid. In the testing process, we rather start from an incompletely known model consisting of an individual with its so far unknown properties. Choose an individual b which satisfies the condition $O_2(b)$ and assume that it also shows the symptoms $O_3, \ldots, O_k (k \leq n)$. Together with the assumption that σ_2 holds of b, the evidence $O_2(b), \ldots, O_k(b)$ now gives us *inductive* reasons for believing that $P(b)$, i.e., that $\{b\}$ is an intended application of K. This belief can be further strengthened by deriving the claims $O_{k+1}(b), \ldots, O_n(b)$ from σ_2 and $P(b)$ and by showing that these claims are true. If some of these predictions turn out to be false, then it follows from σ_2 that after all b does not have P-fobia. In both cases, the law σ_2 will be true of b simply for the reason that we have presupposed it in determining the extension of P in the domain $\{b\}$.

Thus, assuming that, on condition $O_2(b)$, b has been shown to have all the symptoms $O_3, \ldots, O_n$, we have by σ_2 strong support for the claim that $P(b)$. If we now find that $O_1(b)$ is also true, then, by assuming σ_1, the support for $P(b)$ will be even still stronger. Hence, if σ_1 and σ_2 are assumed, we may have very strong evidence for claiming that $\{b\}$ with its real properties is an intended application of K. On the other hand, if we fail in our attempts to show that $O_1(b)$, then we can blame either the assumption $P(b)$ or σ_1. If we keep hold on σ, then the claim that $\{b\}$ is an intended application of K· will be disconfirmed. If we wish to keep the claim that $P(b)$, then our failure in verifying $O_1(b)$ disconfirms σ_1.

As we usually have no final guarantee of the truth of such theoretical statements as $P(b)$ above, the intended theoretical applications in $J(K)$ cannot simply be enumerated, but they have to be characterized by means of an intensional theoretical description. In testing the theoretical claim $J(K) \subseteq M$ there are thus two possible situations. First, we try *to show that a structure to which σ applies is an intended model*. In this case, we argue as follows: Here is a structure z which is a model of M. I believe that z is in fact the structure $z^* \in J$. Therefore, I believe that $z^* \in M$. Secondly, we try *to determine whether the law σ applies to a structure which has been identified, with some degree of certainty, as an intended model.* In this case, we use one part of the theory to appraise another part. In these ways, both the confirmation and the disconfirmation of the theoretical claim $J(K) \subseteq M$ is possible.[24b]

Fifthly, in practice theoretical claims are often tested by means of non-theoretical structures which contain individuals or properties that are analogous with but not identical to those that the theory is supposed to speak about.[25] These 'testing models' are obtained from the full potential models by 'operationalization', i.e., by replacing the theoretical components by their operationalized counterparts. The effect of this operationalization can be described as a mapping of the original core K to another core K' such that the K-theoretical components will be mapped to non-K'-theoretical components. More precisely let us say that o is an *operationalization* of K by K' iff (i) $o: M_p \to M'_p$, (ii) $o(\langle n_1, \ldots, n_m, t_1, \ldots, t_k \rangle) = \langle n_1, \ldots, n_m, t'_1, \ldots, t'_k \rangle$, where $n_1, \ldots, n_m, t'_1, \ldots, t'_k$ are all non-K'-theoretical. The claim

(8) $\bar{o}(J(K)) \in \bar{o}(AT(K))$,

which does not involve any K'-theoretical components, is an *operationalized version* of the theoretical claim (4), i.e., of $J(K) \in AT(K)$. The verification of (8) gives support to (4) to the extent that the values of o 'validly' measure the original theoretical terms.[26,26a]

Sixthly, the status of the assumption that a term is K-theoretical is historically relative and can change when new applications of the core K are found. Even if a theoretical claim of the form (4) may *now* seem to be untestable, it may still have factual content which becomes testable when our knowledge grows. This seems to happen to many dispositional terms which in their 'infancy' are strongly K-theoretical (cf. P-fobia in the above example) but may later become non-K-theoretical. This change may be due to the discovery of new instruments or to the invention of new theories (possibly introducing new theoretical terms) relative to which the measurement of these terms become possible.

(e) *Theoretization and specialization*

The above description of cores K is incomplete for the reason that the different models in M may satisfy some *special laws* and *special constraints*. The formalisms of Sneed (1971) and Stegmüller (1976a) contain devices for defining *expanded cores* which take into account these restrictions. Following a suggestion of Balzer, Sneed (1976) construes these 'expanded cores' as cores which are specializations of the original core. Sets of cores, structured by the specialization relation, define 'core-nets'; together with nets of intended applications these core-nets constitute sets of theory-elements or 'theory-nets'.

Balzer and Sneed (1977) consider three different basic relations between theory-elements: theoretization, specialization, and reduction; they also claim that all the other interesting relations can be defined by means of them.

Let $T' = \langle K', I' \rangle$ be a theory-element which is obtained from another theory-element $T = \langle K, I \rangle$ by adding new theoretical components to the members of M_p. Assume that an element x' of M'_p satisfies the fundamental laws of T' (*i.e.*, $x \in M'$) only if the corresponding element $x = \theta(x')$ of M_p satisfies the fundamental law of T(i.e., $x \in M$). (Here $\theta: M'_p \rightarrow M_p$ is the restriction function.) Then T' is obtained from T by *theoretization*, which is denoted by $T'\tau T$.

This definition is ambiguous, since it does not tell whether some of the K-theoretical terms turn out to be non-K'-theoretical or not. The definition D6 in Balzer and Sneed (1977), p. 200, indicates that *no* such changes occur: M'_p and M_p are supposed to be $k+n$-and $k+m$-matrices ($m \leq n$), which means that $M'_{pp} = M_{pp}$. On the other hand, the authors suggest that one could require that $M'_{pp} \subseteq M$ (*ibid.*, p. 201), which would mean that *all* K-theoretical terms become non-K'-theoretical. In the general case, we should allow that of m such possible changes $j (0 \leq j \leq m)$ will be realized.

A core K' is a specialization of another core K if $M'_p = M_p$, $M'_{pp} = M_{pp}$, $r' = r$, $M' \subseteq M$, and $C' \subseteq C$, i.e., if K' has a stronger fundamental law than K or more restrictive constraints than K. A theory-element T' is a *specialization* of another theory-element T if core K' is a specialization of core K and $I' \subseteq I$. This is denoted by $T'\sigma T$. The condition $T'\sigma T$ guarantees that $A(K') \subseteq A(K)$ and that $I' \in A(K)$ whenever $I \in A(K)$.

Let $\langle L_0, L \rangle$ be a language-pair for core K and $\langle L'_0, L' \rangle$ a language-pair for core K'. If $T'\sigma T$, then L_0 and L'_0, and similarly L and L', are of the same type. But if $T'\tau T$, then the type of L'_0 (resp. L') is an extension of the type of L_0 (resp. L). Specialization does not involve any conceptual change, while theoretization is connected with conceptual enrichment. The logical theory of the definability and the eliminability of new concepts is thus applicable in the study of the theoretization relation τ.

The notions of theoretization and specialization can be generalized to full theory-elements. Thus, let us say that $\langle K', I', J' \rangle$ is a theoretization of $\langle K, I, J \rangle$ iff (i) $\langle K', I' \rangle \tau \langle K, I \rangle$ and (ii) $z \in AT(K')$

entails $\theta(z) \in AT(K)$ for all $z \in \mathrm{Pot}(M'_p)$. It follows from these conditions that the theoretical claim $J' \in AT(K')$ entails the claims $J \in AT(K)$ and $I \in \underline{A}(K)$. Moreover, if θ' is the restriction function θ': $M'_{pp} \to M_{pp}$, then $\bar{\bar{\theta}}'(I') \in A(K)$.

Let $\langle K', I', J' \rangle$ and $\langle K, I, J \rangle$ be full theory-elements such that $\langle K', I' \rangle$ is a specialization of $\langle K, I \rangle$. The definition of σ guarantees that $AT(K') \subseteq AT(K)$. We shall say that $\langle K', I', J' \rangle$ is a specialization of $\langle K, I, J \rangle$ iff (i) $\langle K', I' \rangle \sigma \langle K, I \rangle$ and (ii) $J' = J \cap M'$. Condition (ii) guarantees that the set of the intended theoretical application remains as large as possible in the step from K to K'.

For the notions of reduction ρ and strong reduction $\bar{\rho}$ between theory-elements, and for the corresponding relations of equivalence $\equiv_\rho$ and $\equiv_\rho$, see Section 3(b) below.

(f) *Theory-nets*

Let X be a non-empty finite set, and let $\leq$ be a reflexive and transitive relation on X'. Define a relation $\sim$ on X by

$$x \sim y \quad \text{iff} \quad x \leq y \text{ and } y \leq x.$$

Then $N = \langle X, \leq, \sim \rangle$ is a *net*. The *basic set* of a net N is defined by

$$B(N) = \{z \in X \mid z \leq y \to z \sim y\}.$$

A net may contain several basic elements. A net N is *connected*, if for each two basic elements $x, y \in B(N)$ there is a $z \subset X$ such that $z \leq x$ and $z \leq y$. A net is thus trivially connected if $B(N)$ contains only one element. A net N' is an *expansion* of another net $N (N \sqsubseteq N')$ iff $X \subseteq X'$ and $\leq$ is the restriction of $\leq'$ to $X \times X$. Here N' may contain elements which are 'between' some elements of N; moreover, an element $z \in B(N)$ need not belong to $B(N')$. These possibilities are excluded if N is required to be an *initial part* of N' in the following sense: $y \not\leq x$ for all $x \in N$ and $y \in N'$.

A net $N = \langle X, \leq, \sim \rangle$ is a *theory-net* iff X is a finite set of theory-elements and $I' = I$ implies $K' = K$ for all $\langle K, I \rangle$ and $\langle K', I' \rangle$ in X. To each theory-net there naturally corresponds an associated *core-net* N^* and an *application-net* N^+. If $N = \langle X, \leq, \sim \rangle$ is a theory-net, then the claim of N is the following

(9) $\quad I \in A(K) \quad$ for all $\quad \langle K, I \rangle \in X$

Obviously (9) reduces to (2) if X has only one element. Among the most important special cases of theory-nets are *σ-theory-nets* of the form $N = \langle X, \sigma, =\rangle$, i.e., such that $\leq$ is the specialization relation σ and, hence, $\sim$ is the identity (Sneed, 1976). In this case, $\langle X, \sigma\rangle$ is a partially ordered set of theory-elements. Other examples are of the form $N = \langle X, \tau, \equiv_\tau\rangle$, where τ is the theoretization relation and $x \equiv_\tau y$ iff $x\tau y$ and $y\tau x$, and of the form $N = \langle X, \rho, \equiv_\rho\rangle$ and $N = \langle X, \bar{\rho}, \equiv_{\bar{\rho}}\rangle$ (see Balzer and Sneed, 1977, pp. 210–211).

Following Moulines, a theory-net N is said to be a theory-net relative a scientific community SC and time t iff each $\langle K, I\rangle$ in N is a theory-element relative to SC and t. This pragmatic conception of theory-nets can be used to represent the growth of scientific theories (see Section 3a).

A net $N = \langle X, \leq, \sim\rangle$ is a *full theory-net* iff X is a finite set of full theory-elements such that $J' = J$ entails $I' = I$, and $I' = I$ entails $K' = K$, for all $\langle K, I, J\rangle$ and $\langle K', I', J'\rangle$ in X. The *theoretical claim* of N is

(10) $\quad J \in AT(K) \quad$ for all $\langle K, I, J\rangle \in X$;

this claim (10) entails the claim (9). To a full theory-net N there corresponds a net $N^{\oplus}$ of the theoretical applications J.

(g) *Remarks on the structuralist concept of a theory*

After this extensive survey of the basic concepts of the structuralist conception of a theory, we can now make some methodological and philosophical comments on its general nature.

Let us first note that the structuralist conception is very general: it seems that it can be applied within all fields of empirical science and equally well in mathematics. To distinguish mathematical theories and empirical theories from each other further conditions are needed. Thus, in his definition of 'theories of mathematical physics' Sneed requires that the set I of intended applications should consist of 'physical systems' which are 'of the same kind' and which have 'connected' domains (see Sneed, 1971, pp. 250–260; cf. Stegmüller, 1976a, pp. 163–165).[28] These more or less intuitive requirements are not very clear – here, again, it seems that a more precise definition could be given by talking about the vocabulary associated with a theory. Moreover, it is conceivable that two theories in different fields of science are concerned with precisely the same nontheoretical structures and differ from each other only by virtue of their specific theoretical vocabulary. Thus, we might say that a theory-element $\langle K, I, J \rangle$ (or a theory-net) belongs to the field X if and only if the formulation of the fundamental law (and special laws) and constraints is given by using the theoretical vocabulary of field X.

It was seen, in subsection (a) above, that to a core $K = \langle M_{pp}, M_p, r, M, C \rangle$ one can correlate two languagees L and L^* and the formulations Σ and Γ of M and C in L and L^*, respectively. Hence, to a theory-element $\langle K, I \rangle$ (or $\langle K, I, J \rangle$) one can correlate a pair (or a triple) which involves the formulations – rather than the set-theoretical representations – of the laws and the constraints. It is further interesting to ask whether it is possible to give formulations to the claims $I \in A(K)$ and $J \in AT(K)$, too, in L^*.

The theoretical claim $J \in AT(K)$ can be formulated in L^* by the following sentence:

$$(11) \qquad \forall z \in J(z \models \Sigma) \ \& \ \Gamma^J,$$

where Γ^J is the relativization of Γ with respect to J.[27] Similarly, the claim $I \in A(K)$ can be formulated by

$$(12) \qquad \exists y(y \subseteq M_p \ \& \ \bar{r}(y) = I \ \& \ \forall z \in y(z \models \Sigma) \ \& \ \Gamma^y).$$

Here (11) is a *formulation of the theoretical claim* of the full theory-element $\langle K, I, J \rangle$.[29] Alternatively, we might say, as Harris (1978) essentially suggests, that the set of all the (logical and mathematical) consequences of (11) in L^* is an L^*-theory *based on* Σ and Γ. This proposal gives *a dual-level statement view* of scientific theories: a theory consists of first-order axioms Σ together with some metalinguistic postulates Γ.

Given a full theory-element $\langle K, I, J \rangle$, one can find an L^*-theory corresponding to it, and *vice versa*. It follows that the structuralist conception of theories is essentially equivalent to a suitably emended statement view of theories. It seems, therefore, that Stegmüller, at least in (1976a), has overstated the importance of the question whether a theory consists of statements or of set-theoretical structures.

But a more important difference between the *old* statement view and the Sneedian concept of a theory seems to underlie the notion of an 'intended model' or 'intended application'. According to Stegmüller (1976a), pp. 70–71, the former view is based upon a "fiction" about one "cosmic application" of a theory, while the latter assumes that there are several intended applications for a theory. The importance of this difference can be emphasized by noting that the notion of constraint is dispensable if a core has only one intended application – in this case the structuralist conception of a theory would be essentially equivalent to the old one-level statement view.

In motivating the multi-application approach, there are three different grounds which should be kept separate. One of them could be called 'Wittgensteinian', the second 'instrumentalist', and the third 'pragmatist'.

What could it mean that a theory has one "cosmic application"? The answer is given by Hempel:

> one source of misunderstanding is the view . . . that a hypothesis of the simple form 'Every *P* is a *Q*', such as 'All sodium salts burn yellow', asserts something about a certain limited class of objects only, namely, the class of all *P*'s. This idea involves a confusion of logical and practical considerations: Our interest in the hypothesis may be focussed upon its applicability to that particular class of objects, but the hypothesis nevertheless asserts something about, and indeed imposes restrictions upon, *all* objects . . .. (Hempel, 1965, p. 18.)

In other words, the law 'Every *P* is a *Q*' asserts about *all* objects x a conditional statement that *if* x is *P then* it is also a *Q*. But another way of viewing this law is to say that in it the concept *Q* is applied to certain *intended cases*, namely, to all *P*'s. (More generally, one could say that in this law the concept 'If *P* then *Q*' is applied to some class which includes all *P*'s.) According to this assumption, the range of a generalization is not the whole universe of discourse but rather a pragmatically relativized class of 'intended' cases. If the different *P*'s could be enumerated by $a_1, \ldots, a_n$, then the generalization 'Every *P* is a *Q*' could, on this account, be replaced by the conjunction $Q(a_1) \& \ldots \& Q(a_n)$.

By deliberate inclusion of pragmatic factors within the notion of law, the latter alternative involves a "confusion of logical and practical considerations". This is, in fact, the key idea of the Sneed–Stegmüller approach to theories. Instead of formulating Newton's theory by the *statement*

$$(x)\ (x \text{ is a physical system} \supset x \text{ satisfies Newton's laws}),$$

which applies to the whole universe, they represent it by the *pair* ⟨Newton's laws, Intended applications⟩. This is analogous to replacing the generalization 'Every P is a Q' by the pair $\langle Q, I\rangle$, where I the class of P's. The main virtue of this approach seems to be based on the fact that it can be used in those situations where the class of the intended applications is an 'open class', that is, a class which is defined by the *method of paradigmatic examples*, i.e., by giving a list of some typical cases and admitting into this class other cases which are sufficiently similar to these exemplars (cf. Stegmüller, 1976a, pp. 170–185). As this sort 'openness' of concepts was emphasized by Wittgenstein, one can say that the need for the pragmatic 'non-statement' approach may arise from the 'Wittgensteinian' nature of concept formation in science.

Another motivation for the assumption of multiple intended applications of theories is based upon the idea of theories as *tools*. Reference to theories of pure mathematics is illustrative here. In mathematics, there are two basically different types of theory construction. One of them starts from a particular mathematical structure $\mathfrak{A}$ and tries to find a theory which gives as complete axiomatization as possible for the class $Th(\mathfrak{A})$ of all true sentences about this structure (e.g. axiomatic theories of number theory). Another starts from a set of abstract mathematical assumptions and tries to find as many applications as possible where these assumptions are true (e.g. algebraic theories like group theory).[30] In the former case, the theory has *one intended application* (e.g. the set N of natural numbers), while in the latter case the theory has *several intended applications*. The criteria for appraising these sorts of theories are also different: in the former case, the number of obtained truths about the given structure, and, in the latter case, the number of interesting theoretical and practical applications.[31]

In the Sneedian approach, theories of mathematical physics are thought to be more like group theory than like number theory: such theories contain a mathematical core which is a "tool" or an "instrument" that is hoped to be useful in dealing with certain "chunks of experience" (Sneed, 1971, p. 306). A theoretician creates mathematical cores (Newton's laws, Maxwell's equations, hyperbolic geometry, etc.), and after some initial successes in applying this core to some phenomena attempts will be made to enlarge the scope of these applications to other phenomena.

One cannot deny that there is an instrumentalist flavor in this conception of theories. A realist might indeed argue that *scientific theories should be regarded as analogous to number theory rather than to group theory*: the ultimate aim of theorizing in science is to find the truths about one particular structure, viz. the actual world.[32] Sneed regards his view as "a kind of compromise between an instrumentalist view of theories and a descriptive, or statement view" (Sneed, 1971, p. 306). Moulines says that it would be "unjust to call Sneed's view an instrumentalist one" (Moulines, 1975a, p. 429), and Stegmüller points out that, according to the structuralist conception, "theories are not instruments for creating fictitious pictures of reality, but tools for making empirical claims about reality" (Stegmüller, 1976, p. 163). However, in formulating the 'empirical claim' of a theory–element $T = \langle K, I \rangle$, no function symbols for the K-theoretical components but only existentially quantified function variables are needed. This is one of the reasons why Przełecki (1974) and Tuomela (1978) have concluded that the Sneed–Stegmüller approach better suits an instrumentalist than a full-blown realist.

One difficulty in classifying the position of Sneed and Stegmüller[33] is the fact that their approach does not contain any rules against using the K-theoretical functions – which are *not* used in formulating the 'empirical claim' of $\langle K, I \rangle$ – in the formulation of the 'empirical claim' of *other* theory-elements $\langle K', I' \rangle$, even if their use in the latter connection need not be more testable than in the former. On the basis of this fact, Sneed might rebut the charge of instrumentalism and claim that he takes the theoretical function as seriously as the nontheoretical ones. Pursuing this realist line, I have suggested above that one should add *theoretical claims* of theory-elements to the structuralist formalism. Stegmüller's inclinations are more decidedly instrumentalist, however, which can be best seen in his treatment of theory-dynamics and of the concept of scientific progress (cf. Section 3 below).

Third reason for not accepting the idea that theories have only one intended application is a practical one: the whole universe in its all variety is never the object of one single inquiry. All inquiries concern only some aspects of some parts of the universe, where the focus depends upon the problems and the corresponding 'demands of information'[34] which have given rise to the inquiry. But while inquiries are in this sense 'local', theories are not supposed to be restricted in their applicability to one single research problem. The theoretical importance of these points have been emphasized especially by the pragmatist philosophers of science.

By saying above that one should keep separate the 'Wittgensteinian', 'instrumentalist', and 'pragmatist' motivations for the multi-application approach I mean the following: one can appreciate both the method of paradigmatic examples in the formation of 'open' concepts and the pragmatist emphasis on the local aims of inquiries without committing oneself to an instrumentalist view of theories. This is a philosophical thesis which cannot be proved by any simple argument, but in 3(c) below I try to outline some of my reasons for endorsing it.

3. THE DYNAMICS OF THEORIES

According to Stegmüller, the structuralist conception of theories has five important advantages over the statement view:

> (1) With it a concept corresponding to Kuhn's notion of '*normal science*' can be introduced in an unforced way. (2) With it a *concept of progress* can be introduced which also covers the revolutionary cases. (3) The phenomenon of the immunity of theories to 'recalcitrant experience' can be made clear and understandable. (4) It permits an elegant simplification of what Lakatos intended his *theory of research programmes* to achieve. (5) It removes the danger – and this is perhaps the main advantage – of falling into a rationality monism and, thus, into the *rationality rut* of assuming there could be but *one single* source of scientific rationality. (Stegmüller, 1977, p. 272.)

All these points are related to the dynamic aspect of scientific theorizing.

This section is a commentary on Sneed's and Stegmüller's treatment of theory-dynamics. I shall discuss, in this order, (a) the structuralist explication of Kuhnian normal science, (b) the characterization of scientific revolutions, and (c) the concept of scientific progress.

(a) *Normal science*

Two different strategies can be used in the explication of normal science. First, one defines what is meant by saying that *a person p holds a theory*; then persons who hold the same theory can be said to *belong to the same normal scientific tradition.* This gives an explication of 'Kuhnian' normal science if the persons are required to hold the same *Kuhn-theory* – that is, essentially, a theory in which the class of the intended applications is defined by the method of paradigmatic examples.[35] In the other approach, one employs theory-nets which are relativized to a scientific community *SC* and time t; certain temporal sequences of such nets (relative to the same *SC*) are then characterized as *Kuhnian theory evolution.*[36]

Before stating any formal definitions, let us note that in both of these strategies one has to presuppose the notion of a scientific community as a primitive concept (cf. Section 2(b) above). Therefore, I shall not assume any sharp difference between them.

According to Balzer and Sneed, a quadruple $\langle K_0, I_p, N_p, \mathfrak{N}\rangle$ is a *Kuhn-theory* if and only if (i) $\langle K_0, I_p\rangle$ is a theory-element, (ii) N_p is a σ-theory-net with $B(N_p) = \langle K_0, I_p\rangle$, and (iii) $\mathfrak{N}$ is the set of all σ-theory-nets N such that $N_p \sqsubseteq N$ and $B(N) = \langle K_0, I_0\rangle$ for some $I_0 \supseteq I_p$.

Here K_0 is a core which defines the basic conceptual structure of the theory. I_p is the set of the *paradigmatic applications* of core K_0, i.e., the first successful applications of K_0 by the founder(s) of the theory. N_p is the *paradigmatic σ-theory-net* which indicates how core K_0 is applied to I_p. A *σ-theory-net* N is *admissible* relative to K_0, I_p, and N_p if and only if N is an expansion of N_p and the (unique) basic element of N contains the core K_0 and a set I_0 of intended applications which is an enlargement of I_p. Thus, $\mathfrak{N}$ is the set of all admissible σ-theory-nets relative to K_0, I_p, and N_p.

The set $\mathfrak{N}$ contains all the potential further developments of the theory. The state of the actual development of the theory at time t is a net $N^t \in \mathfrak{N}$. The basic element $\langle K_0, I_0^t\rangle$ of N^t contains all up to t successful (non-theoretical) applications of K_0, not the hoped-for applications. The *historical development* of the Kuhn-theory $\langle K_0, I_p, N_p, \mathfrak{N}\rangle$ can thus be described as a sequence $N^0, N^1, \ldots, N^t, \ldots$, where $N^0 = N_p$ and $N^t \in \mathfrak{N}$ for all t. A step from N^t to N^{t+1} in this development is *cumulative* if N^{t+1} is a proper expansion of N^t. This can happen in three different cases: (i) the core-nets $(N^t)^*$ and $(N^{t+1})^*$ are identical but some of the elements I_i^t of the application-net $(N^t)^*$ are replaced by their proper supersets I^{t+1} in $(N^{t+1})^+$, (ii) $I_0^t = I_0^{t+1}$ but $(N^{t+1})^*$ is a proper expansion of $(N^t)^*$ (i.e., the range of applications remains the same, but new special laws are added), (iii) both the application-net and core-net grow as in (i) and (ii).

Sneed and Balzer suggest that Kuhn-theories correspond precisely to the products of scientific communities during periods of Kuhnian normal science. The historical development of a Kuhn-theory may be only partly cumulative (in the technical sense), but in any case the scientists who are working with this theory at time t are committed to "the use of a certain set of basic concepts" (i.e., core K_0 remains constant) which are "applied to a certain kind of phenomena" (i.e., the nets in $\mathfrak{N}$ are uniquely based). There is also a historical continuity between these scientists and the creator of the theory: they accept the initial paradigm applications of K_0 ($I_p \subseteq I_0^t$) and the paradigmatic theory-net N_p($N_p \sqsubseteq N^t$). If they change any of the elements K_0, I_p, or N_p, they have abandoned to work under the original 'paradigm' or 'disciplinary matrix'[37] and, therefore, do not belong to the same tradition of normal science any more.

In the definition of Kuhn-theories $\langle K_0, I_p, N_p, \mathfrak{N}\rangle$, the status of class $\mathfrak{N}$ is problematic. In the first place, it is a huge class: as Kuhn-theories are supposed to be products of scientific communities, one cannot restrict the admissible theory-nets N e.g. to those which have a true non-theoretical claim. Secondly, $\mathfrak{N}$ seems to be superfluous in the sense that it is definable by means of the three other elements (i.e., K_0, N_p, and I_p) of a Kuhn-theory. For these reasons, one might suggest that $\mathfrak{N}$ should be replaced by a class $\mathfrak{M} \subseteq \mathfrak{N}$ which contains those admissible σ-theory-nets whose claims can be reasonably be expected to be successful. Class $\mathfrak{M}$ thus introduces a *heuristics* to a Kuhn-theory. Let us call the members of $\mathfrak{M}$ *promising* theory-nets. Then we can require that persons who *hold* the Kuhn-theory $\langle K_0, I_P, N_p, \mathfrak{M}\rangle$ at time t agree about the confirmed applications I_0^t of K_0 and the specializations of K_0 in N^t and about the class $\mathfrak{M}^t \subseteq \mathfrak{M}$ of the promising expansions of N_p at t but they may disagree in their *claims* that certain expansions of N^t are (or will be found to be) applicable to sets including I_0^t. This separation of the confirmed and intended applications at t shows that there may exist 'subschools' within a normal scientific tradition. Moreover, a person may hold the same theory at different times even if his beliefs change.[38]

The historical development of Kuhn-theories describes a pattern of the growth of scientific theories. But does it correspond to what Kuhn has called 'normal science'? And does it represent anything which takes place within the actual development of science?[39] Of these two questions, I shall concentrate on the first one.

Let us recall that, for Kuhn, normal science is research carried out by a scientific community which is bounded together by a shared 'disciplinary matrix', i.e., a constellation of group commitments which may include such elements as 'symbolic generalizations', beliefs in particular 'models', and values. Further, they include 'exemplars' or "concrete problem-solutions that students encounter from the start of their scientific education, whether in laboratories, on examinations, or at the ends of chapters in science texts" and "at least some of the technical problem-solutions found in the periodical literature..." (Kuhn, 1970, p. 187). Scientists within a community learn to do their job – to use the symbolic generalizations of a matrix – through the study of exemplars, and the existence of a shared matrix explains the "relative fulness of their professional communication and the relative unaminity of their professional judgments" (*ibid.*, p, 182). Through the matrix, the scientific community acquires "firm answers to questions like the following: What are the fundamental entities of which the universe is composed? How do these interact with each other and with the senses? What questions may legitimately be asked about such entities and what techniques employed in seeking solutions?" (*Ibid.*, pp. 4–5.) As the matrix provides a criterion for choosing problems which can be assumed to have solutions, normal science can be described as an activity of 'puzzle-solving'. On the other hand, as the possession of a matrix involves a commitment to a *Weltanschauung*, the replacement of one matrix by another – a scientific revolution – can be characterized as a 'change of world view'.

In a Kuhn-theory $\langle K_0, I_p, N_p, \mathfrak{M}\rangle$, the elements which correspond to Kuhn's exemplars and disciplinary matrix are I_p and $\langle K_0, I_p, N_p\rangle$, respectively. As a first objection to these identifications, one could point out that it would seem more natural to take the exemplars to be possible models in M_p rather than partial possible models in M_{pp}. In other words, (at least some of) the exemplars should include also the K_0-theoretical functions – for example, the mass and force functions in the case of Newtonian mechanics (cf. Kuhn's examples of exemplars in Kuhn, 1970, pp. 187–190). For this reason, the Kuhn-theory $\langle K_0, I_p, N_p, \mathfrak{M}\rangle$ could be replaced by a quadruple $\langle K_0, J_p, N_p, \mathfrak{M}\rangle$, where $J_p \subseteq M_p$ is the class of the paradigmatic theoretical applications and, further, N_p and the members of $\mathfrak{M}$ are *full* theory-nets (cf. Section 2(f)). This modification of the notion of Kuhn-theory will be assumed below.

As a second objection, let us note that Kuhn's discussion seems to allow for the possibility that the set of exemplars grows over time. Thus, if J_p^t is the set of paradigmatic applications at time t, we require that $J_p \subseteq J_p^t \subseteq J_0^t$. In the same way, net N_p can grow, so that we require $N_p \sqsubseteq N_p^t \sqsubseteq N$ for all N in $\mathfrak{M}$. With this modification, normal science becomes more dynamic than in the Sneedian reconstruction.

A third objection goes in the same direction as the second: the requirement that core K_0 remains constant – or at best is replaced with its specializations – throughout the whole historical development of the theory seems too strict. This requirement excludes from normal science all forms of conceptual change – no matter how trivial they are. For example, one cannot introduce new concepts by means of explicit definitions or make for convenience inessential modifications in the mathematical formalism without breaking the bounds of normal science. It also seems that even more 'radical' conceptual enrichment should be allowed within normal scientific development. Even though it may be reasonable to exclude theoretization (i.e., the introduction of new theoretical concepts) from normal science, one should not rule out the possibility of expanding models with new non-theoretical components. For example, it seems reasonable to expect that at least some periods in the development of modern molecular genetics could be described as normal science. The phenotypes and thus the relevant features of the organisms vary from one application to another. In extending the genetic theory to account for new kinds of hereditary phenomena one therefore may be forced to use in an essential way new non-theoretical concepts.[40]

Continuing this objection, it also seems too strict to require that the fundamental assumptions associated with core K_0 remain entirely unchanged: if the notion of normal science is to be used in the historiography and the sociology of science, we should allow some small changes in the basic laws and constraints of K_0. Arguing for this kind of flexibility, Laudan assumes in his characterization of 'research traditions' (Newtonians, Cartesians, etc.) that at each time there is a set of core assumptions which define a tradition – but this core may change through time (Laudan, 1977, p. 99). We could perhaps go farther and treat even the temporal slices of research traditions as defined by cluster concepts; instead of giving necessary and sufficient conditions for the membership among the Newtonians at time t, we could say that a person supports the Newtonian tradition at t if and only if he satisfies a sufficient number of criteria which are 'semantically relevant' to this classification.

As a fourth objection, one may argue that Kuhn-theories are not able to serve all functions of Kuhn's disciplinary matrixes. It should be noted, however, that nothing told above prevents the formulation of the fundamental law of K_0 to be an existential claim or a statement involving value concepts, so that core K_0 may express ontological commitments and axiological assumptions. But, on the other hand, it is not quite clear to what extent core K_0 can be made to reflect various methodological assumptions concerning proper methods and techniques in 'puzzle-solving'. Perhaps the most crucial question in this direction is the following: does the notion of Kuhn-theory account for the problem generating and problem determining character of disciplinary matrixes?[41] The answer seems to be negative: even though the core K_0 generates questions of the form 'Does K_0 apply to model z?', the Sneedian notion of Kuhn-theory does not explain how such questions turn out to be (or involve) 'puzzles' in Kuhn's sense. The requirement that the set I_0^t of successful applications of K_0 at time t includes as a subset the set I_p of paradigmatic applications is not enough. Following Kuhn's ideas, one should require that the new applications are *similar* to the paradigmatic ones. One should assume that there is a – more or less articulated – relation $\mathfrak{R}$ of similarity or resemblance by means of which proposed applications can be compared to the exemplars.[42] In a formal level, this requirement has been taken into account in my proposal to include the class $\mathfrak{M}$ in Kuhn-theories: one could define the class $\mathfrak{M}^t$ of promising applications at time t as the set of all possible models which are in the relation $\mathfrak{R}$ to some accepted exemplar in I_p^t at time t (and have not yet been tested and ruled out as models of the core K_0).

As a fifth objection, we can point out that a person p may hold, in Sneed's and Stegmüller's sense, two different Kuhn-theories at the same time – even when these theories concern the same non-theoretical structures but conflict with each other in their theoretical part. It is quite possible to believe that the partial possible models can be filled in by theoretical functions in essentially different ways. Instrumentalists have traditionally argued that a scientist can use for his purposes different and conflicting theories.[43] Kuhn explicitly denies such a freedom in the case of 'paradigms': they have a dominant position in scientific communities, and a paradigm change is like a gestalt switch which involves a scientific revolution. In this respect, Sneedian Kuhn-theories and Kuhn's paradigms are not analogous. On the other hand, Kuhn-theories modified so that they involve full theory-nets with intended theoretical applications are more like Kuhnian paradigms – in the sense that a person cannot hold at the same time two such Kuhn-theories which conflict with other.

For Stegmüller, one of the main virtues of the structuralist explication of normal science is the fact that it explains why it is not irrational to assume theories to be immune against refutation by recalcitrant experience. First, a theory need not be 'immunized' against refutation, because "it is the sort of entity of which 'falsified' cannot sensibly be predicated" (Stegmüller, 1976a, p. 13). Secondly, no finite number of unsuccessful attempts to find applications for a core can conclusively show that there are no such applications (cf. *ibid.*, p. 198). Thirdly, if it turns out that a core is not applicable to an intended model *z* (which is not one of the paradigm applications) "the theoretician can decide to drop this element from the set of intended applications" (ibid., p. 174). Because of this immunity, Popper's claim that those working within normal science are uncritical dogmatists is unjustified.

Two comments on these points are in order. First, as Feyerabend (1977) points out, one need not adopt the structuralist view of theories to account for the relative immunity of theories against refutation: this feature has been built e.g. in the 'methodology of research programmes' of Lakatos.[44] Secondly, it is not clear to which extent the structuralist analysis of the situation really supports Kuhn's claim that in normal science negative test results give discredit to the scientist rather than to the theory (cf. Stegmüller, 1976a, p. 13). Kuhn's main ground for this thesis is the assumption that 'paradigms' define legitimate problems and guarantee the existence of their solutions. This is quite different from the idea that a theory can be saved from refutation by making adjustments in its range of applications. Moreover, by dropping an element from the set of the intended applications we in fact change the theory – recall that these applications are a part of theory – and in this way give discredit to the theory rather to the scientist.

As a concluding remark in this subsection, I should like to point out that an explication of normal science is incomplete as long as it does not contain *a theory of belief formation in scientific communities* –

that is, an account of how the members of a community are influenced by the opinions of other members and how they are able to reach a consensus. An interesting theory towards this direction has recently been proposed by Keith Lehrer (1977) who shows that the members of a community bound together by a chain of 'positive mutual respect' can combine their different non-dogmatic degrees of belief to 'consensus probabilities'. The members are even allowed to hold different meaning postulates (i.e., statements which they believe in degree one), so that this account also clarifies the question to what extent communication across different 'paradigms' is possible.

(b) *Scientific revolutions*

Elaborating on Sneed (1971), pp. 297–306, Stegmüller distinguishes between two kinds of scientific revolutions: the transition of 'pretheory' to theory and the 'dislodgment' of one theory by another (Stegmüller, 1976a, pp. 202–219). Stegmüller has emphasized that his notion of 'theory-dislodgment' is not meant as an *explicate* of Kuhn's concept of scientific revolution (Stegmüller, 1978a, p. 72). He has nevertheless argued what revolutions would be like: the dislodged theory should be reducible to the dislodging theory (Stegmüller, 1976a, p. 216).

By scientific revolutions Kuhn means "those non-cumulative developmental episodes in which an older paradigm is replaced in whole or in part by an incompatible new one" (Kuhn, 1970, p. 91). We have already observed that the Sneedian notion of Kuhn-theory does not adequately explain why such paradigm change should be treated as a 'change in world view' and that changes from one Kuhn-theory to another cannot always be called 'revolutionary'.

The emergence of primary theories is obviously a special case of theoretization (see Section 2(e)), and presumably some examples of revolutionary theory-change could be analysed by means of this relation. The main tool which Sneed and Stegmüller use in treating theory-change is the structuralist notion of reduction: a theory-element $T' = \langle K', I' \rangle$ *reduces* to another theory-element $T = \langle K, I \rangle$ if and only if there is a many-one relation R from M'_{pp} to M_{pp} such that the intended applications of T' are correlated with intended applications of T and what T' says about these applications is entailed by what T says about the corresponding applications. A stronger concept of reduction is obtained by requiring the existence of a reducing relation R from M_p' to M_p. These concepts can then be generalized to cover reduction between theory-nets and reduction between Kuhn-theories.[45] They can also be used in illuminating the sense in which rival theories may be said to be 'incommensurable' (Balzer, 1978).

The reduction relation can in principle obtain between theory-elements which employ entirely different conceptual frameworks. Reduction thus defines a 'translation' between these conceptual systems (cf. Balzer and Sneed, 1977, p. 204) – but only on the purely semantical level of potential models or partial potential models. In fact, if L and L' are theoretical languages for cores K and K', respectively, and if T' reduces to T, then L and L' need not be of the same type at all.[45a]

I cannot here discuss problems which are involved in the attempt to use the reduction relations in a (partial) characterization of scientific revolutions.[46] Instead, I make only the general remark that perhaps one should reject the *dichotomy* between normal science and scientific revolutions and replace it by an evaluation of the *degree of radicality* of theory-change. Such evaluation should take into account both the difference between the conceptual frameworks of the theories and the distance between their postulate sets.[47]

(c) *Scientific progress*

According to Stegmüller, normal scientific *progress* can be defined as cumulative development of Kuhn-theories (see Stegmüller, 1976a, p. 220). The emergence of primary theories is always progressive (*ibid.*, p. 212), while theory-dislodgment is progressive if it satisfies – at least approximatively – the reducibility condition (Stegmüller, 1978a, pp. 60–62). Thus, in progressive scientific revolutions "the displaced theory can be partially and approximatively imbedded into the supplanting theory" (*ibid.*, p. 62).

An important feature of this account of scientific progress is the possibility of *progress branching*, both in normal science (set I_0^t can be extended in two different ways either to $I_0^t \cup \{z\}$ or to $I_0^t \cup \{z'\}$) and in revolutionary theory-dislodgment (a theory can be approximately imbeddable in two non-equivalent theories). (See Stegmüller, 1977, pp. 275–277, 284–285.) In such cases, the scientist has to decide on the basis of *value judgments* which way to proceed (*ibid.*, pp. 276, 284).

Stegmüller takes the existence of progress branching to be in sharp contradiction with all *teleological* conceptions of scientific progress (*ibid.*, p. 284), and he thinks that Kuhn is correct in rejecting "a teleological myth of progress" (cf. Kuhn, 1970, pp. 170–173, 206–207), viz. "all variants of a teleological metaphysics explaining progress in terms of 'coming closer and closer to the truth'" (Stegmüller, 1978a, p. 59).

This account of scientific progress is in close agreement with the view that the main virtue of a scientific theory is its *problem-solving ability* (cf. Kuhn, 1970, p. 169, and Laudan, 1977). This view admits an instrumentalist and a realist interpretation, where the former denies, and the latter claims, that the notion of truth is essentially involved in the notion of problem. In giving an account of scientific progress in terms of the problem-solving ability of theories, one can follow Kuhn and *try* to avoid the use of such "metaphors" as "true constitution of nature" (Stegmüller, 1978a, p. 59). But, in spite of his rejection of "teleological metaphysics", Stegmüller's characterization of scientific progress can be claimed to *involve as a special case Popper's problem of verisimilitude.*

To see this, let us see how Stegmüller's definition of progress can be motivated. The class of problems which a Kuhn-theory has solved at time t is represented by the set I_0^t (or, alternatively, by the set J_0^t). A step from $T = \langle K_0, I_0^t \rangle$ to $T' = \langle K_0, I_0^{t+1} \rangle$, where $I_0^t \subseteq I_0^{t+1}$, is progressive, since T' solves more problems than T. A step from $T = \langle K_0, I_0^t \rangle$ to $T' = \langle K_1, I_0^t \rangle$, where K_1 is a specialization of K_0, is progressive, since T' gives a better solution to the same problems as T. Finally, a step from $T = \langle K_0, I_0^t \rangle$ to $T' = \langle K_1, I_1^t \rangle$, where T reduces to T', is progressive, since T' at least solves, at least equally well as T, the same problems as T.[48]

In this account, it is assumed that a 'problem' $z \in M_{pp}$ (alternatively: $z \in M_p$) has been solved by K_0 at time t if and only if the claim that K_0 applies to z is well-confirmed at t (or accepted by the scientific community SC at t). But a well-confirmed claim may be false. It is thus possible that a step from $\langle K, I \rangle$ to $\langle K', I' \rangle$ which exhibits normal scientific progress in Stegmüller's sense leads from a *true* claim $I \in A(K)$ to a *false* claim $I' \in A(K')$ (or from a false claim to another false claim).

In particular, suppose that $I_0^t = \{z\}$. Then one can make normal scientific progress by proposing stronger and stronger laws concerning z, i.e., by specializing the core K_0, and finding empirical confirmation to these laws. But, again, a step from a law σ_1 to law σ_2 may lead us from truth to falsity – or from falsity to another falsity.

Is it reasonable to speak of progress in the cases mentioned above? A positive answer could perhaps be defended by asking: What else could we mean by progress than the introduction of new well-confirmed applications and laws? What other grounds there could be for making claims about progress than the accepted empirical and theoretical evidence? But suppose that at time t we propose to replace a law σ_1 with a stronger law σ_2 and at time $t+1$ we find new evidence which gives us good reasons for rejecting σ_2. Then, instead of still claiming that the step from σ_1 to σ_2 was progressive, it is much more natural to say only that at time t it *seemed* that this step is progressive, while at time $t+1$ it seems otherwise. In other words, we are here employing the notion 'the step from T to T' seems progressive on evidence e'. This seems to presuppose, however, that it is meaningful to distinguish between *real* and *apparent* progress – even though it were the case that we can never know for sure which steps in theory-change are 'really' progressive. Namely, a natural analysis for the claim

(11) The step from T to T' seems progressive on e

can be given by

(12) The claim 'The step from T to T' is progressive' is well-confirmed by e.

In Niiniluoto (1978c), I have shown how the notion of truthlikeness or verisimilitude can be used in defining a non-epistemic notion of 'progress' and its epistemic counterpart 'seems progressive on evidence e'.[49] This account does not involve any kind of "teleological metaphysics": it does not presuppose that in making claims about the progressiveness of scientific developments the scientists already possess the knowledge of the true constitution of the reality, and it does not contain any theses about the future development of science. Moreover, it does not exclude the possibility of 'progress branching' – on the contrary, given a law σ, there are normally several other non-equivalent laws which have a greater degree of truthlikeness than σ.[50] It does not follow, however, that the choice between these alternatives should be made on the basis of value judgments.

One way of utilizing this sort of 'realist' theory of progress in connection with the structuralist theory conception is to define the

value (or degree of adequecy) ***of the solution*** that K gives to a 'problem' $z \in M_p$ by the (estimated) degree of truthlikeness of (the formulation of) the fundamental and special laws of K relative to z.[51] The value of a theory would then depend on the class of problems which it claims to solve and on the value of the given solutions.

NOTES

[1] For a classical exposition of the logistic method, see Church (1956), pp. 1–68. For an excellent account of the development of the Received View and of its problems, see Suppe (1974). See also Tuomela (1973).

[2] See also the articles of Przełecki and Wojcicki in Przełecki and Wojcicki (1977).

[3] See Suppes (1957), Ch. 12, and Suppes (1967a, 1967b, 1969). See also the related work of Herbert Simon, which is most conveniently available in Simon (1977).

[4] This technical suggestion, in the set-theoretic approach, essentially goes back at least to the 1955 dissertation of Adams. A similar suggestion was made independently by Ryszard Wojcicki in 1969; cf. Wojcicki (1973, 1974a).

[5] Sneed's discussion of Ramsey-eliminability was available in a mimeographed form already in 1968. It was commented on by Tuomela in his Stanford dissertation in 1969 (cf. Tuomela, 1973) and by Simon and G. Groen in 1973 (cf. Simon, 1977, Ch. 6.6). See also Swijtink and van Benthem (1978).

[5a] For the Ramsey view of theories, see Ramsey (1931) and Giedymin (1980).

[6] See Stegmüller (1976a), p. *X*.

[7] See Stegmüller (1974, 1975, 1976a, 1976b, 1976c, 1978a, 1978b, 1979).

[8] See also Sneed (1978, 1979), Balzer and Sneed (1978), Sneed and Moulines (1979).

[8a] The structuralist approach has been applied also to arithmetic (Balzer, 1979), geometry (Balzer, 1978b), thermodynamics (Moulines, 1981), biology (Beatty, 1980), classical economics (Händler, 1980a, 1980b; Balzer, 1982a,b), Marxist economics (Diederich and Fulda, 1978; Diederich, 1981), rhytmics (Koster and Schoten, 1982) and Hegel's dialectics (Thagard, 1982). (See also Sneed, 1979.) The concept of approximate reduction is studied in Moulines (1980) and Mayr (1981). See also Pearce (1981a, 1982), Hartkämper and Schmidt (1981).

[9] See Sneed (1976), Stegmüller (1976b), and Kuhn (1976).

[10] For other attempts, see Popper (1963), Suszko (1968), Giedymin (1970), Niiniluoto and Tuomela (1973), Nowak (1975), Scheibe (1976), Törnebohm (1976), Krajewski (1977), Niiniluoto (1978b), Tuomela (1978a), Rantala (1978b) and Przełecki (1978).

[11] For reviews of the Sneed–Stegmüller approach, see Przełecki (1974), Wojcicki (1974a), Moulines (1975a), Feyerabend (1977), Tuomela (1987b), Rantala (1978a, 1980), Pearce (1981a, 1982), Sadovsky (1981), Diederich (1982), Zandvoort (1982a,b) and Sneed (1983).

[12] See Balzer and Sneed (1977). Here Pot(X) is the power set of X.

[13] Stegmüller (1976a) p. 108, avoids this problem by making the more specific assumption that the elements of a matrix are of the form $\langle D, f_1, \ldots, f_n \rangle$, where $D \neq \emptyset$ and each $f_i (i = 1, \ldots, n)$ is a real-valued (partial) function on D.

[14] In a many-sorted structure, one may divide the domain into disjoint sets, such as the

set of nonmathematical objects, the set of real numbers, etc. Many-sorted structures can be reduced, if one wishes, to the ordinary one-sorted structures.

[15] This treatment was used by Carnap in his later work on inductive logic (see Carnap, 1971); it is also a basic tool in modern 'abstract logic' which studies the properties of 'logics' with as few assumptions about syntax as possible. Note also that the characteristic function of $M \subseteq M_p$ is a proposition in the sense of Montague.

[16] Scheibe (1978) makes this point concerning the Sneedian notion of constraint C.

[16a] To say that class M_p and language L have the same type τ_p means simply that all members of M_p are L-structures in the ordinary logical sense. For a more abstract treatment of 'types', see Rantala (1980). The restriction to first-order languages is not essential here. For outlines of a powerful approach which supplements the structuralist program with an abstract logic (which in different applications may be chosen in different ways), see Pearce (1981a), Pearce and Rantala (1983).

[17] If the law M is not an elementary class, it does not have a first-order formulation. If such 'laws' have to be considered in dealing with examples of scientific theories, more powerful (such as higher-order or infinitary) languages are needed.

[18] Harris allows in L^* for quantification over the elements z of M_p and over the elements of the domains of z in M_p, but in the above example this sort of duality in quantification is avoided.

[19] Sneed's criterion of T-theoreticity concerns functions (see Sneed, 1971, pp. 31–33); Stegmüller gives a criterion also for relations (Stegmüller, 1976a, p. 53). It is not clear whether one can speak of theoretical and non-theoretical domains of models – even if in the general definition of cores in Balzer and Sneed (1977) this seems to be presupposed. For discussion about Sneed's criterion, see also Kamlah (1976) and Tuomela (1978b).

[20] A classical argument that the use of such instruments as magnifying glasses, microscopes, etc. is theoretical relative to optics is given by Duhem (1954), pp. 153–158.

[20a] Balzer and Moulines (1980) have made an attempt to give a precise criterion of K-theoreticity without »unclear pragmatic concepts», but in the end they do not get rid of the appeal to »formulas occurring in existing expositions of the theory». They start from the idea that a »method of determination» for a theoretical term P has to be expressible by a set-theoretical formula U(z), where $z \in M_p$, which fixes uniquely the interpretation P^z of P in z, given the other components of z. This requirement is very strong, since by Beth's Theorem it means that P has to be *explicitly definable* by the other terms associated with core K. Further, Balzer and Moulines propose that term P is K-theoretical if and only if each method of determination for P entails the law Σ of core K. Another formulation says that P is K-theoretical if and only if each $z \in M_p$ which satisfies some method of determination U(z) for P is a member of M (i.e., $z \models \Sigma$). However, they note several difficulties with these proposals: first, these definitions have to be restricted to *accepted* methods of determination; secondly, such methods sometimes use only some laws of core K; thirdly, sometimes such methods give only a partial determination of a theoretical function; fourthly, in some cases these methods may rely on other theories than K (e.g., on rigid body mechanics instead of classical particle mechanics). Balzer and Moulines suggest that term P is K-theoretical in the *extended sense* if and only if P is K'-theoretical for all cores K' reducible to K. This suggestion is problematic, however, as Pearce (1981b) points out. Moreover, Balzer and Moulines do not tell whether the »problem of theoretical terms» in the Sneed-Stegmüller sense still exists for their

extended notion of K-theoreticity.

[21] One might, of course, define K-theoreticity by referring, instead of the existing theoretical applications $F_{SC,t}(K)$, to all the possible theoretical applications of K. In this case, one should refer also to the possible expositions of these possible applications. If this makes sense, then the claim that a certain term is K-theoretical would be a strong metascientific hypothesis which is true or false independently of the already known applications of K. However, even in this situation it is still the case that the opinion of the members of SC at time t as to the truth or falsity of this metascientific hypothesis is relative to the existing theoretical applications of K at time t. If the members of SC change their opinion about the status of this hypothesis, then core K will change to another core.

[21a] For this reason, I prefer to call them 'non-K-theoretical' rather than 'K-non-theoretical' as Balzer and Moulines (1980) do.

[21b] It is interesting to note that Balzer has recently allowed that the intended applications I may be any substructures of the potential models in M_p (see Balzer, 1983). This means that I could be a class of theoretical applications in M_p, just as in my suggestion. (Any structure x in M_p is trivially its own substructure in Balzer's sense.) Balzer (1982b, 1983) also realizes that the classical problems of induction or confirmation arise in the structuralist programme as soon as we try to verify the empirical claims of simple theories without any theoretical terms.

[22] Some arguments for the latter alternative have been presented also by Harris (1978).

[23] Cf. Tuomela (1973) and Rantala (1977).

[24] For example, 'double star' is an optics-theoretical but not astronomy-theoretical term in astronomy.

[24a] For Sneed's comments on the first three arguments, see Sneed (1981), pp. 97–98. It seems to me that similar problems that I raise here emerge from Balzer's and Moulines's account of K-theoreticity.

[24b] The idea that theories may be tested by using a part of the theory to confirm theoretical statements entailed by another part has also been discussed by Glymour (1980). One type of such examples could be the following: the basic laws of a theory are employed to fill in the values of theoretical functions, but it is not true a priori that this can be done without violating the constraint C of the theory. Sneed's answer (1981, pp. 98–99) does not touch this problem.

[25] Cf. the distinction between substantive and methodological correspondence rules in Sellars (1963). Cf. also Tuomela (1978b).

[26] For a different treatment of operationalization, see Wojcicki (1973, 1974a).

[26a] I have corrected here a misprint: function o is defined as $o: M_p \to M'_{pp}$. In typical examples of operationalization, the original K-theoretical components are K-theoretical because there are *no* methods for determining their values, and the new core K' does not contain any K'-theoretical terms. A simple case is the following: the concept 'political conviction' is operationalized by 'voting behaviour in the latest elections'.

[27] This means that quantification within members of Γ will be restricted to set I.

[28] For example, the theory of Boolean algebras as applied to switching circuits seems to be, in this sense, a theory of mathematical physics.

[29] For a related way of distinguishing (extralinguistic) theories and their (linguistic) formulations, see Suppe (1974), pp. 204–205.

[30] These assumptions are usually abstracted from the characteristics of certain particular mathematical objects.

[31] For students of mathematics, it is very impressive to see e.g. how easily one can introduce a topological structure into various sets and how much information the basic axioms concerning topological spaces (and their specializations) contain.

[32] This view is a little naive, since it ignores the fact that natural laws have, due to their 'lawlike' character, counterfactual force and thereby refer also to other 'possible worlds' besides the actual one. It is not clear to me whether Sneed's requirement that the domains of the intended applications of physical theories should be connected (Sneed, 1971, p. 255) involves a restriction to the parts of the actual world only.

[33] Another difficulty is the following: it has been argued that the Ramsey-sentence technique which has attracted many instrumentalists is equally well a tool for a scientific realist since it involves the assertion of the existence of certain theoretical entities (cf. Hempel, 1965, p. 216; and Tuomela, 1973). But even if the Ramsey-sentence approach may commit one to accept the existence of certain theoretical *functions*, it is not clear whether the existence of these functions commits one to accept the existence of certain 'real' physical *properties*. If the mass function exists, we can still think *either* that mass is a physical property *or* that masses are just coefficients to be introduced into calculations. (As a historical remark about the Ramsey-technique, let us note that Henri Poincaré used, in his discussion about thermodynamics, the formulation 'There is something which remains constant' for the principle of the conservation of energy; see Poincaré, 1952, p. 127.)

[34] Cf. Levi (1976).

[35] See Sneed (1971), pp. 293–294; Sneed (1976), pp. 130–131; Stegmüller (1976a), pp. 191–195; Balzer and Sneed (1977).

[36] See Moulines (1979) and Stegmüller (1978b).

[37] In the first edition of his *Structure of Scientific Revolutions* (1962) Kuhn used the word 'paradigm' to cover several different sorts of things which a group of scientists may share. In 'Postscript 1969' to the 2nd edition (1970), he distinguishes between two such things which he now calls 'exemplars' and 'disciplinary matrices'. See also Kuhn (1974) and Suppe (1974).

[38] In his slightly different account, Moulines visualizes normal science as a "living net changing and growing in different directions over historical time". He requires essentially that, in a theory-evolution $\{N^t\}$, $t = 0, 1, \ldots$, every theory-element in N^{t+1} must be a specialization of some element in N^t. (This type of requirement will be criticized below.) He does not assume that the paradigmatic theory-element $\langle K_0, I_p\rangle$ is the historically first theory-element in the development of the theory, and he allows that I_p is divided into several 'homogenous' subsets each of which contains its own paradigmatic examples. Cf. also Händler (1982).

[39] These questions are separate: there are those who think that Kuhnian normal science does not exist (see Feyerabend, 1977).

[40] Let K' be obtained from K by adding new non-K-theoretical components to the members of M_{pp} and M_p. Then there may exists between theory-elements $\langle K, I\rangle$ and $\langle K', I'\rangle$ interesting relations which do not seem to be definable by the relations of theoretization, specialization, and reduction. This contradicts the claim of Balzer and Sneed (1977), p. 200. Together with theoretization and reduction relations, this relation may be useful in giving an account of the method of 'idealization and concretization' (cf. Nowak, 1975, and Krajewski, 1977).

[41] For attempts to clarify Kuhn's notion of a 'puzzle' by means of the logic of

questions, see Kleiner (1970) and Girill (1973). Much further work in this direction is needed. Cf. also Niiniluoto (1976).

[42] Cf. Suppe (1974), pp. 135–151.

[43] Cf., for example, Poincaré (1952), p. 163.

[44] See Lakatos (1970); cf. also Laudan (1977). This is connected with the idea that research traditions can be defined by means of 'cluster concepts', so that the successive elements within a tradition bear only a family resemblance to each other. This could be used in the definition of the 'theoretical claims' of Kuhn-theories.

[45] For details, see Sneed (1971), Stegmüller (1976a), Mayr (1976), Sneed (1976), Balzer and Sneed (1977).

[45a] By applying powerful tools from abstract logic, Pearce (1981a, 1982) is able to show that a reduction relation between two cores generates the existence of a translation between the languages of these cores. Therefore, reduction entails some sort of commensurability between theories, so that Stegmüller's idea of giving an account of scientific revolutions by the reduction relation becomes highly problematic.

[46] For interesting comments on this question, see Kuhn (1976).

[47] For attempts to measure theory-distance, see Tuomela (1978a) and Niiniluoto (1978b).

[48] Note that there is no absolute way of counting the number of solved problems, since the solution of several different problems may be reducible to the solution of one and the same problem. This fact is taken into account in the Sneedian concept of reduction.

[49] See also Niiniluoto (1977, 1978b). The basic idea is the following: Let g be a first-order generalization, and let $m(g, C_i)$ be the distance of g from a constituent C_i, where constituents are the strongest consistent generalizations expressible in the language of g. If C_x is the true constituent, then the truthlikeness of g is defined by $m(g, C_x)$. As C_x is normally unknown to us, the most reasonable estimate for the degree of truthlikeness of g on some evidence e seems to be the expected value

$$\mathrm{ver}(g/e) = \sum_i P(C_i/e)m(g, C_i)$$

where the sum is taken over all constituents. A step from g to g' is progressive if and only if $m(g, C_x) < m(g', C_x)$, and this step seems progressive on e if and only if $\mathrm{ver}(g/e) < \mathrm{ver}(g'/e)$.

[50] This is, in fact, not more surprising than the observation that one can travel from one place to another by several different routes. This shows that progress branching is compatible with the view that there is one specified goal of science (cf. Stegmüller, 1978a, p. 59); it is incompatible with this view only if we deny the possibility that the branches later merge together. However, it has been argued, already by William Whewell in 1840, that the most prominent feature of scientific progress is just the merging of different paths in scientific truth-seeking (cf. Niiniluoto, 1978a).

[51] As some attempted solutions may be so misleading that they are worse than the non-existence of a solution, this value might as well be defined by the difference between the truthlikeness of proposed laws and that of a tautology.

BIBLIOGRAPHY

Achinstein, P.: 1968, *Concepts of Science*, Johns Hopkins Press, Baltimore.

Balzer, W.: 1978, 'Incommensurability and Reduction', in Niiniluoto and Tuomela (1978), pp. 313–335. (a)

Balzer, W.: 1978, *Empirische Geometrie und Raum-Zeit-Theorie in mengen-theoretischer Darstellung*, Kronberg, 1978. (b)

Balzer, W.: 1979, 'On the Status of Arithmetic', *Erkenntnis* **14**, 57–85.

Balzer, W.: 1982a, 'A Logical Reconstruction of Pure Exchange Economics', *Erkenntnis* **17**, 23–46.

Balzer, W.: 1982b, 'Empirical Claims in Microeconomics', in W. Balzer, W. Spohn, and W. Stegmüller (eds.), *Philosophy of Economics*, Springer, Berlin, pp. 16–40.

Balzer, W.: 1983, 'Theory and Measurement', *Erkenntnis* **19**, 3–25.

Balzer, W. and Moulines, C.-U.: 1980, 'On Theoreticity', *Synthese* **44**, 467–494.

Balzer, W. and Sneed, J.: 1977, 'Generalized Net Structures of Empirical Theories, I', *Studia Logica* **36**, 195–211.

Balzer, W. and Sneed, J.: 1978, 'Generalized Net Structures of Empirical Theories, Part II', *Studia Logica* **37**, 167–194.

Beatty, J.: 1980, 'Optimal Design Models and the Strategy of Model Building in Evolutionary Biology', *Philosophy of Science* **47**, 532–561.

Benthem, J. van: 1978, 'Ramsey Eliminability', *Studia Logica* **37**, 321–336.

Beth, E., 'Semantics of Physical Theories', in H. Freudenthal (ed.), *The Concept and the Role of the Model in Mathematics and Natural and Social Sciences*, Reidel, Dordrecht, 1961, pp. 48–51.

Bogdan, R.J. (ed.): 1976, *Local Induction*, Reidel, Dordrecht.

Butts, R.E. and Hintikka, J. (eds.): 1977, *Historical and Philosophical Dimensions of Logic, Methodology and Philosophy of Science*, Reidel, Dordrecht.

Carnap, R.: 1971, 'A Basic System of Inductive Logic', in R. Carnap and R.C. Jeffrey (eds.), *Studies in Inductive Logic and Probability*, Vol. 1, University of California Press, Berkeley, pp. 35–165.

Church, A.: 1956, *Introduction to Mathematical Logic*. Princeton University Press, Princeton.

Diederich, W.: 1981, *Strukturalistische Rekonstruktionen*, Fried. Vieweg & Sohn, Braunschweig/Wiesbaden.

Diederich, W.: 1982, 'Stegmüller on the Structuralist Approach in the Philosophy of Science', *Erkenntnis* **17**, 377–398.

Diederich, W. and Fulda, H.F.: 1978, 'Sneed'sche Strukturen in Marx' »Kapital«', *Neue Hefte für Philosophie* 13, 47–80.

Duhem, P.: 1954, *The Aim and Structure of Physical Theory*, Princeton University Press, Princeton.

Feyerabend, P.: 1977, Review of Stegmüller (1973), *The British Journal for the Philosophy of Science* **28**, 351–369.

Fraassen, B. van: 1970, 'On the Extension of Beth's Semantics of Physical Theories', *Philosophy of Science* 37, 325–339.

Frege, G., *Begriffschrift* (1879). English transl. in J. van Heijenoort (ed.): 1967. *From Frege to Gödel: A Sourcebook in Mathematical Logic 1879–1931*, Harvard University Press, Harvard.

Giedymin, J.: 1971, 'The Paradox of Meaning Variance', *The British Journal for the Philosophy of Science* **22** 30–48.

Giedymin, J.: 1980, 'Hamilton's Method in Geometrical Optics and Ramsey's View of Theories', in D.H. Mellor (ed.), *Prospects for Pragmatism*, Cambridge University Press, Cambridge.

Girill, T.R.: 1973, 'The Logic of Scientific Puzzles', *Zeitschrift für allgemeine Wissenschaftstheorie* **4**, 25–40.

Glymour, C.: 1980, *Theory and Evidence*, Princeton University Press, Princeton.

Händler, E.: 1980, 'The Logical Structure of Modern Neoclassical Static Microeconomic Equilibrium Theory', *Erkenntnis* **15**, 33–53. (a)

Händler, E.: 1980, 'The Role of Utility and of Statistical Concepts in Empirical Economic Theories: The Empirical Claims of the Systems of Aggregate Market Supply and Demand Functions Approach', *Erkenntnis* **15**, 129–157. (b)

Händler, E.: 1982, 'The Evolution of Economic Theories. A Formal Approach', *Erkenntnis* **18**, 65–96.

Harris, J.H.: 1978, 'A Semantical Alternative to the Sneed-Stegmüller-Kuhn Conception of Scientific Theories', in Niiniluoto and Tuomela (1978), pp. 184–204.

Hartkämper, A. and Schmidt, H.J. (eds.): 1981, *Structure and Approximation in Physical Theories*, Plenum Press, New York and London.

Heidelberger, M.: 1980, 'Towards a Logical Reconstruction of Revolutionary Change: The Case of Ohm as an Example', *Studies in History and Philosophy of Science* **11**, 103–122.

Hempel, C.G.: 1965, *Aspects of Scientific Explanation*, The Free Press, New York.

Hempel, C.G.: 1970, 'On the "Standard Conception" of Scientific Theories', in M. Radner and S. Winokur (eds.), *Minnesota Studies in the Philosophy of Science*, Vol. IV. University of Minnesota Press, Minneapolis, pp. 142–163.

Hesse, M.: 1974, *The Structure of Scientific Inference*, Macmillan, London.

Hintikka, J. and Tuomela, R.: 1970, 'Towards a General Theory of Auxiliary Concepts and Definability in First-Order Theories', in J. Hintikka and P. Suppes (eds.), *Information and Inference*, Reidel, Dordrecht, pp. 298–330.

Hintikka, J., Gruender, D. and Agazzi, E. (eds.): 1981, *Proceedings of the 1978 Pisa Conference on the History and Philosophy of Science*, D. Reidel, Dordrecht.

Kamlah, A.: 1976, 'An Improved Definition of "Theoretical in a Given Theory"', *Erkenntnis* **10**, 349–359.

Kleiner, S.: 1970, 'Erotetic Logic and the Structure of Scientific Revolution', *The British Journal for the Philosophy of Science* **21**, pp. 149–165.

Koster, J. and Schoten, E.: 1982, 'The Logical Structure of Rhythmics', *Erkenntnis* 18, 269–281.

Krajewski, W.: 1977, *Correspondence Principle and the Growth of Knowledge*, Reidel, Dordrecht and Boston.

Kuhn, T.S.: 1962, *The Structure of Scientific Revolutions*, The University of Chicago Press, Chicago, 2nd ed. 1970.

Kuhn, T.S.: 1974, 'Second Thoughts on Paradigms', in Suppe (1974), pp. 459–482.

Kuhn, T.S.: 1976, 'Theory-Change as Structure Change: Comments on the Sneed Formalism', *Erkenntnis* **10**, 179–199. (Also in Butts and Hintikka, 1977, pp. 289–309.)

Lakatos, I.: 1970, 'Falsification and the Methodology of Scientific Research Programmes', in I. Lakatos and A. Musgrave (eds.), *Criticism and the Growth of Knowledge*, Cambridge University Press, Cambridge, pp. 91–195.

Laudan, L.: 1977, *Progress and Its Problems*, Routledge and Kegan Paul, London.

Lehrer, K.: 1977, 'Social Information', *Monist* **60**.

Levi, I.: 1976, 'Acceptance Revisited', in Bogdan pp. 1–71.

Mayr, D.: 1976, 'Investigations of the Concept of Reduction, I', *Erkenntnis* **10**, 275–294.

Mayr, D.: 1981, 'Investigations of the Concept of Reduction II', *Erkenntnis* 16, 109–130.

Montague, R.: 1961, 'Deterministic Theories', in R. Washburne (ed.), *Decisions, Values, and Groups, II*, Pergamon Press, Oxford, pp. 325–370.

Moulines, C.-U.: 1975, Review of Sneed (1971), *Erkenntnis* **9**, 423–436. [1975a.]

Moulines, C.-U.: 1975, 'A Logical Reconstruction of Simple Equilibrium Thermodynamics', *Erkenntnis* **9**, 101–130. [1975b.]

Moulines, C.-U.: 1976, 'Approximate Application of Empirical Theories: A General Explication', *Erkenntnis* **10**, 201–227.

Moulines, C.-U.: 1979, 'Theory-Nets and the Dynamics of Theories: The Example of Newtonian Mechanics', *Synthese* **41**, 417–439.

Moulines, C.-U.: 1980, 'Intertheoretic Approximation: The Kepler-Newton Case', *Synthese* 45, 387–412.

Moulines, C.-U.: 1981, 'An Example of a Theory-Frame: Equilibrium Thermodynamics', in Hintikka *et al.* (1981), vol. II, pp. 211–238.

Moulines, C.U. and Sneed, J.: 'Suppes' Philosophy of Physics', in R.J. Bogdan (ed.), *Patrick Suppes*, D. Reidel, Dordrecht, 1979, pp. 59–91.

Niiniluoto, I.: 1976, 'Inquiries, Problems, and Questions: Remarks on Local Induction', in Bogdan, pp. 263–296.

Niiniluoto, I.: 1977, 'On the Truthlikeness of Generalizations', in R.E. Butts and J. Hintikka (eds.), *Basic Problems of Methodology and Linguistics*, Reidel, Dordrecht, pp. 121–147.

Niiniluoto, I.: 1978a, 'Notes on Popper as Follower of Whewell and Peirce', *Ajatus* **37**, 272–327.

Niiniluoto, I.: 1978b, 'Truthlikeness: Comments on Recent Discussion', *Synthese* **38**, pp. 281–329.

Niiniluoto, I.: 1978c, 'Verisimilitude, Theory-Change, and Scientific Progress', in Niiniluoto and Tuomela (1978), pp. 243–264.

Niiniluoto, I. and Tuomela, R.: 1973, *Theoretical Concepts and Hypothetico-Inductive Inference*, Reidel, Dordrecht.

Niiniluoto, I. and Tuomela, R. (eds.): 1978, *The Logic and Epistemology of Scientific Change* (Acta Philosophical Fennica **30**), North-Holland, Amsterdam.

Nowak, L.: 1975, 'Relative Truth, the Correspondence Principle and Absolute Truth', *Philosophy of Science* **42**, 187–202.

Pearce, D.: 1981, 'Is There Any Theoretical Justification for a Non-statement View of Theories?', *Synthese* **46**, 1–40. (a)

Pearce, D.: 1981, 'Comments on a Criterion of Theoreticity', *Synthese* 48, 77–86. (b)

Pearce, D.: 1982, 'Logical Properties of the Structuralist Concept of Reduction', *Erkenntnis* 18, 307–333.

Pearce, D. and Rantala, V.: 1983, 'New Foundations for Metascience', *Synthese* 56 (1983), 1–26.

Poincaré, H.: 1952, *Science and Hypothesis*, Dover, New York.

Popper, K.R.: 1963, *Conjectures and Refutations*, Routledge and Kegan Paul, London.

Przełecki, M.: 1969, *The Logic of Empirical Theories*, Routledge and Kegan Paul, London.

Przełecki, M.: 1974, 'A Set-Theoretic versus a Model Theoretic Approach to the Logical Structure of Physical Theories', *Studia Logica* **33**, pp. 91–112.

Przełecki, M.: 1978, 'Commensurable Referents of Incommensurable Terms', in Niiniluoto and Tuomela (1978), pp. 347–365.

Przełecki, M. and Wojcicki, R. (eds.): 1977, *Twenty-Five Years of Logical Methodology in Poland*, Reidel, Dordrecht.

Ramsey, F.P.: 1931, 'Theories', in *The Foundations of Mathematics and Other Logical Essays*, Kegan Paul, London.

Rantala, V.: 1977, *Aspects of Definability* (Acta Philosophica Fennica **29**), North-Holland, Amsterdam.

Rantala, V.: 1978a, 'The Old and New Logic of Metascience', *Synthese* **39**, 233–247.

Rantala, V.: 1978b, 'Correspondence and Non-Standard Models: A Case Study', in Niiniluoto and Tuomela (1978), pp. 366–378.

Rantala, V.: 1980, 'On the Logical Basis of the Structuralist Philosophy of Science', *Erkenntnis* **15**, 269–286.

Sadovsky, V.N.: 1981, 'Logic and the Theory of Scientific Change', in Hintikka *et al.* (1981), vol. I, pp. 49–62.

Scheibe, E.: 1973, *The Logical Analysis of Quantum Mechanics*, Pergamon Press, Oxford.

Scheibe, E.: 1976, 'Conditions of Progress and the Comparability of Theories', in R.S. Cohen *et al.* (eds.), *Essays in Memory of Imre Lakatos*, Reidel, Dordrecht.

Scheibe, E.: 1978, 'On the Structure of Physical Theories', in Niiniluoto and Tuomela (1978), pp. 205–224.

Sellars, W.: 1963, *Science, Perception and Reality*, Routledge and Kegan Paul, London.

Shapere, D.: 1974, 'Scientific Theories and Their Domains', in Suppe (1974), pp. 518–570.

Simon, H.A.: 1977, *Models of Discovery*, Reidel, Dordrecht.

Sneed, J.: 1971, *The Logical Structure of Mathematical Physics*, Reidel, Dordrecht.

Sneed, J., 1976, 'Philosophical Problems in the Empirical Science of Science: A Formal Approach', *Erkenntnis* **10**, pp. 115–146. (Also in Butts and Hintikka, 1977, pp. 245–268.)

Sneed, J.: 1978, 'Invariance Principles and Theoretization', in Niiniluoto and Tuomela (1978), pp. 130–178.

Sneed, J.: 1979, 'Introduction', in *The Logical Structure of Mathematical Physics*, 2nd revised edition, D. Reidel, Dordrecht, pp. xvii–xxv.

Sneed, J.: 1981, 'Comments on Bechler, Niiniluoto and Sadovsky', in Hintikka *et al.* (1981), vol. I, pp. 49–62.

Sneed, J.: 1983, 'Structuralism and Scientific Realism', *Erkenntnis* **19**, 345–370.

Stegmüller, W.: 1970, *Theorie und Erfahrung*, Springer-Verlag, Berlin-Heidelberg-New York.

Stegmüller, W.: 1973, *Theorienstrukturen und Theoriendynamik*, Springer-Verlag, Berlin–Heidelberg–New York.

Stegmüller, W.: 1974, 'Theorienstruktur und logisches Verständnis', in W. Diederich (ed.), *Theorien der Wissenschaftsgeschichte*, Suhrkamp, Frankfurt am Main, pp. 167–209. English transl. 'Logical Understanding and the Dynamics of Theories', in W. Stegmüller, 1977, *Collected Papers*, II, Reidel, Dordrecht, pp. 150–176.

Stegmüller, W.: 1975, 'Structures and Dynamics of Theories: Some Reflections on J.D.

Sneed and T.S. Kuhn', *Erkenntnis* **9**, 75–100. Also in *Collected Papers*, II, pp. 177–202.

Stegmüller, W.: 1976a, *The Structure and Dynamics of Theories*, Springer–Verlag, New York–Heidelberg–Berlin.

Stegmüller, W.: 1976b, 'Accidental ("Non-Substantial") Theory Change and Theory Dislodgment', *Erkenntnis* **10**, 147–178. Also in Butts and Hintikka (1977), pp. 269–288.

Stegmüller, W.: 1976, 'Normale Wissenschaft und wissenschaftliche Revolutionen – Kritische Betrachtungen zur Kontroverse zwischen Karl Popper und Thomas Kuhn', *Wissenschaft und Weltbild* **29**, 169–180. [1976c.]

Stegmüller, W.: 1978a, 'A Combined Approach to the Dynamics of Theories', *Theory and Decision* **9**, 39–75.

Stegmüller, W.: 1978, 'The Structuralist View: Survey, Recent Developments and Answers to Some Criticisms', in Niiniluoto and Tuomela (1978), pp. 113–129.

Stegmüller, W.: *The Structuralist View of Theories: A Possible Analogue of the Bourbaki Programme in Physical Science*, Springer-Verlag, Berlin, Heidelberg, New York, 1979.

Suppe, F.: 1972, 'What's Wrong with the Received View on the Structure of Scientific Theories', *Philosophy of Science* **39**, 1–19.

Suppe, F. (ed.): 1974, *The Structure of Scientific Theories*, The University of Illinois Press, Urbana.

Suppes, P.: 1957, *Introduction to Logic*, Van Nostrand, New York.

Suppes, P.: 1967, 'What is a Scientific Theory', in S. Morgenbesser (ed.), *Philosophy of Science Today*, Basic Books, New York, pp. 55–67. [1967a.]

Suppes, P.: 1967, *Set-Theoretic Structures in Science*, mimeographed, Stanford University, Stanford, [1967b.]

Suppes, P.: 1969, *Studies in the Methodology and Foundations of Science*, Reidel, Dordrecht.

Suszko, R.: 1968, 'Formal Logic and the Development of Knowledge', in I. Lakatos and A. Musgrave (eds.), *Problems in the Philosophy of Science*, North-Holland, Amsterdam, pp. 210–222.

Swijtink, Z.: 1976, 'Eliminability in a Cardinal', *Studia Logica* **35**, 73–89.

Thagard, P.: 1982, 'Hegel, Science, and Set Theory', *Erkenntnis* 18, 397–410.

Toulmin, S.: 1972, *Human Understanding*, Vol. I, Princeton Unviersity Press, Princeton.

Tuomela, R.: 1973, *Theoretical Concepts*, Springer, Vienna.

Tuomela, R.: 1978a, 'Scientific Change and Approximation', in Niiniluoto and Tuomela (1978), pp. 265–297.

Tuomela, R.: 1978b, 'On the Structuralist Approach to the Structure and Dynamics of Theories', *Synthese* **39**, 211–231.

Törnebohm, H.: 1976, 'On Piecemeal Knowledge-Formation', in Bogdan, pp. 297–318.

Wojcicki, R.: 1973, 'Basic Concepts of Formal Methodology of Empirical Sciences', *Ajatus* **35**, 168–196.

Wojcicki, R.: 1974a, 'Set-Theoretic Representations of Empirical Phenomena', *Journal of Philosophical Logic* **3**, 337–343.

Wojcicki, R., 1974b, 'Comments on Przełecki', *Studia Logica* **33**, 105–107.

Zandvoort, H.: 1982a, 'Comments on the Notion »Empirical Claim of a Specialization

Theory Net» within the Structuralist Conception of Theories', *Erkenntnis* 18, 25–38.
Zandvoort, H.: 1982b, 'An Extension of Sneed's Reconstruction of Classical Particle Mechanics to Complex Applications, and an Alternative Approach to Special Force Laws', *Erkenntnis* 18, 39–64.

Chapter 7

TRUTHLIKENESS, REALISM, AND PROGRESSIVE THEORY-CHANGE

1. Introduction: Sarton's Definition of Science

In his influential book *The Study of the History of Science* (1936), George Sarton – the first Professor of History of Science at Harvard University and the long-time Editor of *Isis* – formulated a definition of science, dating from his earlier work in 1913:

> »*Definition.* Science is systematized positive knowledge, or what has been taken as such at different ages and in different places.
> *Theorem.* The acquisition and systematization of positive knowledge are the only human activities which are truly cumulative and progressive.
> *Corollary.* The history of science is the only history which can illustrate the progress of mankind. In fact, progress has no definite and unquestionable meaning in other fields than the field of science.» (Sarton, 1957, p. 5.)

In other words, Sarton claims here that scientific progress means the cumulation of systematized positive knowledge. Further, the notion of knowledge is associated by him with the notions of accuracy and truth: science is »search after truth», and

> »our merit is in every case proportional to the ratio of the amount of truth which we discover to the amount discoverable.» (*Ibid.*, p. 18.)

At first sight, Sarton seems to be a typical representative of the traditional empiricist school which views scientific development as piecemeal accumulation of truths. However, his position is in fact more complicated – it contains tensions which also illuminate the reasons why the traditional conception of science was sharply criticized and re-evaluated in the 1960's and 70's. As a result of this recent discussion, it has been questioned whether the notion of progress has any »definite and unquestionable meaning» even in the field of science.[1]

As a historian of science, Sarton was fully aware that »knowledge is not something dead and static, but something fluid, alive, and moving» (*ibid.*, p. 5). He makes a contrast between historical knowledge, which »grows slowly and precariously» and tends asymptotically »towards completeness» without ever reaching its goal (*ibid.*, p. 7), and scientific knowledge the growth of which is »unpredictable, luxuriant, and sometimes explosive in its intensity and destructiveness» (*ibid.*, p. 8). Indeed, Sarton argues, any branch of natural science »may be completely revolutionized at any time by a discovery necessitating a radically new approach to the subject».[2] While we may »safely assume» that chemistry after five hundred years will be as unlike the present as the chemistry five hundred years ago, similar revolutions in the methods, scope, fundamental notions, and contents are not to be expected in »literary subjects», such as Latin grammar, Greek literature, and the history of the eighteenth century (*ibid.*, p. 8).[3]

Sarton's conception of science thus contains three basic theses:

(1) Science, as systematic pursuit of knowledge, is *progressive.*

(2) The growth of science is *cumulative.*

(3) The development of science involves conceptual, theoretical, and methodological *revolutions.*

But how is it possible to combine the theses (1), (2), and (3)? If the progress of science is guaranteed by its cumulative nature, what does progress mean in connection with revolutions? How could science be cumulative and contain destructive explosions at the same time?

2. Cumulation and Revolution in Scientific Change

Many models of scientific change can be viewed as attempts to solve the conflict between theses (1), (2), and (3). In most cases, this is done by accepting two of them and by rejecting or reinterpreting the remaining third one.

The *accumulation-of-truths* view of progress, associated with naive forms of empiricism and rationalism, accepts theses (1) and (2) as unproblematic. On the other hand, thesis (3) is taken to be a claim about extraordinarily rapid periods of cumulation: the so-called 'revolutions' are simply such phases of development where a novel idea or theory breaks through and stimulates further research in a new direction. The main arguments against this view are of two types: first, the underlying epistemology is not well-founded, and secondly, the history of science does not support it. The development of science is at least not entirely cumulative, since science proceeds also by correcting and rejecting earlier results and by replacing old theories with conceptually different theoretical frameworks.[4]

One attempt at 'saving' the accumulation model is based upon the counter-inductive inference that revolutionary developments belong only to the infancy of a scientific discipline and will eventually fade away: after an initial period with radical theoretical changes, a point of *maturity* is reached – and the rest is just linear growth and refinement within a fixed framework.[5] However, this thesis would not relieve us from the task of analyzing scientific development before maturity. Its supporters have also failed to give any clear criteria for judging when the point of maturity has actually been reached. Moreover, historians of science (like Sarton and Kuhn) have usually followed the inductive strategy of predicting revolutions in the future on the basis of past ones.

The representatives of the various types of *scientific realism* accept thesis (1) in the strong form which takes science to be progressive as a *cognitive* (i.e., knowledge-seeking) enterprise. But when scientific theories are interpreted realistically, most revolutions in science have to be understood as changes where one false theory is replaced by another false theory. Therefore, critical realists – who reject the accumulation model and accept theses (1) and (3) – are bound to reject thesis (2) and replace it with a claim that the 'cognitive virtues' of scientific theories increase or may increase within the development of science.[6] The 'epistemic utility' which measures such virtues is usually taken to be *truthlikeness* (verisimilitude), essentially a combination of truth and information, so that scientific progress is viewed as 'approach to the truth' or towards better and more comprehensive representations of reality (e.g., Peirce, Engels, Lenin, Popper, Putnam, Nowak).[7] For Sellars, the primary epistemic utility is *explanatory coherence.*[8]

The realist view is in principle compatible with many different kinds of inter-theory relations which may obtain between two (realistically interpreted) successive scientific theories T_1 and T_2. Among these relations, we have various explications for the *reduction* relation from T_1 to T_2 and the relation of *correspondence* between T_1 and T_2.[9]

Popper's falsificationism is based upon the method of refuting bold conjectures: science is, and should be, in the state of *permanent revolution.*[10] However, Popper tries to combine all the theses (1), (2), and (3) by giving an essentially cumulative explication for the notion of verisimilitude. According to his qualitative definition, theory T' is closer to the truth than theory T only if T' entails at least all the true sentences entailed by T. Popper (1975) argues that »striking progress» in science is always revolutionary in the sense that a new theory should contradict its predecessor. However, in contrast with political revolutions, scientific revolutions can be assessed rationally since they have to be »conservative» in the sense that the new theory »must always be able to explain fully the success of its predecessor». It is precisely this cumulativist trend in Popper's thought which has lead his account of scientific progress to serious troubles (cf. Miller, 1974; Grünbaum, 1976; Popper, 1979, Appendix 2).

Feyerabend, who accepts the idea of permanent revolutions, has radicalized the Popperian position to the view that science is an »essentially anarchistic enterprise». Slightly paradoxically, this leads him to an extremely cumulative view of science, since for him knowledge is »an ever increasing ocean of mutually incompatible (and perhaps even incommensurable) alternatives», where »nothing is ever settled, no view can ever be omitted from a comprehensive account» (Feyerabend, 1975, p. 30). At the same time, he denies the thesis (1) in its Sartonian form, since he does not any more accept a demarcation between science and other human activities.

Kuhn's account of science is a compromise between cumulative and non-cumulative views: while paradigm-based *normal science* is cumulative, periods of normal science end up with a crisis and with a *revolution* where the old paradigm is replaced by a new one. According to Kuhn, different paradigms are »incommensurable», and later theories cannot be said to be closer to the truth than earlier ones. It is nevertheless possible to maintain that revolutions are progressive – in the sense that the new paradigm preserves a »relatively large part» of the problem-solving ability of its predecessor and also resolves »some outstanding and generally recognized problem that can be met in no other way» (Kuhn, 1962, p. 169). The progress is not cumulative, however, since some problems which the old paradigm solved »must be banished», because they are not meaningful any more within the new paradigm. (This situation in paradigm change is sometimes called a 'Kuhn loss'.) Thus, Kuhn's compromise between theses (2) and (3) leads him to re-evaluate the thesis (1): science is not any more characterized in cognitive terms which presuppose the notion of truth, but rather theories are evaluated as instruments for problem-solving.

Lakatos's methodology of *scientific research programmes* is a similar compromise as Kuhn's model, but Lakatos associates his notions of progressive and degenerative programmes with the idea of verisimilitude.[11] Lakatos accepts the cumulative idea that a theory T' superseding another theory T should, besides having corroborated excess content relative to T, contain »all the unrefuted content» of T (Lakatos, 1970, p. 116). On the other hand, Laudan's account of *research traditions* emphasizes the claim that science is a problem-solving rather than a knowledge-seeking activity. Even though science is not cumulative even as a procedure of problem-solving, mainly because of the so-called Kuhn losses (Laudan, 1977, pp. 149–150), Laudan suggests that progress can be defined by an index expressing the problem-solving ability of a theory: the weighted number of the empirical problems solved by a theory minus the weighted number of the anomalies and the conceptual problems associated with it.[12]

For the scientific realists, theories are primarily attempts to give informative and lawlike descriptions of (some aspect of) reality. For the *instrumentalists,* theories are conceptual tools for performing certain tasks (prediction, control, systematization, problem-solving, etc.), so that scientific progress is definable by the achievements of theories relative to these tasks.[13] Most instrumentalists are prepared to accept thesis (3), since there are revolutionary changes on the level of scientific theories. But since they refuse to give a realistic interpretation to theoretical terms, they are able to combine this view with modified theses (1) and (2): science is propressive and (possibly) cumulative on the level of the achievements of theories. A classical expression of this position was given by Duhem[14] in 1908:

> »Scientific progress has often been compared to a mounting tide; applied to the evolution of physical theories, this comparison seems to us very appropriate, and it may be pursued in further detail.
> Whoever casts a brief glance at the waves striking a beach does not see the tide mount; he sees a wave rise, run, uncurl itself, and cover a narrow strip of sand,

then withdraw by leaving dry the terrain which it had seemed to conquer; a new wave follows, sometimes going a little farther than the preceding one, but also sometimes not even reaching the sea shell made wet by the former wave. But under this superficial to-and-fro motion, another movement is produced, deeper, slower, imperceptible to the casual observer; it is a progressive movement continuing steadily in the same direction and by virtue of it the sea constantly rises. The going and coming of the waves is the faithful image of those attempts at explanation which arise only to be crumbled, which advance only to retreat; underneath there continues the slow and constant progress whose flow steadily conquers new lands, and guarantees to physical doctrines the continuity of a tradition.» (Duhem, 1954, pp. 38–39.)

The characterizations of scientific progress in terms of problem-solving (Kuhn, Laudan) or effectivity in prediction and control (Rescher) are variations of this Duhemian theme.

In Niiniluoto (1979b), I characterized as *realist* those accounts of scientific progress which employ the realist or non-epistemic notion of truth (cf. Chapter 5.). It is interesting to note that, as regards the notion of truth, the instrumentalists can be divided into two groups. One of them is prepared to assign a truth value to observational statements but not to theoretical statements – and thus represents a sort of minimal empirical realism.[15] The other group of anti-realists (pragmatists, verificationists, and intuitionists) rejects the notion of truth as correspondence between language and world and replaces the semantic concept of truth with some epistemic surrogate.[16] While the first group may try to give a minimally 'realist' account of progress by relating the notion of progress to the class of true observational statements entailed by a theory, no realist theory of progress is available to the latter group.

Most realists and instrumentalists who deny the cumulative nature of scientific progress still accept the idea that there is some sort of *continuity* even within revolutionary theory-change from T_1 to T_2. The source of this continuity has been searched among the *sentences* belonging to T_1 and T_2,[17] the common *referents* of the theoretical terms in T_1 and T_2,[18] the *models* or successful *applications* of T_1 and T_2,[19] transtheoretically legitimate *problems* that both T_1 and T_2 try to solve,[20] successful *praxis*,[21] or *reasons* connecting ideas in T_2 to their ancestors in T_1.[22] These different accounts indicate the direction where models of scientific progress will most likely develop in the future. They replace the sharp dichotomy between non-revolutionary cumulation and non-cumulative revolution with a more flexible account where theory change may exhibit continuity to varying degrees. And as 'continuous' theory-change contains cumulation as a special case, which may or may not obtain, the realist continuity models preserve as much as seems feasible of the original Sartonian theses (1), (2), and (3).

3. Remarks on Laudan's Problem-Solving Model

In a recent paper on scientific change, Larry Laudan makes the following statement:

> »Given the notorious difficulties with notions of 'approximate truth' – at both the semantic and epistemic levels – it is implausible that characterizations of scientific progress which view evolution towards greater truthlikeness as the central aim of science will allow one to represent science as a rational activity.« (Laudan, 1981b, pp. 144–145.)

Laudan is certainly justified in his criticism of those philosophers who have made their case for realism without adequate semantical and epistemological tools. But, while I agree that the notion of 'approximate truth' is difficult, I also happen to think that a satisfactory theory of degrees of truthlikeness and their estimation is already available. Before I defend this claim in the next sections, some comments on Laudan's own problem-solving model of scientific change are given in this section.

In my view, the recent debates about the nature of 'problems' and their 'solutions' (cf. Koertge, 1981; Mellor, 1981; Laudan, 1981c; Sarkar, 1981; Finocciaro, 1981; Leplin, 1981; Laudan, 1982) indicate that at present the theory of truthlikeness – even though still in progress – is in fact better developed and understood than the problem-solving account of science. However, erotetic logic would give a valuable tool for analyzing the relevant notions – such as various kinds of questions, the criteria for a reply being a potential answer to a question, the distinction between potential and correct answers, the distinction between complete and partial answers, etc. Most of the work in this direction still remains to be done.[23] As most scientific questions are either truth-using or truth-seeking or both[24], I am also convinced that the problem-solving account of scientific change ultimately needs – rather than makes unnecessary – the concepts of truth and truthlikeness.

Following Mellor and Gutting, Sarkar (1981) has argued that inconsistent theories create a trouble to Laudan's model, since they entail every statement and, hence, are bound to have at least as high problem-solving ability as any other rival theory. Laudan (1982) correctly points out that Sarkar ignores here the fact that inconsistency is a serious conceptual problem which strongly reduces the problem-solving ability of a theory. But Laudan (1981c) and Finocciaro (1981) also claim that an inconsistent theory will »always be confronted by as many anomalies as solved problems«. This is hard to combine with Laudan's conception of anomaly: according to him, a problem is anomalous for a theory T if it is not solved by T but is solved by another theory T' (Laudan, 1977, p. 17). If problem-solving means that a »statement of the problem« is »deduced« from a theory (*ibid.*, p. 25), then an inconsistent theory solves all problems. This observation has the following consequence: as soon as an inconsistent theory is introduced, there are no »unsolved« empirical problems any more, so that all the problems that a theory T does not solve are after all anomalies for it. Hence, Laudan's central distinction between unsolved and anomalous problems in the appraisal of theories disappears.

Laudan's other objection to inconsistent theories suggests a different view about problems. He says that the »*raison d'être* of science» is to put us »in a position to anticipate nature» – and this is precisely where inconsistent theories fail (Laudan, 1982, p. 80). In this case, a problem is not represented by a statement which is believed to be true, but rather by a question 'Is P true or not?'. This predictive problem is solved by a theory T which either entails P or entails $\sim P$ but not both. This account of problems would have the desirable consequence that there are problems unsolved by an inconsistent theory. However, it is unfortunately incompatible with Laudan's (1981c) claim that »to speak of those problems whose solutions can be credited to a theory is to refer to those *already observed* states of affairs which the theory entails». A further difficulty for Laudan is the following: if a theory gets credit for its predictions only after their correctness has been observed, then the aim of science is to make *true* predictions[25] – which would make science a truth-seeking enterprise. Alternatively, if we disregard issues about the truth value of predictions, then non-scientific predictive devices (horoscopes, chrystal balls, cards, etc.) could have a very high problem-solving ability.[26]

Laudan's restriction of problems to »already observed states of affairs» can be motivated by a wish to avoid such paradoxes that Grünbaum presented against Popper (cf. Sarkar, 1981). Laudan also emphasizes that, in contrast to the always infinite class of the deductive consequences of a theory, the class of solved problems is always finite. This claim is hard to evaluate, however, until the individuation of problems is analyzed in more detail. For example, Newton's theory is able to solve the problem concerning the orbit of Mars around the sun. It follows that it can solve, precisely in the same way, the corresponding problem for any planet (real or imagined) relative to any other (real or imagined) star. Do we say that Newton's theory gets credit only for those planets whose orbits actually have been presented as problems? Or do we say that Newton's theory solves an infinite number of problems belonging to the same problem type? If we count this problem type as *one* solved problem, then it does not make sense to associate this problem to any »already observed state of affairs».

Similar troubles can be raised for problems like 'What is the position of Mars within interval $(0, t)$?'. If an answer is given as a function of time t, then the theory has given a (correct) description of the position of Mars in an *infinite* number of points. Did it solve one problem or an infinite number of problems?

I don't think that these comments raise any unsolvable problems for the problem-solving approach. However, I don't see any reason to expect that it is easier to define a measure for the problem-solving ability of a theory than to succeed in the notoriously hard task of defining the degree of confirmation of a theory within inductive logic.

4. Truthlikeness and its Estimation

In this section, I shall outline a general theory of truthlikeness. In the next section, I develop some of its consequences for a realist view of scientific progress.

Let **S** be a set of alternative *states of nature*. Each element s in **S** represents a possible state in which the world (or rather some fragment of the world) may be. Let L be a language which is sufficiently rich for describing the states s in **S**.[27] The statement in L which claims that the state s obtains in the world is denoted by the same symbol 's'. Thus,

$$(1)\quad \begin{aligned} &\vdash \bigvee_{s \in \mathbf{S}} s \\ &\vdash \sim (s \,\&\, s') \text{ for all } s, s' \in \mathbf{S},\ s \neq s'. \end{aligned}$$

Conditions (1) guarantee that there is one and only one element of **S** which is *true* (in the ordinary correspondence-theoretical sense); let us denote it by $s_\star$. In other words, statement $s_\star$ expresses what the state of the world actually is.

Set **S** defines a *cognitive problem* of the following form: Which of the states s in **S** is the true state $s_\star$? The strongest *potential answers* to this problem are just the elements s of **S**. All the other relevant answers can be expressed as disjunctions of elements from **S**. Let us define

$$(2)\quad D(\mathbf{S}) = \left\{ \bigvee_{s \in \mathbf{I}} s \mid \mathbf{I} \subseteq \mathbf{S} \right\}.$$

Hence, a statement h of L belongs to $D(\mathbf{S})$ if and only if

$$(3)\quad \vdash h \equiv \bigvee_{s \in \mathbf{I}_h} s \quad \text{for some } \mathbf{I}_h \subseteq \mathbf{S}.$$

The representation (3) is called the **S**-normal form of h. The elements h of $D(\mathbf{S})$ are potential *partial answers* to the cognitive problem associated with **S**. A partial answer h is true if and only if $s_\star \in \mathbf{I}_h$, i.e., $s_\star$ belongs to its normal form. The information content of a partial answer h depends on the width $|\mathbf{I}_h|$ of its normal form. If all elements s of **S** are regarded as equally informative, the *content* of h may be defined by

$$(4)\quad B(h) = \frac{1 - |\mathbf{I}_h|/|\mathbf{S}|}{1 - 1/|\mathbf{S}|}$$

so that $B(s) = 1$ for each $s \in \mathbf{S}$ and $B(h) = 0$ for a tautology h. Otherwise the content of h may be measured by the standard cont-measure of the semantic information theory (cf. e.g. Niiniluoto and Tuomela, 1973).

As examples of cognitive problems the following cases may be mentioned:

(a) **S** consists of the *state descriptions* of a first-order language,
(b) **S** consists of the *structure descriptions* of a first-order language,
(c) **S** consists of the *constituents* of a first-order language,
(d) **S** consists of the *constituents* of a higher-order language,
(e) **S** consists of the *basic causal laws* of a language with the operation of nomic necessity,
(f) **S** consists of *points* in a real-valued interval $[a, b] \subseteq \mathbf{R}$ (or in a region $C \subseteq \mathbf{R}^n$),
(g) **S** consists of all real-valued *polynomials* of degree n ($n \in \mathbf{N}$)
(h) **S** consists of a class of continuous *functions* of the type $f: \mathbf{R}^n \rightarrow \mathbf{R}$.
(i) **S** consists of a class of *probability measures* on a given algebra of sets.

Of these situations, (a) is relevant for cases where we are interested in singular sentences, (b) for numerical statements about class occupancy, (c) and (d) for qualitative generalizations and theories,[28] (e) for nomic generalizations and theories,[29] (f) for point and interval estimation, (g) and (h) for quantitative laws and theories, and (i) for probabilistic laws.[30]

Many scientific problems are *open* in the sense that we are not able to enumerate all of their potential answers. For example, usually we cannot give all the alternative theories that may be used in answering an explanatory question of the form 'Why p?'. However, the basic method of approaching open questions is to transform them to *closed* questions with a known set of potential answers, i.e., to cognitive problems in our sense above. (See Niiniluoto, 1976.) In particular, a question can be 'closed' by fixing the language L in which its answers are supposed to be formulated. If the formulation of the problem turns out to be insufficient or inadequate, we have to proceed to consider *deeper* cognitive problems by extending the language L with new vocabulary. It seems to me that, in this sense, all the interesting scientific problems can in principle be reduced to cognitive problems of the types (a) – (i).

The next move in the theory of truthlikeness is to introduce a metric d in the set **S**. If **S** does not already have a natural topological structure, this can be done by utilizing the syntactic structure of the elements s of **S**. I shall illustrate the basic idea only in cases (c), (f), and (h).

Case (c). If L is a monadic first-order language with a finite vocabulary, its constituents s_i are statements of the form

$$(5) \quad \bigwedge_{j \in \mathbf{CT}_i} (\exists x)\, Ct_j(x) \;\&\; (x)\left[\bigvee_{j \in \mathbf{CT}_i} Ct_j(x)\right]$$

where $Ct_j(x)$ are Carnapian Q-predicates. This means that predicates $Ct_j(x)$,

$j = 1,\ldots, K$, constitute a complete classification system with K mutually exclusive cells. Constituent (5) claims that cells $Ct_j \in \mathbf{CT}_i$ are non-empty and other cells are empty. The simplest measure for the distance $d(s_i, s_k)$ between two constituents s_i and s_k of form (5) is due to W.K. Clifford (1877):

$$(6) \quad d(s_i, s_k) = |\mathbf{CT}_i \Delta \mathbf{CT}_k|/K,$$

i.e., the normalized cardinality of the symmetric difference of sets $\mathbf{CT}_i$ and $\mathbf{CT}_k$. Definition (6) can be modified by taking into account the distances between Q-predicates. These ideas can also be generalized to arbitrary first-order languages with polyadic predicates.[31]

Case (f). The distance between two points $x, y \in \mathbf{R}$ is the absolute value of their difference or the square of their difference:

$$(7) \quad \begin{aligned} d(x, y) &= |x-y| \\ d(x, y) &= (x-y)^2. \end{aligned}$$

If $x, y \in \mathbf{R}^n$, where $x = (x_1, \ldots, x_n)$ and $y = (y_1, \ldots, y_n)$, their distance may be defined by

$$(8) \quad d^k(x, y) = \left(\sum_{i=1}^{n} |x_i - y_i|^k \right)^{\frac{1}{k}}$$

Here the choice k=1 gives the *city-block distance* and k=2 the *Euclidean distance* between x and y.

Case (h). The distance between two n-place functions $f: C \to \mathbf{R}$ and $g: C \to \mathbf{R}$, where $C \subseteq \mathbf{R}^n$, can be defined by

$$(9) \quad d^k(f, g) = (\int_C |f(x) - g(x)|^k \, dx)^{\frac{1}{k}} \; (k=1,2)$$

or by

$$(10) \; d(f, g) = \sup_{x \in C} |f(x) - g(x)|.$$

When a metric $d: \mathbf{S} \to \mathbf{R}$ has been introduced to set $\mathbf{S}$, we may say that the value $d(s, s_\star)$ expresses the *distance* of s *from the truth* $s_\star$. The *degree of truthlikeness* of s (relative to $s_\star$) can then be defined by one of the following functions:

$$(11) \; \begin{aligned} M(s, s_\star) &= 1 - d(s, s_\star) \\ M(s, s_\star) &= 1/d(s, s_\star). \end{aligned}$$

The next step is to extend the function $d: \mathbf{S} \to \mathbf{R}$ to a function $d: D(\mathbf{S}) \to \mathbf{R}$. Let $h \in D(\mathbf{S})$. The simplest strategy is to define the distance of h from s

by some function G of the distances $d(s', s)$ for states s' in the normal form of h:

(12) $d(h, s) = G(\{d(s', s) \mid s' \in \mathbf{I}_h\})$.

Then $d(h, s_\star)$ is the distance of h from the truth, and the degree $M(h, s_\star)$ of truthlikeness of h is defined as in (11). Further, h is *closer to the truth* than h' if and only if $M(h, s_\star) > M(h', s_\star)$, and h is *approximately true* if its degree of truthlikeness is sufficiently high.

The form of function G in (12) is a matter of dispute. One possibility is to choose G so that the degree of truthlikeness of h is given by

(13) $M(h, s_\star) = \gamma m_{\min}(h, s_\star) + (1-\gamma) B(h) m_{\max}(h, s_\star)$,

where $0 < \gamma < 1$, $B(h)$ is defined by (4), and

$$m_{\min}(h, s_\star) = 1 - \min_{s \in \mathbf{I}_h} d(s, s_\star)$$

$$m_{\max}(h, s_\star) = 1 - \max_{s \in \mathbf{I}_h} d(s, s_\star)$$

(cf. Niiniluoto, 1978). Hence, $M(s, s_\star) = 1 - d(s, s_\star)$ for $s \in \mathbf{S}$, and $M(h, s_\star) = \gamma$ for a tautology h. Thus, s is closer to the truth than a tautology if and only if $d(s, s_\star) < 1 - \gamma$.

In the cognitive problem associated with $\mathbf{S}$ the true state $s_\star$ is unknown. Let e be a statement of the available (observational and theoretical) *evidence* which is relevant to the task of finding $s_\star$. Assume that a probability measure P, given e, is defined on set $\mathbf{S}$. In other words, $P(s/e)$ is the probability given e that s is the true state $s_\star$. (If $\mathbf{S}$ is an infinite subset of $\mathbf{R}^n$, P is a n-dimensional probability density on $\mathbf{S}$.) Here P is an epistemic probability measure, and the values $P(s/e)$ are obtained by Bayesian conditionalization from prior probabilities, specified by some system of inductive logic or simply expressing the initial epistemic state of the investigator. To *estimate* the degree of truthlikeness of a hypothesis h in $D(\mathbf{S})$ we calculate the expected value of $M(h, s)$, where the possible values of s are weighted with the probabilities $P(s/e)$:

(14) $ver(h/e) = \sum_{s \in \mathbf{S}} P(s/e) M(h, s)$.

If $\mathbf{S} \subseteq \mathbf{R}^n$, (14) is replaced by the integral

$$\int_{\mathbf{S}} P(s/e) M(h, s)\, ds.$$

By (4), (13), and (14), $ver(h/e) = \gamma$ for a tautology h. If $e \vdash s_0$ for some s_0 in $\mathbf{S}$, then (14) reduces to $ver(h/e) = M(h, s_0)$. Similarly, if $P(s_0/e) \to 1$ for some s_0 in $\mathbf{S}$, when the size of the evidence e increases, then

$ver(h/e) \to M(h, s_0)$.

In the same situation,

(15) $ver(h/e) \to 1$ iff $\vdash h \equiv s_0$.[32]

If follows that

(16) $ver(h/e) \to M(h, s_\star)$ iff $s_0 = s_\star$.

For problems of type (c), the condition $s_0 = s_\star$ holds if and only if evidence e is both true (i.e., does not eliminate $s_\star$) and non-misleading (i.e., exhibits all the qualitative variety relative to L there is in the world).[33] More generally, if $P(s_i/e) \to p_i$ for some $i = 1,\ldots, m$, then

(17) $ver(h/e) \to \sum_{i=1}^{m} p_i M(h, s_i)$.

In this case, the disjunction $s_1 \vee \ldots \vee s_m$ has a relatively high expected verisimilitude.[34] Similarly, if $P(s/e)$ is a probability density which heavily concentrates in region $C_0 \subseteq \mathbf{R}^n$, i.e.

$$\int_{C_0} P(s/e)\, ds$$

is high, then the claim that $s_0 \in C_0$ has a relatively high degree of expected verisimilitude.

5. Progress as Increase of Truthlikeness

Assume now that h_1 and h_2 are two theories in $D(\mathbf{S})$. Then the step h_1/h_2 (i.e., the step from h_1 to h_2) is *progressive* if and only if $M(h_1, s_\star) < M(h_2, s_\star)$. Definition (12) entails then the following theorem (cf. Niiniluoto, 1979b, p. 251).

Thm. 1. (i) If $h \neq s_\star$, then $h/s_\star$ is progressive.[36]
(ii) If h is false, then $h/(h \vee s_\star)$ is progressive.
(iii) If h_1 and h_2 are true and if $h_2 \vdash h_1$ and $h_1 \nvdash h_2$, then h_1/h_2 is progressive.
(iv) If h_1 and h_2 are both false and $h_2 \vdash h_1$, then h_1/h_2 may be either progressive or non-progressive.
(v) If h_1 is true and h_2 is false, then h_1/h_2 may be progressive.

Here (i) says that the step to $s_\star$ from any other theory h in $D(\mathbf{S})$ is progressive. According to (ii), the step from a false theory h to the strongest true theory entailed by h is progressive. Among true theories, but not among false theories, increase in logical strength entails progress (by (iii) and (iv). Finally, (v) follows from the fact that sufficiently informative false theories may have a higher degree of truthlikeness than a less informative true theory.

Let us say that step h_1/h_2 *seems progressive on evidence e* if and only if $ver(h_1/e) < ver(h_2/e)$. Then, corresponding to Theorem 1, we have:

Thm. 2. Assume that $P(s_0/e)$ is sufficiently close to one.
(i) If $h \neq s_0$, then h/s_0 seems progressive on e.
(ii) If $s_0 \notin \mathbf{I}_h$, then $h/(h \vee s_0)$ seems progressive on e.
(iii) If $s_0 \in \mathbf{I}_{h_1}$ and $s_0 \in \mathbf{I}_{h_2}$ and if $h_2 \vdash h_1$ and $h_1 \nvdash h_2$, then h_1/h_2 seems progressive on e.
(iv) If $s_0 \in \mathbf{I}_{h_1}$ and $s_0 \notin \mathbf{I}_{h_2}$, then h_1/h_2 may seem progressive on e.

As a corollary from Thm. 2 (iii), we obtain the following result.

Thm. 3. Assume that $P(s_0/e\&h_1)$ is sufficiently close to one. If $s_0 \in \mathbf{I}_{h_1}$ and $s_0 \in \mathbf{I}_{h_2}$ and if $h_2 \vdash h_1$ and $h_1 \nvdash h_2$, then h_1/h_2 seems progressive on $e\&h_1$.

Thm. 3 shows that, in the case where we have accepted a theory h_1 in $D(\mathbf{S})$ as part of the evidence, a step from h_1 to a logically stronger theory h_2 may seem progressive on the same evidential situation. In some cases, such a theory h_2 may be even incompatible with h_1 (see Niiniluoto, 1979b, p. 253). On the other hand, if we have accepted an element s_0 of $\mathbf{S}$ as our evidence, then no step from s_0 to any other theory h in $D(\mathbf{S})$ seems progressive on s_0, since

$$ver(h/s_0) = M(h, s_0) < 1 = M(s_0, s_0) = ver(s_0/s_0).$$

The only way of making further progress is by *conceptual enrichment*, i.e., by introducing new descriptive vocabulary to the language – and thereby replacing the original problem $\mathbf{S}$ with a deeper problem $\mathbf{S}'$ such that $s \in D(\mathbf{S}')$ for each $s \in \mathbf{S}$ (cf. *ibid.*, p. 254).

If we have accepted a theory as evidence, it is possible to make *retrospective* comparative judgments of verisimilitude concerning historically earlier theories. This idea is illustrated by the following examples. (See also Niiniluoto, 1984b.)

Example 1. When a light ray passes from one medium to a denser medium, it bends towards the perpendicular drawn at the point of incidence. Let α be the angle of incidence and β the angle of refraction (see Fig. 1). Since Ptolemy's *Optica,* it was known that, for two given media, β depends in some regular way on α. The problem of refraction was discussed by many medieval scientists (Alhazen, Grosseteste, Roger Bacon, Witelo). Ptolemy's data were sometimes taken to suggest that $\alpha/\beta = \text{constant}$. For light passing from air to water, we have roughly

$$(18)\ \beta = \frac{2}{3}\alpha.$$

In fact, a more accurate formula which fits Ptolemy's data is

$$(19)\ \beta = \frac{33}{45}\alpha - \frac{1}{400}\alpha^2$$

(cf. Cohen and Drabkin, 1948, pp. 271–281). Grosseteste suggested that the refracted ray divides into two equal parts the angle between the perpendicular and the imagined straight continuation of the incident ray. His law of refraction thus takes the form

$$(20)\ \beta = \frac{1}{2}\alpha.$$

The sine law of Snell (1620) has later been accepted as giving the correct relation between α and β. In Newton's *Opticks*, it is presented as Axiom V: »The Sine of Incidence is either accurately or very nearly in a given Ratio to the Sine of Refraction». In particular, for the case of air and water Newton gives the formula

$$\frac{\sin\alpha}{\sin\beta} = \frac{4}{3},$$

which can be written in the form

$$(21)\ \beta = \arcsin\left(\frac{3}{4}\sin\alpha\right).$$

If we have accepted law (21) as true, then the metrics (9) and (10) give the result that the false laws (20), (18), and (19) – in this order – are increasingly better approximations to the truth.

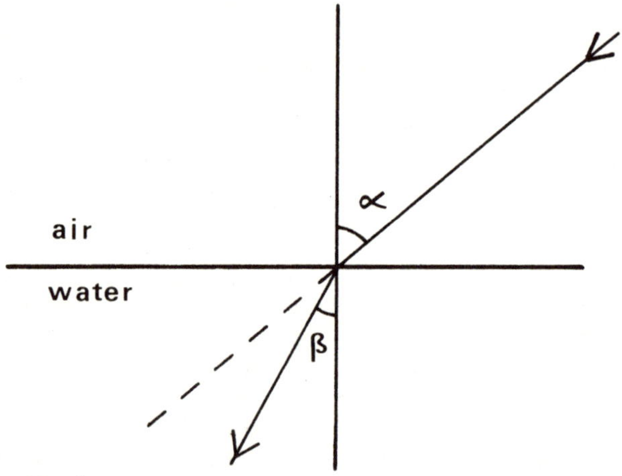

Fig. 1.

Example 2. According to van der Waals's law,

$$\left(p + \frac{a}{V^2}\right)(V-b) = RT,$$

where p is the pressure of a gas, V is its volume, T its absolute temperature, R is the gas constant, a expresses the intermolecular attractive forces, and b the finite size of the gas molecules. When this equation is solved with respect to p, we obtain

$$(22)\ p = \frac{RT}{V-b} - \frac{a}{V^2}.$$

If we assume that (22) is true, then the following claims can be seen to be increasingly truthlike attempts at formulating the gas law:

(23) $p \propto 1/V$ (Boyle)

(24) $p = RT/V$ (Boyle-Mariotte)

$$(25)\ p = \frac{RT}{V} - \frac{a}{V^2}.$$

Here the »ideal gas law» (24) is in the relation of *correspondence* to the laws (25) and (22); similarly (25) is in the relation of correspondence to (22). Namely, from equation (22) we may deduce the following *idealizational laws:*

(26) If $b = 0$, then $p = \dfrac{RT}{V} - \dfrac{a}{V^2}$.

(27) If $a = 0$ and $b = 0$, then $p = RT/V$.

It can also be seen from (22) that, when V is sufficiently large in relation to a and b, the law (24) is approximately true (in the sense of Niiniluoto, 1982b).

As further illustrations of progressive theory-change we give the following two examples (cf. Niiniluoto, 1979b, pp. 254–255), where **S** consists of the constituents of a monadic first-order language L with K Q-predicates (i.e., case (c)). Further, e^c_n describes a sample of n individuals exemplifying c different Q-predicates in the set $\mathbf{CT}_c$, i.e., $|\mathbf{CT}_c| = c$.

Example 3. Let s_1 be a constituent which is compatible with evidence e^c_n, so that $\mathbf{CT}_c$ is a proper subset of $\mathbf{CT}_1$. Assume that a new individual a is found to exemplify a Q-predicate $Ct_0 \notin \mathbf{CT}_1$. (See Fig. 2.) Then s_1 is incompatible with the new evidence $e^{c+1}_{n+1} = e^n_c \,\&\, Ct_0(a)$. Let s_2 be the constituent with $\mathbf{CT}_2 = \mathbf{CT}_1 \cup \{Ct_0\}$. Then the step s_1/s_2 does not seem progressive on e^c_n but seems progressive on e^{c+1}_{n+1}. Let s_3 be the constituent with $\mathbf{CT}_3 = \mathbf{CT}_c \cup \{Ct_0\}$. Then the step s_2/s_3 seems progressive on e^{c+1}_{n+1}.

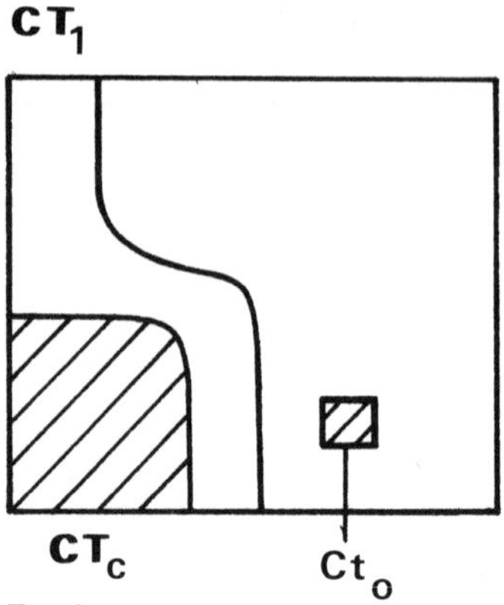

Fig. 2.

Example 4. Let h be a generalization in L which claims that certain b Q-predicates are empty. Thus, h is a disjunction of 2^{K-b} constituents. Assume that evidence e_n^c exemplifies d of these b Q-predicates. (See Fig. 3.) Let s_0 be the constituent which claims that precisely the Q-predicates in $\mathbf{CT}_c$ are non-empty. If n is sufficiently large, formulae (6), (13), and (14) give the result:

$$(28)\ ver(h/e_n^c) \simeq M(h, s_0)$$
$$= \gamma\left(1-\frac{d}{K}\right) + (1-\gamma)\cdot\frac{2^K-2^{K-b}}{2^K-1}\cdot\frac{b-d}{K}.$$

Hence, $ver(h/e_n^c)$ decreases when d grows and b is fixed, and increases when b grows and d is fixed. Let h' be the generalization which revises h simply by weakening it so that it becomes consistent with evidence e_n^c; thus h' claims that $b-d$ Q-predicates are empty. Then, if γ is not very small, step h/h' seems progressive on e_n^c. However, this step is by no means the most

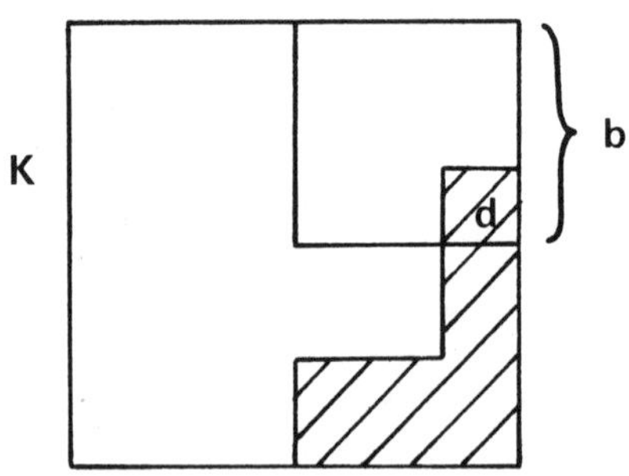

Fig. 3.

progressive, if $b+c-d<K$. Formula (28) will have its maximal value, if h is chosen as the maximally bold generalization compatible with e_n^c. This generalization – which claims that $K-c$ Q-predicates are empty – both revises the original h and has 'excess content' relative to it. Thus, both examples 3 and 4 are related to the Lakatosian conception of progress[35] – without presupposing cumulativity.

We have assumed above that the rival theories h_1 and h_2 belong to the same set $D(\mathbf{S})$, so that they can be expressed as disjunctive statements in the same language L associated with **S**. This assumption may seem problematic in cases where h_1 and h_2 are *incommensurable* theories which have been formulated in different conceptual frameworks. To see how such situations can be handled, let us note first that the comparison of the truthlikeness of two theories h_1 and h_2 is interesting only if h_1 and h_2 are *rival* theories. It may be that statements of the form 'Newton's celestial mechanics is more truthlike than Darwin's theory of evolution' could be shown to be meaningful, but I doubt that. For all purposes that I can imagine, it is sufficient to define the relation

h_1 is more truthlike than h_2

for theories which are rival solutions to a cognitive problem. But this condition means that there has to be a set **S**, and a language L associated with **S**, such that h_1 and h_2 belong to $D(\mathbf{S})$. If h_i was originally a potential solution to problem $\mathbf{S}_i$ $(i=1,2)$, then the common problem **S** for h_1 and h_2 will be deeper than $\mathbf{S}_1$ and $\mathbf{S}_2$, and the language L associated with **S** will be a common extension of the languages L_1 and L_2 (for $\mathbf{S}_1$ and $\mathbf{S}_2$, respectively). Further, if languages L_1 and L_2 are 'incommensurable' in the sense that they presuppose different, or even conflicting, principles as 'analytic truths' or 'meaning postulates' (say, MP_1 and MP_2, respectively), then in language L the relevant comparison has to be made between $h_1\&MP_1$ and $h_2\&MP_2$ (cf. Niiniluoto, 1980, pp. 445–446).

In the next two sections, I shall consider how my account of truthlikeness is able to stand up to the criticism that has recently been presented against the realist views of scientific progress.

6. Realism and Truth

To define scientific progress in terms of increasing truthlikeness amounts to the claim that science is primarily a cognitive enterprise which aims at finding highly truthlike information about reality. This definition gives us a methodological norm for advancing science: prefer those theories which can be claimed to be more truthlike than their rivals. It does not entail that all the steps in the actual historical development of science have been and will be progressive.[36] It is also compatible with different views about the end of scientific progress: some realists think that a 'final theory' will eventually be reached, while some regard such a final theory only as an ideal towards which we shall indefinitely approach without ever reaching it.[37]

L.J. Cohen (1980) has argued that the aim of science should not be truth or truthlikeness *simpliciter* but rather truthlikeness with evidence which justifies belief in truthlikeness. If we modify the classical conception of knowledge by saying that *knowledge* means justified truthlike belief, Cohen's point can be expressed as follows: the aim of science is not truthlikeness but knowledge.[38] In my view, this argument has a bite against those theorists of truthlikeness who have studied the absolute logical (non-epistemic) notion without saying anything about the epistemic problem concerning the estimation of truthlikeness.[39] My distinction between real and estimated truthlikeness (i.e., measures M and ver), and the corresponding distinction between real and estimated progress, give a sufficient reply to Cohen. A rational scientist whose aim is to find highly truthlike theories, and who therefore chooses truthlikeness as his epistemic utility, will try to maximize the expected utility of his choices of theories. This procedure is in fact equivalent to the decision to prefer theories h which have a higher value of estimated verisimilitude $ver(h/e)$ than their rivals, given evidence e (cf. Niiniluoto, 1979b, 1982a).

Our aim in science is to find theories which are correct solutions to cognitive problems. In proposing a theory h, we hope that its real degree of truthlikeness (i.e., $M(h, s_\star)$) is high. But since there are no infallible methods for proving this, the best we can do to realize this aim is to choose h so that $ver(h/e)$ is high. As evidence e may be either false or true but misleading, the fact that $ver(h/e)$ is high is no *proof* that h is in fact highly truthlike. Whether we are *rationally justified* to claim that h is highly truthlike, depends essentially on the rationality of the epistemic probabilities that occur in formula (14) for $ver(h/e)$. In this sense, the problem of estimating degrees of truthlikeness is as difficult as the traditional problem of induction (cf. Niiniluoto, 1982a).

Some realists have tried to prove that science approaches to the truth by a sort of natural necessity.[40] The strongest results in this direction show that, on certain conditions, our empirical procedures lead *with probability one* to approximately true results (cf. Chapter 5). This is sufficient to confute cognitive skepticism which views truth and truthlikeness »utopian» as the aims of science (cf. Laudan, 1981a,c). On the other hand, these results do not guarantee that all truths are accessible to human beings and will be revealed within the history of science. Whether this is the case or not cannot be settled *a priori* by philosophers.[41] In particular, there are no convincing proofs that the predicates 'true' and 'limit of scientific inquiry', when applied to scientific theories, are coextensive.[42]

These observations may clarify the somewhat obscure dispute whether truth is »radically epistemic». Putnam (1980b, 1981, 1982) has suggested that Kant can be interpreted

»as rejecting the idea of truth as correspondence (to a mind-independent reality) and as saying that the only sort of truth we can have an idea of or use for is *assertibility* (by creatures with our rational natures) *under optimal conditions* (as determined by our sensible natures). Truth becomes a radically epistemic notion.» (Putnam, 1982, p. 146.)

Apart from the implausibility of the historical claim that American pragmatism (i.e., identification of truth with warranted assertibility) was invented already by Kant,[43] Putnam seems to confuse two different questions:

(29) Is the world ready-made? Is there a built-in structure in the world?

and

(30) Is the correspondence theory of truth more satisfactory (as a theory of truth) than the pragmatist theories?

Putnam's »metaphysical realists» answer both of these questions positively, »internal realists» negatively. In my view, a reasonable scientific realist should answer (29) negatively and (30) positively.

According to sentence 1.1 in Wittgenstein's *Tractatus*, »the world is the totality of facts, not of things». If these facts are assumed to be in the world without any conceptualizing activity of human beings, this sentence is an expression of metaphysical realism – or the ontological Myth of the Given.[44] Against this position, it can be argued that facts are semi-linguistic entities: as structured wholes, they presuppose a conceptual framework. The existence of these frameworks in turn presupposes human minds. In this sense, facts are always man-made. This conclusion does not lead to idealism, however, since conceptual frameworks *alone* are not sufficient to determine any facts. There exists something independently of us which can be »carved up» differently by different conceptual systems. This something – call it THE WORLD – is not man-made, and it is not a collection of facts. Still, it is not completely amorphous mass which can moulded in any way we like. Rather it is a collection of *potential facts* which become actual when we impose some concepts upon it. Indeed, when a conceptual framework L is given, there is a *uniquely* determined structure W_L which is the fact that L picks out from THE WORLD.[45] Another way of expressing this view is to follow Goodman (1960) by saying that »there are many ways the world is, and every true description captures one of them». While the mystic holds that there is a way which is *the* way THE WORLD is, and this structure of THE WORLD is to be grasped by some concept-free intuition, a reasonable realist should deny this assumption but admit that each true description tells us *a* way THE WORLD is.

Rorty has argued that

»the notion of 'the world' as used in a phrase like 'different conceptual schemas carve up the world differently' must be the notion of something *completely* unspecified and unspecifiable – the thing-in-itself, in fact.» (Rorty, 1972, p. 663.)

If THE WORLD is taken to be Kantian *Dinge an sich,* we ignore Peirce's great insight: we are able to be in causal interaction with THE WORLD, and the knowledge which is produced through this interaction is – more or less truthful – knowledge about THE WORLD, not about something, like Kantian phenomena, which stands between us and THE WORLD.

From the negative answer to (29) – which I share with Putnam – one cannot infer anything about the theories of truth. Putnam fails to appreciate the following fact: the notion of truth as 'correspondence with reality' is applicable quite independently of the question whether the reality in question is man-made or not. Our statements can be in correspondence with such concrete and abstract creations of human mind as artefacts, works of art, language, numbers, legal orders in different societies, etc.[46] In Tarski's version of the correspondence theory, truth is a relation between sentences h in a language L and any structures of the type W_L discussed above. If h is true in W_L, we may derivatively say that h is true of THE WORLD as well. Which sentences of L are true in W_L is not determined by the conceptual framework L but by THE WORLD. There is nothing 'metaphysical' in this account of truth – for example, it does not presuppose 'essential' properties (cf. Putnam, 1982), and it does not try to correlate the sentence h with THE WORLD *directly*. And there is nothing 'epistemic' in this account either: it only tells what is means for a sentence h to be true, when a structure W_L and the interpretation of the vocabulary of L in the domain of W_L is given. When the conceptual system L is given, the structure W_L is normally unknown to us – and that precisely is the reason why we have cognitive problems for scientists to solve.[47]

Horwich (1982) claims against Putnam and Dummett that »truth may exist beyond our capability to recognize it». However, he associates this »semantical realism» with the redundancy theory of truth, and argues against those »metaphysical realists» who take the concept of truth to involve »a primitive non-epistemic idea – for example, 'correspondence with reality' – not entirely captured by a Tarski-style disquotation schema (»p» is true iff p)». His main complaint about »metaphysical realism» is that is makes truth »radically independent of our practice». It seems that Horwich fails to distinguish two concepts of independence: according to the correspondence theory, truth is *conceptually* independent of epistemic notions, but this is compatible with the existence of many kinds of *factual* dependencies between truth and our knowledge-seeking procedures. In other words, while truth is non-epistemic, our scientific procedures may have the property that they yield results which approach to the truth at least in the long run.[48]

7. Approximate Truth and Success

In his recent criticism of »convergent realism», Laudan includes among »urgent items on the realists' agenda» the following two tasks: first, to articulate »a coherent account of approximate truth which entails that approximately true theories will, across the range where we can test them, be successful predictors»; secondly, to give a »criterion that would *epistemically* warrant the ascription of approximate truth to a theory» (Laudan, 1981a, p. 32). It seems to me that the account of truthlikeness, outlined in Section 4, gives a satisfactory answer to Laudan's challenge.

Laudan's basic strategy is to point out difficulties with two principles:

(T1) If a theory is approximately true, then it will be empirically successful.

(T2) If a theory is empirically successful, then it is probably approximately true.

He assumes that a realist accepts, besides (T1) and (T2), the view that a theory is not approximately true if its central theoretical terms fail to refer (*ibid.*, p. 33), i.e.,

(T3) If a theory is approximately true, then it is genuinely referential.

Then Laudan argues that there is a »plethora of theories» (such as the aether theory) which were empirically successful and »by present lights» non-referential (*ibid.*, p. 33). By (T3), such theories are not approximately true; hence, they violate (T2) and block the »upward path» from empirical success to approximate truth. Against the »downward path» from approximate truth to empirical success Laudan does not present any counter-examples. He notes that there are genuinely referential theories which were not empirically successful (such as 19th century atomism) (*ibid.*, p. 24). Further, he argues that, while the principle

(T1′) If a theory is true, then it will be successful

is »self-evident» (*ibid.*, p. 30), the realists have so far failed to give a notion of approximate truth which would justify the principle (T1).

To answer Laudan's criticism, let us first note that our definition of function *ver* gives us an epistemic criterion for the ascription of degrees of truthlikeness and approximate truth to a theory – and thus solves the problems concerning the 'upward path'. As all our knowledge about matters of fact is fallible, this criterion is not infallible. Thus, it is possible that $ver(h/e)$ is high but $ver(h/e\&e')$, where e' is new evidence, is not high, i.e., it may be rational to claim on evidence e that h is approximately true but not rational any more on evidence $e\&e'$. There is nothing strange in J.C. Maxwell's remark that the aether *was* »better confirmed than any other theoretical entity in natural philosophy» (Laudan, 1981a, p. 27). By our present lights, the status of the aether may quite well seem different than in the nineteenth century. To ask for an infallible criterion which would prove the approximate truth of a scientific theory is to demand too much of the realists.

But – the critic of realism may continue – if a theory may be empirically successful without really being approximately true, how could the approximate truth of a theory *explain* its empirical success? And how could one claim that the success of science would be a »miracle» (cf. Putnam, 1975) unless its theories were approximately true? These questions can be answered only by analyzing in more detail the notion of 'empirical success'.

It is easy to see that condition (T1), as formulated above, cannot be generally valid. The same holds even of condition (T1′) – which Laudan finds »self-evident». Assume that h is a theory which is formulated in a purely theoretical language L_T. Then, depending on the factual structure of the world W_{L_T}, theory h may be highly truthlike or true even though it does not have any empirical consequences! Assume then that h is formulated in a language L which contains an empirical sublanguage L_0. Then the following holds:

(T4) If h is true, then all its deductive consequences in L_0 are true.

On the other hand, the following principle is not valid

(T5) If h is true, then all its deductive consequences in L_0 are highly truthlike,

since some of the true consequences of h are uniformative. For the same reason, the following principle fails:

(T6) If h is highly truthlike, then all its deductive consequences in L_0 are highly truthlike.

The failure of (T5) and (T6) does not block the 'downward path' from approximate truth to empirical success, however. It only shows that this success typically concerns *the most informative empirical predictions* derivable from a theory: it need not be the case that most or many of these predictions are strictly true, since in fact they all may be false, but nevertheless many of them cannot be at a great distance from the truth. Typical illustrations of this situation are the following (cf. Niiniluoto, 1982a). If a quantitative law $y = f(x)$ is approximately true within range $C \subseteq \mathbf{R}$, given the metric (10), then *all* of its predictions of the form $y_a = f(x_a)$, when $x_a \in C$, are close to the true value $f_\star(x_a)$. (Here y_a and x_a are the values of quantities y and x for an object a.) If a law $y = g(x)$ is approximately true within $C \subseteq \mathbf{R}$, given the metric (9), then its predictions of the form $y_a = g(x_a)$ will *in the average* be close to the true values $f(x_a)$, for objects a such that $x_a \in C$. (see Fig. 4.)

These observations can be extended to theories which contain theoretical terms. Laudan demands the realists to give an argument to the effect

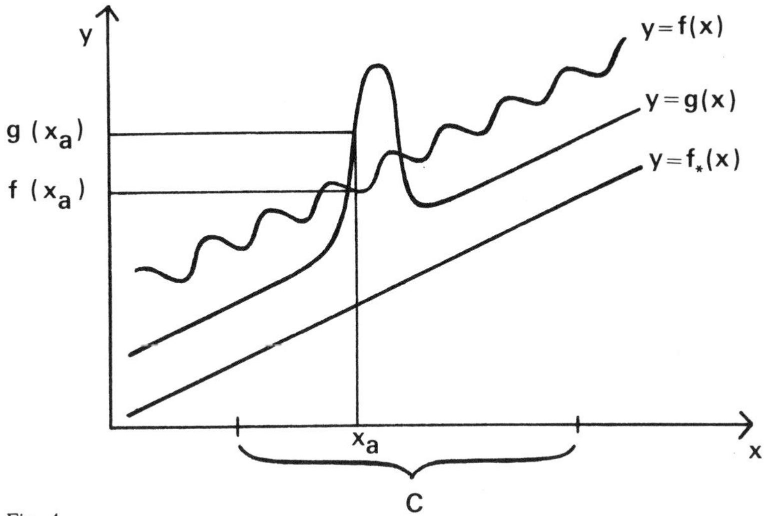

Fig. 4.

that »greater correspondence at the level of unobservable claims is more likely than not to reveal itself in greater accuracy at the experimental level» (Laudan, 1981a, p. 35), but again this is a too strong requirement. If h is a theory in language $L = L_T \cup L_0$, and h_T is the purely theoretical part of h in L_T, then the question whether an increase in the truthlikeness of h_T has any effects upon the empirical consequences of h in L_0 depends essentially upon the connections between the theoretical and empirical terms in h. If these connections are completely wrong, it does not 'help' theory h even if its theoretical part h_T would be maximally truthlike in L_T. However, in this case the theory h would not be highly truthlike in the whole language L, since its empirical part would be far from the truth. Similarly, if h in L does not have empirical consequences, it is empirically uninformative – and cannot be highly truthlike in the whole language L. It follows that a highly truthlike theory h in $L = L_T \cup L_0$ has to be at least to some extent empirically successful.

A similar remark can be made about purely observational 'theories': they cannot be highly truthlike in the whole language L. For example, Hesse thinks that the Bayesian theory is partial »toward empirical rather than theoretical content» (Hesse, 1981, p. 17). It is true that an informative theory h in L is always less *probable* than its observational consequences e in L_0, but the estimated *truthlikeness* of h may be larger than that of e.

If a theory h is empirically successful to a limited degree only, its success can hardly be explained by claiming that it is approximately true. Nevertheless it can be argued that even in this case the relative success of h can be explained by the assumption that h is truthlike to some degree.

Take as an example the phlogiston theory which replaced the Ancient doctrine of fire as one of the elements and correctly treated combustion and calcination as processes involving the same principles. This theory explains a number of empirical generalizations which tell that pieces of certain stuffs $A_1, \ldots, A_n$ will burn in certain conditions C:

(31) $\forall x(A_i x \,\&\, Cx \rightarrow Bx) \qquad (i = 1, \ldots, n)$

and that certain other stuffs $D_1, \ldots, D_k$ do not burn in similar conditions:

(32) $\forall x(D_j x \,\&\, Cx \rightarrow \sim Bx) \quad (j = 1, \ldots, k)$

by assuming that all A-stuffs possess a theoretical property F, and all D-stuffs lack this property F, and F is responsible for the process of burning:

(33) $\bigwedge_{i=1}^{n} \forall x(A_i x \rightarrow Fx) \,\&\, \bigwedge_{j=1}^{k} \forall x(B_j x \rightarrow \sim Fx) \,\&\, \forall x(Cx \rightarrow (Fx \leftrightarrow Bx)).$

Note that here (31) and (32) are deducible from (33) no matter what theoretical property F is taken to be. Technically speaking, this means that (31) and (32) are derivable from the *Ramsey-sentence* of (33) which says that there is a property F such that (33) holds. Moreover, by present lights, this Ramsey-sentence is *true:* while the phlogistonists interpreted 'Fx' as 'x contains phlogiston', for modern chemists 'Fx' means 'x has the capability of combining with oxygen'. It follows that the phlogiston theory has some truthlikeness – and was progressive relative to its predecessors. This theory went wrong in its attempted definition of combustion in terms of phlogiston:

(34) $\forall x(Bx \leftrightarrow \exists y(Fy \,\&\, Sxy))$

where 'Sxy' means that y leaves x or separates from x, when the correct theory can be expressed by

(35) $\forall x(Bx \leftrightarrow \exists y(Oy \,\&\, Uxy))$

where 'Oy' means that y is oxygen and 'Uxy' means that y unites with x. The Ramsey-sentence of the conjunction of (33) and (34) is false. The mistake in (34) is so radical that no charitable interpretation would justify the treatment of 'F' as a referring term. Thus, the phlogiston theory was not genuinely referential, but still it had some empirical success and some truthlikeness.

This account of the phlogiston theory does not violate principle (T3). However, there are examples of scientific theories which might be regarded – depending on our theory of reference – as counterexamples to it. The Boyle-Mariotte law (24) is approximately true for a large range of the values of its variables (cf. Example 2 in Section 5). Many physicists have nevertheless interpreted it as a theory about »ideal gases» which do not really exist. In this case, this theory is not genuinely referential. In order to be able to treat the sequence of laws in Example 2 as exhibiting continuous progress, a realist has to construe all of them as being about »real gases» or else he has to define a notion of ontological convergence for a sequence of various kinds of ideal gases.

Laudan (1981a) presents also a number of objections to the view that later theories should retain appropriate portions of the »laws and mechanisms» of earlier theories. I don't want to make any historical claims about the frequency in which this 'retentionist' strategy is in fact followed by scientists. It is clear that this strategy is not necessary for scientific progress. However, our function *ver* helps us to understand why it is often very reasonable. Krajewski claims that the great advantage of the Correspondence Principle is the following: if the new theory contains the old theory as a counterfactual special case, it

> »has its empirical support not only in the results of new, specific ... experiments but also in the results of the whole experience confirming the old theory». (Krajewski, 1977, p. 1.)

It is difficult to justify this idea on the basis of the traditional concept of confirmation. Assume that the new theory h_2 contains a special case h_0 which is close to the old theory h_1 but inconsistent with h_1. Then if evidence e highly confirms h_1, it cannot highly confirm h_0. If we claim that e confirms h_0 (and thereby h_2), it has to be admitted that e did not confirm h_1 in the first place. The notion of estimated verisimilitude behaves in a nicer way in this situation, since it generally satisfies the following principle:

(36) If $ver(h_1/e)$ is high and h_1 is close to h_0, then $ver(h_0/e)$ is high.[49]

And if h_2 entails h_0, which seems highly truthlike on evidence e, there is a good chance that h_2 itself is highly truthlike.

8. Concluding Remarks

Dilworth (1981) has recently presented an account of scientific progress which takes seriously Kuhn's idea of theory change as analogous to a Gestalt switch. He ends up with a definition which can readily be understood on the realist basis – even though Dilworth himself mistakenly thinks that his view is in sharp contradiction with »empiricism» and »Popperianism». He assumes that there is something common to which rival theories refer, and compares theories with respect to their accuracy, simplicity, and scope. It is obvious that our definition of progress in terms of the function *ver* takes into account such factors as accuracy (between theory and observations) and scope.[50] It may also reflect considerations of simplicity through the probability measure P: in some systems of epistemic probabilities, there is a built-in preference in the long run for the simplest hypothesis compatible with the evidence.

It remains to be studied more precisely how the concept of truthlikeness is related to other epistemic and methodological notions which might be taken to be relevant to the aims of science.[51] The definition of scientific progress in terms of increasing truthlikeness expresses the realist conviction that information about the reality is the *basic* aim of science. This is compatible with the introduction of other 'utilities' as independent but secondary goals of scientific inquiry. Moreover, there may be special fields of science (e.g. the engineering or technological sciences[52]) where a particularly strong weight on some of these secondary utilities (e.g. practical applicability and computability) can be motivated by the intended use of their results.

Notes

1 For a survey of problems concerning the notion of scientific progress, see Chapter 5 above.

2 According to Cohen (1976), the notion of 'revolution' was first applied to science in the 18th century by Turgot and Bailly.

3 It seems that Sarton underestimates here the possibilities of revolutions within the humanities: think, for example, about the 'Chomskyan revolution' in linguistics or the radical re-evaluation of mediaeval science in this century! For attempts to apply Kuhn's notions of 'normal science' and 'revolution' to the human and social sciences, see the articles and bibliography in Gutting (1980). For progress in the humanitics, see Davenport (1980).

4 Convincing evidence for this is given in Kuhn (1962). See also Chapter 5.

5 »With each scientific revolution the odds against there ever being another increase.» (Levin, 1979, p. 420.) Heisenberg's concept of a 'closed theory' might be mentioned as an attempt to characterize mature science (cf. Böhme *et al.*, 1978; Böhme, 1980): closed theories are claimed to be »classical» in the sense that they are »valid for all time» and even survive scientific revolutions. Krajewski (1977) suggests that mature sciences follow the principle of correspondence in their development.

6 On the other hand, I do not include among these realist those pragmatist philosophers who define truth in epistemic terms – and thereby make the thesis about the convergence of knowledge »towards the truth» an analytic truth. (Cf. Putnam, 1980, 1981, and Shapere, 1980, for epistemic conceptions of truth.)

7 Cf. Chapter 5. Also Sarton can be included in this group of realists, but his characterization of our »merit» (quoted in Section 1) illustrates the difficulties with presystematic notions of truthlikeness: his »ratio» is not well-defined, since the amount of truth which is »discoverable» is infinite.

8 Cf. Pitt's (1981) exposition of Sellars's ideas.

9 For inter-theory relations, see for example Bunge (1970), Strauss (1970), Sklar (1967), Nickles (1973), Yoshida (1977), Stegmüller (1977). For the correspondence principle, see Krajewski (1977), Nowak (1980), Rosenberg (1980), and Niiniluoto (1984b). See also Example 2 of Section 5 below.

10 Popper mistakenly attributes Trotsky's principle 'Revolution in permanence' to Karl Marx (Popper, 1975, p. 83).

11 For example, Lakatos says that his »brand of methodological instrumentalism» is consistent with realism: »I hold that the succession of progressive research programmes constantly superseding each other is likely to produce theories with everincreasing verisimilitude» (Lakatos, 1971, p. 175). For this reason, I find it misleading to characterize Lakatos's project as an attempt to »provide a theory of objectivity without a representational theory of truth» (cf. Hacking, 1981, p. 129), or to claim that Lakatos is »explicit in divorcing progress from truth» (cf. Leplin, 1981, p. 270). It is correct to say, however, that Lakatos did not see how the concept of verisimilitude and his own methodological concept of progress can be systematically linked with each other. This link is provided by my notion of estimated truthlikeness (see Niiniluoto, 1979b, pp. 254–255, and Section 5 below).

12 The idea that science is a problem-solving activity is central to Popper as well. In contrast with Popper's realist view, Laudan wants to separate questions about truth from problem-solving. Some comments on Laudan's model are given in Section 3 below. See also Chapter 11.

13 Toulmin's instrumentalist model, where the development of knowledge is viewed as an evolutionary process, does not make any distinction between progressive and non-progressive changes in science. See Chapter 4.

14 Duhem's position is complex, because he tries to combine an instrumentalist view of theories with the idea that one task of theories is to »classify» experimental laws. He further thinks that there is a »natural classification» of these laws towards which scientific theories converge. (I am grateful to Mr. Veli Verronen for challenging my earlier account of Duhem.) As Duhem explicitly speaks of the »correspondence» between the classification given by a theory and the transcendental natural classification (Duhem, 1954, p. 335), his view may be taken to represent a peculiar type of convergent realism. Duhem's position can be criticized on the following grounds. It seems reasonable to say that a theory of electricity has the function of grouping together all laws concerning electric phenomena. But what other reason for this grouping there might exist except the fact that *there is electricity* which these laws talk about? But this existential claim is not accepted in Duhem's instrumentalism.

15 This view was supported by some logical empiricists. Also van Fraassen's (1980) »empiricist alternative to scientific realism» is close to this position. Cf. also Hesse (1977, 1981).

16 Laudan's position concerning progress comes close to this second group, even though he endorses the claim that scientific theories have truth values (cf. Laudan, 1981c, 1982), since at the same time he regards it as an »utopian» task to try to find out these truth values.

17 For example, Post's (1971) *General Correspondence Principle* requires that an approximate or qualified version $T^{\star}_{1Q}$ of some subset T_{1Q} of T_1 is derivable in T_2 (cf. Koertge, 1972). For related suggestions, see also Harris (1975), Watkins (1975, 1978), Scheibe (1976) and Krajewski (1977). This idea includes also the case where T_2 *approximately explains* T_1 (cf. Schaffner, 1967; Nickles, 1973; Scheibe, 1973; Tuomela, 1979). A related idea is to measure the amount of *meaning variance* in theory change by representing the 'analytic component' of a theory T_i by a set $\mathbf{MP}_i$ of meaning postulates and by evaluating the distance between sets $\mathbf{MP}_1$ and $\mathbf{MP}_2$ (cf. Giedymin, 1970; Suppe, 1974; Tuomela, 1979).

18 See, e.g., Scheffler (1967), Putnam (1975), Levin (1979), Przelecki (1979), Papineau (1979), Nola (1980, 1981), Smith (1981). The general idea is to develop a theory of reference which allows us to say the following: T_1 and T_2 refer to the same entities but express different claims about them. (The alternative is to say that T_1 and T_2 speak about different things – which makes them incommensurable with each other.) Cf. also Field (1973), Earman and Fine (1977), and Jardine (1978). In Sellars's account, an expression X in T_1 will have a counterpart Y in T_2 which has the same 'role' or 'function' in T_2 as X has in T_1, but the 'substantive correspondence rule'

linking X to Y also implies that the entities X denotes in T_1 do not exist (cf. Pitt, 1981, pp. 108–110). (It is not clear to me how Sellars would handle, e.g., the Copernican Revolution within his model of theory change.)

19 The Adams-Suppes-Sneed concept of reduction between models (cf. Stegmüller, 1977) and its approximate versions (cf. Moulines, 1980; Mayr, 1981) are developments of this idea. See also Rantala's (1979) model-theoretic account of the correspondence relation and Pearce (1981).

20 Cf. Kuhn (1970) and Laudan (1977). For discussion of the legitimacy of problems relative to theories, see Leplin (1981).

21 Cf. Rescher (1980).

22 Cf. Shapere (1982), who argues that a »chain-of-reasoning connection» is sufficient to defend the idea that electrons can be »transtheoretical» – »something about which we can have competing theories without assuming that there is either a common meaning or a common reference to the term 'electron'». This methodological account of continuity seems problematic, however. For example, observations concerning an eclipse may serve as reasons for an astronomical theory about planets or for a physical theory of relativity – so that »chains of reasoning» may get mixed with each other in a rather disturbing manner.

23 For attempts to apply some ideas from the logic of questions to science, see Niiniluoto (1976) and Hintikka (1981). Cf. also Rescher (1979).

24 See my discussion in Chapter 5 above and in Chapter 11 below.

25 Laudan (1982) gives, as an example of a truth-independent characterization of the aims of science, the requirement that theories should entail »novel facts». However, a *fact* is a state of affairs described by a *true* sentence. Note also that, at least in some cases, Laudan (1982) speaks of situations where theory T entails P and theory T' entails $\sim P$, but only the former gets credit for the solution of a problem.

26 Such devices solve all empirical problems, have no anomalies, and at least are not associated with the same conceptual problems as inconsistent theories.

27 In some cases, the relevant states s cannot be expressed by any single sentences of language L. For example, the structures for a first-order language with relations are limits of structures of depth d, when $d \to \infty$, where a structure of depth d is described by a constituent with the quantificational depth d. For details, see Niiniluoto (1979a).

28 See Niiniluoto (1978) and Oddie (1979, 1982).

29 See Niiniluoto (1984a).

30 For the treatment of (g) and (h), see Niiniluoto (1982b). For (i), see Rosenkrantz (1980).

31 For discussion about alternative definitions, see Niiniluoto (1978, 1982c) and Oddie (1981).

32 Hintikka's inductive logic which is designed for situations of type (c) satisfies the principle that $P(s_0/e) \to 1$ for the constituent s_0 of L which is qualitatively similar to the evidence e. Cf. Niiniluoto and Tuomela (1973).

33 This condition has an interesting connection to Peirce's notion of a 'likely' or 'verisimilar' theory (see Niiniluoto, 1978b, p. 250).

34 This situation will obtain, if constituents s_i are expressed in a richer language than evidence e. It is thus related to the Quinean problem concerning the »underdetermination of theories by observations». (See note 41 below.)

35 See Lakatos (1970), p. 116. See also Worrall (1976, 1978) and Watkins (1975, 1978). Worrall summarizes the Lakatosian position as follows: »According to the standards laid down by the methodology of scientific research programmes, a shift from one theory to the next constitutes scientific progress if the new theory not only explains the facts it was introduced to explain but makes extra predictions as well, some of which are empirically confirmed» (Worrall, 1976, p. 143). Further, the new theory should predict facts which are *novel* in the sense that they were »not used in the construction of the theory». The Lakatos-Zahar-Worrall conditions of progress seem

to be essentially the same as William Whewell's requirements for new hypotheses. Writing in 1840, Whewell demanded that a hypothesis should predict facts which – instead of being absolutely new – were not »contemplated in the formation of our hypotheses». In my view, the simplest definition of the novelty of a fact *e* relative to a theory *h* is the following: either *e* itself is unknown (to the scientific community) or it is unknown that *h* explains *e*. See Niiniluoto (1983).

36 For example, we have seen above that Grosseteste's law of refraction is farther from the truth than Ptolemy's.

37 Cf. Chapter 5. Pitt interprets Sellars's account of science to require that »we finally reach a completed science» or »final theory» *FT*, but he finds this aim »ill-conceived» (Pitt, 1981, pp. 134–137). He argues against Sellars that *FT* cannot be a theory at all, since it does not have any »external subject matter» (*ibid.*, p. 134). I don't find this convincing: if the final theory *FT* will be reached in the development of science, then *FT* will be, for some *n*, the *nth* member T_n of a sequence of theories

$$(\star) \quad F_{CS}, T_1, T_2, \ldots, T_{n-1}, T_n$$

where F_{CS} is the framework of common sense and each theory T_i externally explains its predecessor T_{i-1}. Therefore, the 'external subject matter' of *FT* is simply its predecessor T_{n-1}. If *FT* is *not* reached except as an ideal limit, so that the sequence $(\star)$ is infinite, *FT* would not have a predecessor any more. But even in this case one might perhaps say that the external subject matter of *FT* is given by all theories preceding it in $(\star)$.

38 Cohen's third argument against the »verisimilitudinist» programme is based upon his distinction between verisimilitude and legisimilitude. For different replies to this issue, see Oddie (1982) and Niiniluoto (1984a).

39 Harris (1975) defines a relation $V <_C W$ (where *V* and *W* are theories) which can be interpreted in the epistemic sense by choosing *C* as the accepted evidence. With this exception, my theory for the estimation of truthlikeness seems to be the only proposal in the literature so far. (See also Rosenkrantz, 1980.)

40 This seems to be a common assumption of Marxism-Leninism and the Sellarsian realism. Jardine (1980) has investigated general conditions which would guarantee the convergence of science towards »absolute truth».

41 This point can be illustrated by Quine's thesis of the »underdetermination» of theories by observational data. Boyd (1973) has argued that underdetermination does not exist, while Newton-Smith (1980) defends the possibility that there may be incompatible theories which »fit all actual and possible observations». In such cases, Newton-Smith favours the »arrogance response»: if *P* is »empirically undecidable», then the world is indeterminate relative to *P*. Arguments of this type ignore the fact that our capability for observation is influenced by new instruments and theories – and therefore it increases with the progress of science. Even if something is empirically undecidable *today*, we cannot know that it will still be empirically undecidable tomorrow. Hence, we are never in a position to give the »arrogance response». Hesse (1981) defends a form of the underdetermination thesis which is so weak that, in my view, it does not challenge realism at all. If »uniquely true theories cannot be known to be attained by science in any finite time», it does not follow that »realism is in practice an epistemologically and scientifically empty doctrine» (*ibid.*, p. 9): realism is »non-empty» as soon as one can support non-tautologous disjunctions of theoretical hypotheses.

42 Putnam has proposed a model-theoretic argument to show that an »epistemically ideal» theory could not be false: if a theory T_I satisfies all desirable epistemic conditions and all »theoretical and operational constraints», the claim that it is false is unintelligible (see 'Realism and Reason' in Putnam, 1978, and Putnam, 1980a). This argument fails for several reasons, since Putnam makes a number of questionable assumptions. (For criticism, see Merrill, 1980; Tuomela, 1980; Glymour, 1982; Pearce and Rantala, 1982.)

Putnam presupposes that the ideal theory T_I is a *first-order* theory which entails *singular atomic sentences* about macroscopic objects; that T_I is *uniquely* determined by the constraints; that T_I may be semantically closed in the sense of containing a theory of reference for its own terms; and that the intended interpretation(s) of T_I have to be chosen among the models of T_I.

43 At best Kant may be regarded as a grandfather of pragmatism, since he claimed – by the traditional *diallelus* argument – that there is no criterion for material truth. However, he makes it clear that material truth consists of the »adequacy» of an idea for its object – which is a classical formulation of the correspondence theory of truth. Moreover, he clearly distinguishes the »objective» notion of truth from various epistemic attitudes of – problematic, assertoric and apodeictic – »holding-to-be-true». See Kant's *Logic* (containing the lectures that he gave twice a year from 1755 to 1796).

44 The connection between metaphysical realism and Sellars's notion of the Myth of the Given is discussed in Tuomela (1980).

45 A simple example may clarify this point. Suppose that the conceptual system L contains the primitive terms A ('human being'), B ('male') and C ('female'), where B and C are defined independently of each other. Then the structure W_L has the form $\langle X, A^X, B^X, C^X\rangle$, where X is the domain of objects, $A^X \subseteq X$ is the class of human beings, B^X is the class of males, C^X is the class of females. Moreover, it is the case that $A^X \subseteq B^X \cup C^X$ and $B^X \cap C^X = \emptyset$.

46 See, for example, Niiniluoto (1981) and Chapter 9 below.

47 Putnam points out that there are many different ways of putting the signs of a language and the things in a given set in »correspondence with each other». This is in fact the starting point of Tarski's theory where the truth of a sentence in a structure is defined relative to each interpretation of the language. In my view, such interpretations are human conventions, established through our practices and 'language games', rather than any physical or causal relations. However, I am not convinced that a 'consensus theory of reference' has to lead to a Duhemian view of scientific progress, as Jardine (1978) argues.

48 This is not more surprising than the analogous fact that the notion of 'female' is conceptually independent of the concept of 'male', but yet some women can be approached by men who employ the right sorts of procedures.

49 The distance between theories can be defined as a generalization of the notion of truthlikeness. See Tuomela (1979) and Niiniluoto (1978).

50 The connection between the notions of truthlikeness and explanatory power can be illustrated by the following result: if theory T is maximally truthlike in language L and if $\mathbf{C}$ is the class of constituents of the observational sublanguage L_o of L, then the explanatory power of T with respect to $\mathbf{C}$ is also maximally great – assuming that explanatory power is measured by $\text{syst}_i(T, \mathbf{C})$, $i = 1, 2, 3$, defined in Niiniluoto and Tuomela (1973), pp. 91–92.

51 For discussion in this direction, see M. Sintonen, *A Pragmatic Theory of Explanation,* forthcoming in *Acta Philosophica Fennica* in 1984.

52 See Chapter 12 below.

Bibliography

Böhme, G., Daele, W. van den, and Hohlfeld, R., 'Finalisierung Revisited', in G. Böhme *et al., Starnberg Studien 1: Die Gesellschaftliche Orientierung des wissenschaftlichen Fortschritts,* Suhrkamp, Frankfurt, 1978, pp. 195–250.

Böhme, G., 'On the Possibility of »Closed Theories»', *Studies in the History and Philosophy of Science* **11** (1980), 163–172.

Boyd, R., 'Realism, Underdetermination, and a Causal Theory of Evidence', *Nous* **7** (1973), 1–12.
Boyd, R., 'Scientific Realism and Naturalistic Epistemology', in P.D. Asquith and R. Giere (eds.), *PSA 1980*, vol. 2, Philosophy of Science Association, East Lansing, 1981, pp. 613–662.
Bunge, M., 'Problems Concerning Intertheory Relations', in P. Weingartner and G. Zecha (eds.), *Induction, Physics, and Ethics,* D. Reidel, Dordrecht, 1970, pp. 285–315.
Cohen, I.B., 'The Eighteen-Century Origins of the Concept of Scientific Revolution', *Journal of the History of Ideas* **37** (1976), 257–288.
Cohen, L.J., 'What Has Science to do with Truth?', *Synthese* **45** (1980), 489–510.
Cohen, M.R. and Drabkin, I.E. (eds.), *Source Books in Greek Science,* Harvard University Press, Cambridge, Mass., 1948.
Davenport, E., 'Progress in Literary Study', in P.D. Asquith and R.N. Giere (eds.), *PSA 1980*, vol. 1, Philosophy of Science Association, East Lansing, 1980, pp. 141–148.
Dilworth, C., *Scientific Progress: A Study Concerning the Nature of the Relation Between Successive Scientific Theories,* D. Reidel, Dordrecht, 1981.
Duhem, P., *The Aim and Structure of Physical Theory,* Princeton University Press, Princeton, 1954. (Atheneum, New York, 1962.)
Earman, J. and Fine, A., 'Against Indeterminacy', *The Journal of Philosophy* **74** (1977), 535–538.
Field, H., 'Theory Change and the Indeterminacy of Reference', *The Journal of Philosophy* **70** (1973), 462–481.
Feyerabend, P., *Against Method,* New Left Books, London, 1975.
Finocciaro, M.A., 'Remarks on Truth, Problem-Solving, and Methodology', *Studies in History and Philosophy of Science* **12** (1981), 261–268.
Fraassen, B. van, *The Scientific Image,* Clarendon Press, Oxford, 1980.
Giedymin, J., 'The Paradox of Meaning Variance', *The British Journal for the Philosophy of Science* **21** (1970), 257–268.
Glymour, C., 'Conceptual Scheming, or, Confessions of a Metaphysical Realist', *Synthese* **51** (1982), 169–180.
Goodman, N., 'The Way the World Is', *Review of Metaphysics* **14** (1960), 48–56. (Reprinted in *Problems and Projects,* Bobbs-Merrill, Indianapolis, 1972, pp. 24–32.)
Grünbaum, A., 'Can a Theory Answer More Questions than one of its Rivals?', *The British Journal for the Philosophy of Science* **27** (1976), 1–23.
Gutting, G. (ed.), *Paradigms and Revolutions: Applications and Appraisals of Thomas Kuhn's Philosophy of Science,* University of Notre Dame Press, Notre Dame, 1980.
Hacking, I., 'Lakatos's Philosophy of Science', in Hacking (1981), pp. 128–143.
Hacking, I. (ed.), *Scientific Revolutions,* Oxford University Press, Oxford, 1981.
Harris, J.H., 'On Comparing Theories', *Synthese* **32** (1975), 29–76.
Hesse, M., 'Truth and the Growth of Scientific Knowledge', in F. Suppe and P.D. Asquith (eds.), *PSA 1976,* vol. 2, Philosophy of Science Association, East Lansing, 1977, pp. 261–280.
Hesse, M., 'The Hunt for Scientific Reason', in P.D. Asquith and R. Giere (eds.), *PSA 1980,* vol. 2, Philosophy of Science Association, East Lansing, 1981, pp. 3–22.
Hintikka, J., 'On the Logic of an Interrogative Model of Scientific Inquiry', *Synthese* **47** (1981), 69–84.
Horwich, P., 'Three Forms of Realism', *Synthese* **51** (1982), 181–201.
Jardine, N., '»Realistic« Realism and the Progress of Science', in C. Hookway and P. Pettit (eds.), *Action and Interpretation,* Cambridge University Press, Cambridge, 1978.

Jardine, N., 'The Possibility of Absolutism', in D.H. Mellor (ed.), *Science, Belief and Behaviour: Essays in Honour of R.B. Braithwaite,* Cambridge University Press, Cambridge, 1980, pp. 23–42.

Kant, I., *Logic* (ed. by G.B. Jäsche in 1800), Bobbs-Merrill, Indianapolis, 1974.

Koertge, N., 'Theory Change in Science', in G. Pearce and P. Maynard (eds.), *Conceptual Change,* D. Reidel, Dordrecht, 1973, pp. 164–195.

Krajewski, W., *Correspondence Principle and the Growth of Knowledge,* D. Reidel, Dordrecht, 1977.

Kuhn, T.S., *The Structure of Scientific Revolutions,* The University of Chicago Press, Chicago, 1962. (2nd ed. 1970.)

Lakatos, I., 'Falsification and the Methodology of Scientific Research Programmes', in I. Lakatos and A. Musgrave (eds.), *Criticism and the Growth of Knowledge,* Cambridge University Press, Cambridge, 1970, pp. 91–195.

Lakatos, I., 'Reply to Critics', in R.C. Buck and R.S. Cohen (eds.), *PSA 1970,* D. Reidel, Dordrecht, 1971, pp. 174–182.

Laudan, L., *Progress and Its Problems,* Routledge and Kegan Paul, London, 1977.

Laudan, L., 'A Confutation of Convergent Realism', *Philosophy of Science* **48** (1981), 19–49. (a)

Laudan, L., 'A Problem-Solving Approach to Scientific Progress', in Hacking (1981), pp. 144–155. (b)

Laudan, L., 'The Philosophy of *Progress...*', in P. Asquith and I. Hacking (eds.), *PSA 1978,* vol. 2, Philosophy of Science Association, East Lansing, 1981. (c)

Laudan, L., 'Problems, Truth, and Consistency', *Studies in History and Philosophy of Science* **13** (1982), 73–80.

Leplin, J., 'Truth and Scientific Progress', *Studies in History and Philosophy of Science* **12** (1981), 269–292.

Levin, M., 'On Theory-Change and Meaning-Change', *Philosophy of Science* **46** (1979), 407–425.

Mayr, D., 'Investigations of the Concept of Reduction II', *Erkenntnis* **16** (1981), 109–130.

Merrill, G.H., 'The Model-Theoretical Argument Against Realism', *Philosophy of Science* **47** (1980), 69–81.

Miller, D., 'Popper's Qualitative Theory of Verisimilitude', *The British Journal for the Philosophy of Science* **25** (1974), 166–177.

Moulines, C.U., 'Intertheoretic Approximation: The Kepler-Newton Case', *Synthese* **45** (1980), 387–412.

Newton-Smith, W., 'The Underdetermination of Theory by Data', in R. Hilpinen (ed.), *Rationality in Science,* D. Reidel, Dordrecht, 1980, pp. 91–110.

Nickles, T., 'Two Concepts of Intertheoretic Reduction', *The Journal of Philosophy* **70** (1973), 181–201.

Niiniluoto, I., 'Inquiries, Problems, and Questions: Remarks on Local Induction', in R.J. Bogdan (ed.), *Local Induction,* D. Reidel, Dordrecht, 1976, pp. 263–296.

Niiniluoto, I., 'Truthlikeness: Comments on Recent Discussion', *Synthese* **38** (1978), 281–330.

Niiniluoto, I., 'Truthlikeness in First-Order Languages', in J. Hintikka, I. Niiniluoto, and E. Saarinen (eds.), *Essays on Mathematical and Philosophical Logic,* D. Reidel, Dordrecht, 1979, pp. 437–458. (a)

Niiniluoto, I., 'Verisimilitude, Theory-Change, and Scientific Progress', in Niiniluoto and Tuomela (1979), pp. 243–264. (b)

Niiniluoto, I., 'Language, Norms, and Truth', in I. Pörn (ed.), *Essays in Philosophical Analysis* (Acta Philosophica Fennica **32**), Helsinki, 1981, pp. 168–189.

Niiniluoto, I., 'What Shall We Do With Verisimilitude?', *Philosophy of Science* **49** (1982), 181–197. (a)

Niiniluoto, I., 'Truthlikeness for Quantitative Statements', in P.D. Asquith and R.N.

Giere (eds.), *PSA 1982*, vol. 1, Philosophy of Science Association, East Lansing, 1982, pp. 208–161. (b)

Niiniluoto, I., 'On Explicating Verisimilitude: A Reply to Oddie', *The British Journal for the Philosophy of Science* **33** (1982), 290–296. (c)

Niiniluoto, I., 'Novel Facts and Bayesianism', *The British Journal for the Philosophy of Science* **34** (1983), 375–379.

Niiniluoto, I., 'Verisimilitude vs. Legisimilitude', in I. Niiniluoto and J. Zygmunt (eds.), *Proceedings of the Finnish-Polish-Soviet Logic Conference 1981*, Studia Logica XLII, 2/3, Ossolineum, Wroclaw, 1984. (a)

Niiniluoto, I., 'Theories, Approximations, and Idealizations', forthcoming in *Logic Methodology, and Philosophy of Science* VII, North-Holland, Amsterdam, 1984. (b)

Niiniluoto, I. and Tuomela, R., *Theoretical Concepts and Hypothetico-Inductive Inference*, D. Reidel, Dordrecht, 1973.

Niiniluoto, I. and Tuomela, R. (eds.), *The Logic and Epistemology of Scientific Change*, Acta Philosophica Fennica **30**:2–4, North-Holland, Amsterdam, 1979.

Nola, R., '»Paradigms Lost, or the World Regained» – An Excursion into Realism and Idealism in Science', *Synthese* **45** (1980), 317–350.

Nola, R., 'Fixing the Reference of Theoretical Terms', *Philosophy of Science* **47** (1980), 505–531.

Nowak, L., *The Structure of Idealization: Towards a Systematic Interpretation of the Marxian Idea of Science*, D. Reidel, Dordrecht, 1980.

Oddie, G., 'Verisimilitude and Distance in Logical Space', in Niiniluoto and Tuomela (1979), pp. 243–264.

Oddie, G., 'Verisimilitude Reviewed', *The British Journal for the Philosophy of Science* **32** (1981), 237–265.

Oddie, G., 'Cohen on Verisimilitude and Natural Necessity', *Synthese* **51** (1982), 355–379.

Papineau, D., *Theory and Meaning*, Oxford University Press, Oxford, 1979.

Pearce, D., 'Is There Any Theoretical Justification for a Non-statement View of Theories?', *Synthese* **46** (1981), 1–40.

Pearce, D. and Rantala, V., 'Realism and Formal Semantics', *Synthese* **52** (1982), 39 53.

Pitt, J.C., *Pictures, Images, and Conceptual Change: An Analysis of Wilfrid Sellars' Philosophy of Science*, D. Reidel, Dordrecht, 1981.

Popper, K., 'The Rationality of Scientific Revolutions', in R. Harré (ed.), *Problems of Scientific Revolution*, Oxford University Press, Oxford, 1975, pp. 72–101. (Also in Hacking, 1981, pp. 80–106.)

Popper, K., *Objective Knowledge*, Oxford University Press, Oxford, 1972. Revised edition 1979.

Post, H.R., 'Correspondence, Invariance and Heuristics: In Praise of Conservative Induction', *Studies in History and Philosophy of Science* **2** (1971), 213–255.

Przelecki, M., 'Commensurable Referents of Incommensurable Theories', in Niiniluoto and Tuomela (1979), pp. 347–365.

Putnam, H., *Mind, Language and Reality: Philosophical Papers*, vol. 2, Cambridge University Press, Cambridge, 1975.

Putnam, H., *Meaning and the Moral Sciences*, Routledge and Kegan Paul, London, 1978.

Putnam, H., 'Models and Reality', *The Journal of Symbolic Logic* **45** (1980), 464–482. (a)

Putnam, H., 'How to be an Internal Realist and a Transcendental Idealist (at the Same Time)', in R. Haller and W. Grassl (eds.), *Language, Logic, and Philosophy*, Hölder-Pichler-Tempsley, Vienna, 1980, pp. 100–108. (b)

Putnam, H., *Reason, Truth, and History*, Cambridge University Press, Cambridge, 1981.

Putnam, H., 'Why There Isn't a Ready-Made World', *Synthese* **51** (1982), 141–167.

Radnitzky, G. and Andersson, G. (eds.), *Progress and Rationality in Science,* D. Reidel, Dordrecht, 1978.

Rantala, V., 'Correspondence and Non-Standard Models: A Case Study', in Niiniluoto and Tuomela (1979), pp. 366–378.

Rescher, N., 'Some Issues Regarding the Completeness of Science and the Limits of Scientific Knowledge', in G. Radnitzky and G. Andersson (eds.), *The Structure and Development of Science,* D. Reidel, Dordrecht, 1979, pp. 19–40.

Rescher, N., 'Conceptual Schemes', in P.A. French, T.E. Vehling, Jr. and H.E. Wettstein (eds.), *Midwest Studies in Philosophy V: Studies in Epistemology,* University of Minnesota Press, Minneapolis, 1980, pp. 323–345.

Rorty, R., 'The World Well Lost', *The Journal of Philosophy* **69** (1972), 649–665.

Rosenberg, J., 'Coupling, Retheoretization, and the Correspondence Principle', *Synthese* **45** (1980), 351–386.

Rosenkrantz, R., 'Measuring Truthlikeness', *Synthese* **45** (1980), 463–488.

Sarkar, H., 'Truth, Problem-Solving and Methodology', *Studies in History and Philosophy of Science* **12** (1981), 61–74.

Sarton, G., *The Study of the History of Science,* Harvard University Press, Harvard, 1936. (Also Dover, New York, 1957.)

Schaffner, K., 'Approaches to Reduction', *Philosophy of Science* **34** (1967), 137–147.

Scheffler, I., *Science and Subjectivity,* Bobbs-Merrill, Indianapolis, 1967.

Scheibe, E., 'The Approximative Explanation and the Development of Physics', in P. Suppes *et al.* (eds.), *Logic, Methodology and Philosophy of Science* IV, North-Holland, Amsterdam, 1973, pp. 931–942.

Scheibe, E., 'Conditions of Progress and Comparability of Theories', in R.S. Cohen *et al.* (eds.), *Essays in Memory of Imre Lakatos,* D. Reidel, Dordrecht, 1976, pp. 547-568.

Shapere, D., 'The Character of Theory-Change', in T. Nickles (ed.), *Scientific Discovery, Logic, and Rationality,* D. Reidel, Dordrecht, 1980, pp. 61–102.

Shapere, D., 'Reason, Reference, and the Quest for Knowledge', *Philosophy of Science* **49** (1982), 1–23.

Sklar, L., 'Types of Intertheoretic Reduction', *The British Journal for the Philosophy of Science* **18** (1967), 109–124.

Smith, P., *Realism and the Progress of Science,* Cambridge University Press, Cambridge, 1981.

Stegmüller, W., *The Structure and Dynamics of Theories,* Springer-Verlag, Berlin, 1976.

Strauss, M., 'Intertheory Relations', in P. Weingartner and G. Zecha (eds.), *Induction, Physics, and Ethics,* D. Reidel, Dordrecht, 1970, pp. 220–284.

Suppe, F., 'The Search for Philosophic Understanding of Scientific Theories', in F. Suppe (ed.), *The Structure of Scientific Theories,* University of Illinois Press, Urbana, 1974, pp. 1–241.

Tuomela, R., 'Scientific Change and Approximation', in Niiniluoto and Tuomela (1979), pp. 265–297.

Tuomela, R., 'Metaphysical versus Internal Realism', *Reports from the Department of Philosophy, University of Helsinki* N:o 8/1980.

Watkins, J., 'Metaphysics and the Advancement of Science', *The British Journal for the Philosophy of Science* **26** (1975), 91–121.

Watkins, J., 'Corroboration and the Problem of Content-Comparison', in Radnitzky and Andersson (1978), pp. 339–378.

Worrall, J., 'Thomas Young and the »Refutation» of Newtonian Optics', in C. Howson (ed.), *Method and Appraisal in the Physical Sciences,* Cambridge University Press, Cambridge, 1976, pp. 107–179.

Worrall, J., 'The Ways in Which the Methodology of Scientific Research Programmes Improves on Popper's Methodology', in Radnitzky and Andersson (1978), pp. 45–70.

Yoshida, R.M., *Reduction in the Physical Sciences,* Dalhousie University Press, Halifax, 1977.

CHAPTER 8

THE GROWTH OF KNOWLEDGE IN MATHEMATICS

In his paper 'History of Mathematics: Why and How' for the 1978 International Congress of Mathematicians in Helsinki, André Weil makes the following statement:

> »Some universities have established chairs for 'the history and philosophy of mathematics'; it is hard for me to imagine what those two subjects can have in common.» (Weil, 1980, p. 230.)

This view may be contrasted with Imre Lakatos's claim:

> »The history of mathematics has been distorted by false philosophies even more than has the history of science.» (Lakatos, 1978, p. 43.)

While it is hard for me to imagine that an eminent mathematician like Weil would lack the imagination required for seeing connections between the history and philosophy of mathematics, I find it more illuminating to regard his statement as a philosophical thesis about such connections – and, moreover, as a thesis which Lakatos would have regarded as false.

One of the aims of this paper is to illustrate how and why the historians of mathematics have to adopt – consciously or unconsciously – opinions about methodological issues within the historiography of science in general[1] and about the nature of mathematics in particular[2]. I shall approach these issues by relating them to the debates about the growth of knowledge within the recent philosophy of science.

1. GROWTH OF KNOWLEDGE: QUANTITATIVE PERSPECTIVE

In an address to a meeting of natural scientists in 1899, Ludwig Boltzmann gives a vivid characterization of scientific progress during the 19th century:

> In earlier centuries, science advanced steadily but slowly through the work of the most select minds, just as an old town constantly grows through new buildings put up by industrious and enterprising citizens. In contrast, our present century of steam and telegraphy has set its seal of nervous and precipitate activity on scientific progress too. Especially the development of natural science in recent times resembles rather that of a modern American town which in a few decades grows from a village into a city of millions.
>
> The consequence of this vast and rapidly growing extent of our positive knowledge was a division of labour in science right down to the minutest detail, almost reminiscent of a modern factory where one person does nothing but measure carbon filaments, while another cuts them, a thid welds them in and so on. Such a division of labour certainly helps greatly to promote rapid progress

> in science and is indeed indispensable for it; but just as certainly it harbours great dangers. For we lose the overview of the whole, required for any mental activity aiming at discovering something essentially new or even just essentially new combinations of old ideas.[3]

Boltzmann is here referring to a feature which has later been studied in quantitative terms by historians and sociologists of science.[4] Henry Adams estimated in 1904 that the amount of scientific knowledge grows in a geometrical series doubling itself within every 50 years. Later studies have confirmed that since the end of the 18th century the total volume of science – as measured by the number of scientific disciplines, scientists, books, articles or journals – has approximately followed the exponential pattern of growth. The doubling time is even shorter than Adams thought: usually from 10 to 15 years, depending on the chosen indicator.

For example, the number of mathematicians in the U.S.A. doubles in 8 years and the number of publications in mathematics in 10 years. This growth is more rapid than in the natural sciences, where the corresponding times are 12 and 15 years, respectively. This means, in practice, that during the decade between 1970 and 1980 an equal amount of mathematical works were published as in all times before 1970 – quite a startling piece of information for those who finished their studies in mathematics around the year 1970.

2. Birth of New Branches in Mathematics

The extraordinary speed in the growth of mathematics is very clearly visible in the continuous birth of new relatively independent branches of mathematics. From this perspective, Boltzmann's two similes – science as a city which grows by building new suburbs and by repairing and reorganizing the central areas, and science as a factory which is based on the division of labour and produces new products with refined tools and instruments – are very appropriate to mathematics.

The classification of the branches of mathematics is to some extent a conventional matter. Special prominence can be given to those classifications which have been accepted by the community of mathematicians themselves. The classification scheme which was used in *Mathematical Reviews* in 1940 contained 11 headlines:

History
Foundations
Algebra
Number Theory
Group Theory
Analysis
Topology
Geometry
Numerical and Graphical Methods
Mechanics
Mathematical Physics.

The AMS (American Mathematical Society) Subject Classification Scheme 1979/80, which is used in *Mathematical Reviews* and in *Zentralblatt für Mathematik*, contains already 60 main headlines:

00–XX General
01–XX History and Biography
03–XX Mathematical Logic and Foundations
04–XX Set Theory
05–XX Combinatorics
06–XX Order, Lattices, Ordered Algebraic Structures
08–XX General Mathematical Systems
10–XX Number Theory
12–XX Algebraic Number Theory, Field Theory and Polynomials
13–XX Commutative Rings and Algebras
14–XX Algebraic Geometry
15–XX Linear and Multilinear Algebra; Matrix Theory (finite and infinite)
16–XX Associative Rings and Algebras
17–XX Nonassociative Rings and Algebras
18–XX Category Theory, Homological Algebra
20–XX Group Theory and Generalizations
22–XX Topological Groups, Lie Groups
26–XX Real Functions
28–XX Measure and Integration
30–XX Functions of a Complex Variable
31–XX Potential Theory
32–XX Several Complex Variables and Analytic Spaces
33–XX Special Functions
34–XX Ordinary Differential Equations
35–XX Partial Differential Equations
39–XX Finite Differences and Functional Equations
40–XX Sequences, Series, Summability
41–XX Approximations and Expansions
42–XX Fourier Analysis
43–XX Abstract Harmonic Analysis
44–XX Integral Transforms, Operational Calculus
45–XX Integral Equations
46–XX Functional Analysis
47–XX Operator Theory
49–XX Calculus of Variations and Optimal Control; Optimization
51–XX Geometry
52–XX Convex Sets and Related Geometric Topics
53–XX Differential Geometry
54–XX General Topology
55–XX Algebraic Topology
57–XX Manifolds and Cell Complexes
58–XX Global Analysis, Analysis of Manifolds
60–XX Probability Theory and Stochastic Processes
62–XX Statistics
65–XX Numerical Analysis
68–XX Computer Science (including Automata)
70–XX Mechanics of Particles and Systems
73–XX Mechanics of Solids
76–XX Fluid Mechanics
78–XX Optics, Electromagnetic Theory

80–XX Classical Thermodynamics, Heat Transfer
81–XX Quantum Mechanics
82–XX Statistical Physics, Structure of Matter
83–XX Relativity
85–XX Astronomy and Astrophysics
86–XX Geophysics
90–XX Economics, Operations Research, Programming, Games
92–XX Biology and Behavioral Sciences
93–XX Systems, Control
94–XX Information and Communication, Circuits

Each of these 60 main branches is divided into several subbranches. For example, Probability Theory and Stochastic Processes contains 10 subareas:

60Axx Foundations of probability theory
60Bxx Probability theory on algebraic and topological structures
60C05 Combinatorial probability
60D05 Geometric probability; stochastic geometry; random sets
60Exx Distribution theory
60Fxx Limit theorems
60Gxx Stochastic processes
60Hxx Stochastic analysis
60Jxx Markov processes
60Kxx Special processes.

Further, each of these subbranches is divided into several subsubbranches. For example, in the above list the title 'Limit theorems' contains the following areas:

60Fxx Limit theorems
60F05 Central limit and other weak theorems
60F10 Large deviations
60F15 Strong theorems
60F17 Functional limit theorems; invariance theorems
60F20 Zero-one law
60F25 L^p-limit theorems
60F99 None of the above, but in this section.

The exposition of the complete Classification Scheme includes 47 printed pages.

3. Integration of Knowledge

The exponential growth of mathematical publications and the branching structure of the birth of new special fields in mathematics are illustrations of *divergent* growth of knowledge. As William Whewell pointed out in 1840, this kind of development in physics is balanced by the *integrative* growth of the contents of knowledge: a new theory may contain as its special cases earlier results, just like Newton's mechanics contains Galileo's laws of free fall and Kepler's planetary laws. (See Chapters 3 and 4 above.) In this sense, Whewell notes, the development of science resembles the flow of a river which unites to itself a whole net of branches. This idea of integrative growth may be considered as the leading principle concerning the development of mathematical knowledge.

The strive for more *general* and *abstract* results is a characteristic feature of mathematical theory formation. Most fields of mathematical research start originally from the treatment of specific problems and proceed to more complicated general situations by eliminating some restrictive assumption. This problem-oriented stage includes also the development of new powerful techniques. With them it becomes possible to prove theorems about large problem areas, and these theorems will then be generalized and applied to new situations. The result of this development is finally an abstract axiomatic theory which is interesting by its own right – independently of the specific problems from which it started and which now appear as simple exercises in textbooks. As illustrative examples of this pattern of pre-axiomatic development one may mention the development of geometry from practical problems of surveying and division of lands to Euclid's system and later to abstract linear algebra[5] and the change of probability theory from a calculus associated with games of chance to an independent branch of abstract measure theory.[6]

Some fields of mathematics are born already on a high level of abstraction even though they do not directly build upon any earlier area of research. Cantor's set theory is a case in point.[7] Sometimes a new branch is generated as an offspring of a well-developed mathematical theory – just by changing some of the axiomatic assumptions (e.g., non-Euclidean geometries, non-Archimedean fields). It is more common, however, that a new mathematical theory is invented to combine on a higher level two or more earlier fields. A typical example is algebraic topology, which introduces to its objects both an algebraic and a topological structure at the same time. Other examples are given in Figure 1: its first step of abstraction is the creation of the concept of natural numbers, the next is the invention of arithmetic as a theory of natural numbers, then the invention of algebra for the study of laws of symbolic calculation, the creation of abstract algebra for the study of axiomatically defined algebraic structures (groups, fields, etc.), the birth of universal algebra for the abstract treatment of the properties and relations of these structures (isomorphism etc.), the combination of universal algebra with formal logic in model theory, and finally the invention of abstract logic to study the model-theoretic properties of arbitrary logics.[8]

This way of abstraction and unification does not seem to have any end point. It is of course possible to turn backwards and pay attention to the specializations and applications of the already gained mathematical theories. However, in their strive to abstractness, mathematicians are guided by the belief that the general theories are not only stronger or more powerful but they are often simpler, easier to prove, and more 'beautiful' than their special cases. For example, the step from two-dimensional to n-dimensional spaces (for arbitrary n) is irresistible for a good mathematician who

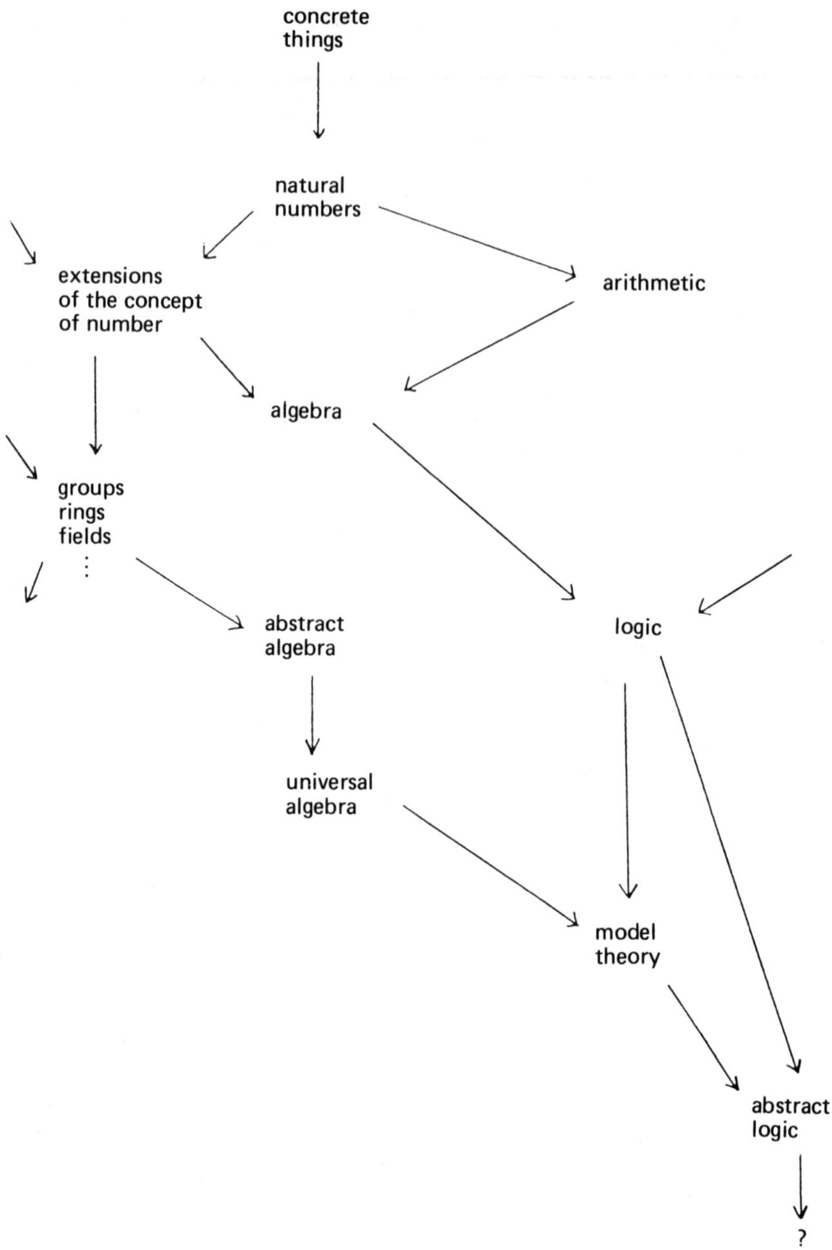

Fig. 1. Integrative growth in mathematics.

seeks informative, intellectually economical, generally applicable and elegant results.

4. Explanation of the Development of Mathematics

Beth (1959) has calculated that the number of publications in mathematical logic remained very steady and low (about one study in ten years) until a sudden increase occurred in the middle of the 19th century (31 studies in 1869–77, and 100 in 1878–87). Beth's explanation for this fact is that before 1847 the development of logic was determined only by the influence of Aristotle, and only after that year the frequency of publication started to depend on the immediately preceding works and their number. The fact that the new algebraic treatment of logic by de Morgan and Boole – Leibniz's earlier ideas excluded – was born not earlier than 1847 may be explained by the late invention of the idea of abstract, non-interpreted mathematical systems in the early 19th century (e.g. non-Euclidean geometry, group theory, theory of algebraic invariants).

These explanations of the development of logic are *internalist* in the sense that they refer to 'internal' factors – such as new ideas which break through and stimulate further research, the stage of development in the neighbouring fields of mathematics, and the opinions among the investigators about the nature of mathematics. On the other hand, *externalist* explanations view the development of mathematics as a response to 'external' demands and challenge from commerce, production, military armament, or other scientific disciplines. As standard examples of such external factors the following are well-known: the influence of surveying to geometry, commerce to arithmetics, astronomy to trigonometry, physics to differential calculus, gambling institution to probability theory, insurance to mathematical risk theory, and the 2nd World War to operations research.

External factors can no doubt be used to explain the quantitative growth of research activity in some special fields of mathematics. However, no social 'order' can guarantee that the mathematicians are able to give answers to the externally determined problems – as is shown by the relatively late invention of the mathematical theory of probability in the 17th century.[9] To explain the developments of the *content* of mathematical knowledge, reference to internal factors is needed – in particular, to the strive for abstraction and generalization, the influences across different branches of mathematics, and the existence of research traditions in mathematics.

Raimo Lehti (1971) has suggested that the development of mathematics could be compared to the wandering of a flock of lemmings. He says that there may be some »deeper» reasons for the direction that lemmings choose, but the study of this direction and the search for its reasons is a »purely empirical question»; therefore »any restricting presupposition of the real essence of mathematics» may be harmful for the attempt to find the answer to this question.

It seems to me that there is more internal dynamic in the development of mathematics (viewed as a group behaviour of mathematicians) than in the wandering of flocks of lemmings. The existence of simultaneous discoveries in mathematics[10] suggests that the advance of mathematics may lead to a point where certain discoveries become possible – and then are fairly soon realized. Moreover, the existence of mathematical 'schools', manifesting themselves as successful research traditions[11] or as opposition against some fields of research[12], shows that mathematicians themselves often have strong opinions about the directions in which mathematics should be advanced.

It also seems that to find out where the mathematicians through times are wandering is not a *purely* empirical question: to distinguish mathematicians from natural scientists presupposes already some preliminary view on mathematics. For example, many ancient and mediaeval men regarded astronomy as a branch of mathematics. The AMS Classification Scheme includes such areas as astrophysics and geophysics. Similar problems of classification arise in all historiography of science: for example, the decision to include or exclude mediaeval alchemy from the history of science depends on the conception of science that the historian has adopted. It is fair to require that such a conception should not be too restrictive, and one should not accept it dogmatically. The understanding that we now have about the 'essence' of mathematics will be deepened by the future development of mathematics itself – and this process will force us to re-evaluate the historical course of mathematics.

5. History and Philosophy of Mathematics

The remarks of Section 4 help us to answer the question posed by Weil: What can history and philosophy of mathematics have in common?

First, mathematicians themselves have normally a more or less articulated conception of the criteria of good and bad mathematics. This conception, which may contain assumptions about the existence of mathematical objects, the nature of acceptable proofs in mathematics, or the criteria of progress in mathematics, can be called the *implicit philosophy of mathematics* of the researcher. Mathematicians who study same problems and share the same implicit philosophy of mathematics cultivate 'research programmes' (Lakatos, 1976) or 'research traditions' (Laudan, 1977) – such as Pythagorean geometry, classical theory of probability, Leibnizian differential calculus, Cantorian set theory, and the various trends in constructive mathematics.[13] A historian of mathematics is entitled to take account of such implicit conceptions which have guided mathematical research. For example, it is hard to imagine how the history of geometry could be understood without considering discussions about the nature of the axiomatic method in the ancient times and in the 19th century or how the history of probability theory could be understood without paying attention to changes in the philosophical interpretation of the concept of probability.

Secondly, when a historian interprets an implicit conception of mathematics he has to use his own knowledge about the philosophy of mathematics. The less explicit the latter is, the easier it is for him to distort the facts that he is studying. The same phenomenon is visible in the historiography in science where the historians often have interpreted their data through their own implicit view about science.[14] In particular, the assumption that the history of mathematics is independent of the philosophy of mathematics is itself a philosophical presupposition which will in many cases mislead its supporters.

Lakatos has colourfully expressed the theory-ladenness of historical reconstructions, in his 1966 essay 'Cauchy and the Continuum':

> »The history of mathematics has been distorted by false philosophies even more than has the history of science. It is still regarded by many as an accumulation of eternal truths; false theories or theorems are banished to the dark limbo of prehistory or recorded as regrettable mistakes, of interest only to curiosity collectors. According to some historians of mathematics the 'proper' history of mathematics starts with those works which conform to the standards they regard as being ultimate. Others descend to the prehistoric ages only to pick out shining fragments of eternal truth from among the rubbish. Both miss some of the most exciting patterns of conjectures and refutations in the history of mathematical thought. Still worse, interesting inconsistent theories are distorted into 'correct' but uninteresting precursors of up-to-date theories. Efforts to save the authority of the giants of the past by giving them a polished modern look have gone further than one would imagine.»

6. Progress in Mathematics

One way of approaching general principles about mathematics is to try to answer the question concerning the criteria of *progress* in mathematical research.

One type of answer gives emphasis to the theoretical and practical applications of mathematical theories. Thus, David Hilbert says that the »acid test» of a new mathematical theory is »its ability to solve problems which, though known for a long time, the theory was not expressly designed to solve».[15] Kurt Gödel argues that the new axioms of infinity in set theory have to be evaluated »inductively» by their »success» or »fruitfulness in consequences», while Hermann Weyl puts special emphasis on consequences relevant to physics.

Michael Hallett (1979), pp. 9–10, tries to define progress by the problem-solving ability of a mathematical theory:

> »I propose that the setting up of a new mathematical theory T'_m constitutes progress with respect to its predecessor T_m if T'_m is used in the solution of at least one problem P which T_m did not solve [up to the time of T'_m's intro-

duction], provided that P is not of T'_m's own making and that the statement solving P was not used in the construction of T'_m.»

According to Hallett, Cantor's set theory became progressive, when Mittag-Leffler used it in 1884 in the theory of analytic functions. Hallett calls his definition 'Hilbert's criterion' and compares it with the 'methodology of scientific research programmes' (MSRP) of Lakatos, Zahar, and Worrall: »MSRP states that a consequence of T' provides support for T' providing it is true and was not used in the construction of T'» (p. 10).

One problem with Hallett's criterion is that he does not give any analysis of the meaning of 'problems' and their 'solution' in mathematics. He writes as if any derivation of a consequence from a theory would constitute a solution, but this is much too general to be adequate. Suppose, for example, that I propose an inconsistent theory T and derive the continuum hypothesis CH from it.[17] Or suppose that I propose a new axiom A to set theory and it turns out that A is logically equivalent to CH. In both cases Hallett's criterion entails that progress has been made. This shows that Hallett's criterion is not *sufficient* for progress.

It can be argued that this criterion is not *necessary*, either. Discovery of new proofs for old theorems, construction of new mathematical entities, and discovery of theorems about such new entities – these typical instances of mathematical progress are not covered by Hallett's criterion. It seems odd to deny that Cantor's discovery of transfinite numbers and his successful proofs of many of their properties constituted progress already before any application of this theory outside set theory.

Incidentally, we may ask why Hallett does not directly apply the MSRP criterion of progress to mathematics. In other words, we could say that a mathematical theory T' is progressive relative to theory T if T' entails some *true* theorem B such that B is not entailed by T and B was not used in the construction of T' – where T' is at least as 'beautiful' as T.[18] (Here T may be an empty theory.) With this definition, we could also make a distinction between real and presumed progress (cf. Chapter 4 above), not made by Hallett: T' can be *claimed to be* progressive relative to T when it is known that T' entails the true theorem B, but T' *is* progressive even if this entailment holds but is unknown.

An attempt to combine the notions of progress and truth in mathematics will be denied by those 'antirealists' who do not admit any role for a semantic conception of truth in mathematics.[19] On the other hand, realists – ranging from platonists to those constructivists[20] who place mathematical entities to the man-made 'world 3' – can make a distinction between two cases:

(a) *Theories with a standard model*, i.e., theories which speak about a specific mathematical structure (which is unique up isomorphism). For example, the standard model of arithmetic is the triple $\mathbf{N} = \langle N, S, 0 \rangle$, where N is the set of natural numbers, S is the successor function (i.e., $S(x) = x + 1$ for all $x \in N$) and 0 is zero. The so-called complete arithmetic

$Th(\mathbf{N}) = \{h \mid \mathbf{N} \models h\}$ consists of all arithmetical statements h which are true (in the Tarskian sense) in structure **N**. Gödel's Incompleteness Theorem tells that no recursive axiom system $A \subseteq Th(\mathbf{N})$ can comprise all the truths in the complete arithmetic. Let $A_1 \subseteq Th(\mathbf{N})$ and $A_2 \subseteq Th(\mathbf{N})$ be two axiom systems for arithmetic, and let $Cn(A_i)$ be the set of the deductive consequences of A_i $(i = 1,2)$. Then theory A_2 may be said to be *closer to the truth* than theory A_1 at least if $Cn(A_1)$ is a proper subset of $Cn(A_2)$.[21] In this case, A_2 is 'closer' to the complete arithmetic than A_1, since it entails more truths about natural numbers than A_1.

(b) *Theories without a standard model.* Group theory is not a description of any particular mathematical structure, but it is rather a theory of groups. Progress in group theory may of course mean that new, earlier unknown consequences are derived from the axioms for groups.[22] However, the growth of group theory to a larger extent consists in the study of special kinds of groups, obtained by adding new assumptions, or in the exploration of the applications of this theory. This process may be described as the articulation of the group-theoretic axioms rather than as 'approach to the truth'. It closely resembles the formal account of Kuhn's 'normal science' that Sneed, Stegmüller, and Moulines have proposed for natural science.[23]

Both in cases (a) and (b) mathematical advance may also follow an indirect way. Information about structure **N** can be obtained by embedding it to some larger structure (e.g., the set of real numbers or complex numbers) and by solving questions concerning this larger structure. Group theory may be embedded to universal algebra. Further, modern logic and model theory provide methods for studying the properties of theories. With this step to the 'metalevel', important insight has been obtained e.g. about results which are true but not provable in Peano arithmetic.[24]

7. Is the Growth of Mathematics Cumulative?

Since the end of 1950's, a number of objections have been directed against the view that scientific knowledge grows by *cumulation,* i.e., by adding new established results upon the earlier ones. Popper's (1963) falsificationism emphasizes the trial-and-error procedure of presenting bold conjectures and criticizing and refuting them. Kuhn (1962) and Feyerabend (1975) have discussed theoretical changes involved in the overthrow of conceptual frameworks. For Kuhn, only 'normal science', which is bound with a fixed 'paradigm' or 'disciplinary matrix', is cumulative, while 'scientific revolutions' involve the replacement of 'incommensurable' frameworks with each other. Lakatos (1976) describes science as the competition between long-range 'research programmes' which are identified by fixed 'hard core' assumptions surrounded by a 'protective belt' of more easily changed auxiliary assumptions. The rejection of a degenerative research program for a more progressive one corresponds then to Kuhnian scientific revolutions.

In small mathematical communities, the research activity may have characteristics which Kuhn attributes to normal science: a dominant line of research prescribes limits to good and bad mathematics and offers well-defined 'puzzles' for its members to solve. It was also noted above that the development of certain areas of mathematics (such as group theory) structurally resembles normal science – it consists of attempts to develop the theory and find to it new applications without challenging its basic assumptions. Moreover, concrete problem solutions function as 'exemplars' for further mathematical research just as Kuhn's description of normal science implies.

But does the development of mathematics contain also 'revolutions'? Has the recent discussion about the fallible but self-corrective nature of scientific inquiry or about the non-linearity of the development of scientific knowledge anything to do with such a formal science as mathematics? Is mathematics after all, with its strict methods of proving theorems, a paradigmatic example of the cumulative pattern of growth?

The answer to these questions depends essentially on the aspect of 'mathematical knowledge' that we choose to consider. In this respect, it is important to distinguish the *results* (proofs, theorems) of mathematical research, the mathematical *research process* (discovery of theorems and proofs), and the methodological and philosophical *background assumptions* which guide the work in mathematics.

In most cases, the mathematicians publish only their results in their final form, without indicating the steps by which they have been reached. Luckily enough, *mathematical heuristics* has been studied in interesting ways already from the method of analysis and synthesis in Greek geometry (Hintikka and Remes, 1974) to the psychology and methods of invention (Hadamard, 1954). Indication of progress in this field is provided by the computer program AM which, given 115 basic concepts in elementary set theory, was able to discover the concept of natural number and propose as hypotheses the fundamental theorem of algebra and Goldbach's conjecture (Simon *et al.*, 1981).

George Polya (1945), p. 117, argues that

> »many mathematical results were found by induction first and proved later. Mathematics presented with rigor is a systematic deductive science but mathematics in the making is an experimental inductive science».

Lakatos generalizes this point to all mathematics which is not presented in a strictly 'Euclidean' form, but he replaces Polya's induction with the Popperian method of conjectures and refutations:

> »Informal, quasi-empirical, mathematics does not grow through a monotonous increase of the number of indubitably established theorems but through the incessant improvement of guesses by speculation and criticism, by the logic of proofs and refutations.» (Lakatos, 1976, p. 5.)

According to Lakatos, a typical pattern of mathematical discovery contains the following steps:

(1) Primitive conjecture.
(2) Attempted proof.
(3) Global counterexamples to the conjecture appear.
(4) The proof is re-examined, a false lemma is spotted, and this lemma is built into a condition in the improved conjecture.

As illuminating examples of this pattern, Lakatos considers attempts to generalize Euler's theorem to arbitrary polyhedra (Lakatos, 1976) and the limit theorems concerning uniformly continuous functions (Lakatos, 1978, Ch. 3).[25]

It seems to me that Lakatos is successful in showing that mathematical knowledge formation involves interesting dynamic aspects which should be studied further. On the other hand, he makes sweeping generalizations on the basis of rather limited material. To mention only one example, the pre-euclidean period of probability theory does not seem to correspond to his description of the development of 'quasi-empirical' mathematics. Moreover, a supporter of the cumulative model for mathematical knowledge may restrict his thesis to the established theorems: most mathematical results – perhaps after an explorative stage following Lakatos's pattern one or two rounds – are indubitable and will not change any more in future research. They may become special cases of more general theorems, and simpler and more elegant ways of proving them will be found – but this does not change their final status. Even great mathematicians may occasionally make mistakes, but otherwise the cumulative model is valid – or so it can be claimed.

A different kind of objection against the cumulative model can be based on the observation that the proofs of mathematical theorems often have to use set-theoretical assumptions (such as the Axiom of Choice AC or the Continuum Hypothesis CH) which are problematic in their validity. If these auxiliary assumptions become rejected, also the theorems have to be rejected – unless new proofs are found for them without these assumptions. Mathematicians have invented a clever way of handling this problem: many books attach a special mark as a warning to those theorems B which rely on disputed principles (say, AC). Then the real theorem which is proved is not B itself but rather the claim 'If AC then B'. Hilary Putnam has given a pertinent name *if-thenism* to this strategy (see Musgrave, 1977). It can now be claimed that mathematics is cumulative on the level of such conditional if-then-statements.

8. Incommensurability and Mathematics

Among the most interesting situations in the growth of mathematical knowledge are new theories which lead to a re-evaluation of old results. One example is the attempt to replace set theory with category theory as the foundation of mathematics (see MacLane, 1971). A more restricted example is the theory of integration: some functions which are not Riemann-integrable are integrable in Lebesgue's sense. Thus, conceptual change and revision occurs in mathematics – just as in Kuhnian revolutions – and they are associated with the problem of *meaning variance:* while the terms 'mass' has a different meaning in classical and relativistic mechanics, the term 'integral' has a different meaning in Riemann's and Lebesgue's theories.

There seems to be an important difference between these two cases, however. The theories of integration of Riemann and Lebesgue are *incommensurable* in the strong sense that they speak of different matters: one of Riemann-integration, the other of Lebesgue-integration. Hence, there is no meaning-preserving translation between these theories. Moreover, claims of the form 'function f is Lebesgue-integrable but not Riemann-integrable' are not inconsistent. On the other hand, the idea that classical mechanics and relativity theory were incommensurable in the same sense – one speaking of 'Newtonian mass', the other of 'relativistic mass' – seems odd: the claim 'the Newtonian mass of particle x is m_1 and its relativistic mass m_2' does not seem meaningful, for if Einstein is right then mass *is* relativistic and there is no such thing as 'Newtonian mass'. For this reason, it seems natural to think that material particles and bodies have a quantitative physical property, mass, *about which* Newton's and Einstein's theories make different, mutually conflicting statements.[26] No corresponding assumption can be made in the theory of integration.

It should be noted that not all conceptual change in mathematics is 'revolutionary'. Take, for example, three definitions of natural numbers as the equivalence classes of the relation 'has the same power' (Frege, Russell), as set-theoretical constructions

$$\begin{aligned}0 &= \emptyset\\ 1 &= \{\emptyset\} = \{0\}\\ 2 &= \{\{\emptyset\}\} = \{1\}\\ &\vdots\\ n &= \{n-1\}\\ &\vdots\end{aligned}$$

(Zermelo), or as constructions

$$\begin{aligned}0 &= \emptyset\\ 1 &= \{\emptyset\} = \{0\}\\ 2 &= \{\emptyset, \{\emptyset\}\} = \{0,1\}\\ &\vdots\\ n &= \{0,\ldots, n-1\}\\ &\vdots\end{aligned}$$

(von Neumann). All these structures of natural numbers are isomorphic, so that mathematicians regard them as essentially equal to each other. Similarly, the definition of real numbers by Dedekind cuts or by the limits of converging Cauchy sequences give essentially the same result. Instead, the non-standard models of analysis, which contain infinitesimals, are not isomorphic to the structure **R** of real numbers. Also the structure of 'constructive real numbers', obtained by considering effectively converging Cauchy sequences, differs essentially from **R**.[27]

It is interesting to note that the theory of integration is a cumulative enterprise just for the reason that the different theories of integral are incommensurable: all results concerning Riemann's integral remain valid in spite of the new theory of Lebesgue, because strictly speaking these theories do not contradict each other.[28] On the other hand, many mathematicians may feel that constructivistic and intuitionistic analysis, classical analysis, and non-standard analysis are in some sense rivals to each other. If platonists are right in thinking that real numbers exist independently of human mind, then these theories can be understood as mutually incompatible descriptions of the nature of real numbers (cf. the rival theories of mass). If constructivists are right in claiming that mathematical objects are created by man, these theories can be understood as incommensurable accounts of different things (cf. the theories of integral). Therefore, the possibility of non-cumulative revolutions in mathematics seems to presuppose that platonism is valid at least to some degree.

The remarks above concern the mathematical research process and its results. They should be distinguished from changes in the ontological and methodological background assumptions. The distinction between pure and applied geometry in the 19th century and the rejection of the traditional view of geometrical axioms as self-evident truths about the space – these changes constituted a fundament shift in the prevailed conception of geometry, even though the set of theorems in Euclidean geometry remained unchanged. Similarly, constructivist mathematics is based upon views about mathematical proofs which are radically different from classical ideas. In this sense, there may exist revolutionary developments within the *philosophy* of mathematics which, in the Kuhnian fashion, are changes in the 'world view' of the mathematicians.

9. Future Prospects

In the end of the 18th century, some leading mathematicians (Lagrange, d'Alembert) presented doubts that the progress of their field will not continue for long, since the most important mathematical discoveries have already been made.[29] This prediction could not have been farther from the truth: the early 19th century was a period of vigorous activity (e.g., abstract algebra, non-Euclidean geometry), and since then the growth of mathematical knowledge has even accelerated.

Rescher (1978) has presented plausible arguments to support the view that science will grow in the future without a fixed upper bound – even if the growth will become slower. This conclusion is even easier to defend in the case of mathematics than in natural science. Mathematicians can base their work upon the results of earlier generations, and they find new tools to help their work.[30] Mathematical research is not so much dependent on the expensive high-power technology as the natural sciences. What is more, the subject matter of mathematics – man-made abstract structures and theories about them – is in principle extendible *ad libitum,* so that there is no fear of a time when all interesting mathematical problems have been solved.

NOTES

1 See, for example, Sarton (1957) and Kuhn (1977).

2 For the philosophical foundations of mathematics, see Beth (1959), Benacerraf and Putnam (1964).

3 See Boltzmann (1974), p. 76.

4 See de Solla Price (1963) and Rescher (1978). Already Friedrich Engels, in his *Dialectics of Nature* (1873–83), proposed a parabolic law saying that the volume of science is proportional to the square of time (cf. Rescher, 1978, p. 75).

5 See Knorr (1975).

6 See Todhunter (1865) and Maistrov (1974).

7 The axiomatic foundations of set theory is of course still an intensively studied field.

8 For a short history of algebra, see Struik (1967). For model theory and abstract logic, see Henkin *et al.* (1974), Lindström (1969), and Barwise (1974).

9 Cf. Hacking (1975) and the discussion in Hintikka *et al.* (1981).

10 See Kuhn (1977), Chapter 4. The most famous of them is the independent discovery of non-Euclidean geometry by Gauss, Bolyai and Lobachevsky in the early 19th century.

11 The Finnish school in the theory of analytic functions is a good example.

12 Gauss never ventured to publish his discoveries on non-Euclidean geometry; Cantor's set theory has been attacked since its birth in the 1870's.

13 For the existence of 'research programmes' in mathematics, see Lakatos (1978), Howson (1979) and Hallett (1979).

14 For arguments against Baconian, presupposition-free historiography of science, see Agassi (1963) and Howson (1976).

15 See Benacerraf and Putnam (1964), p. 150. This principle may be compared with Whewell's hypothetico-deductive account of scientific inference (cf. Chapter 3 above).

16 See Benacerraf and Putnam (1964), p. 265.

17 Note that, according to Hallett (1979), p. 22, theories may be progressive on the basis of their heuristic role – they need not be logically indispensable for the solutions that they generate.

18 Mathematicians often think that they have a built-in capability of recognizing important results in mathematics without any formal criteria of progress. The mathematician »may not be able to define what is a mathematical idea, but he likes to think that when he smells one he knows it» (Weil, 1980, p. 230). It is also claimed that the mathematicians have an »intuitive certainty» (Hadamard, 1954) on mathematical progress based upon esthetic criteria. The Finnish mathematician Rolf Nevanlinna used to emphasize that »beautiful» mathematical structures have had an almost »predestined» significance also in relation to applications – combining thus the ideas of esthetic intuition and predictive success.

19 Cf. the discussion in Dummett (1977, 1978) and Wright (1980).

20 For Popper's conception of 'world 3', see Chapters 3 and 9.

21 If neither $Cn(A_1) \subseteq Cn(A_2)$ nor $Cn(A_2) \subseteq Cn(A_1)$ holds, the definition of truthlikeness requires a measure of the *information content* for arithmetical statements.

22 For an attempt to measure the increase of information within *deductive* inference, see Hintikka (1970).

23 See Moulines (1979) and Chapter 5 above.

24 For interesting recent results in this direction, see Paris and Harrington (1977).

25 Fine (1981) proposes a 'refinement' to Lakatos's treatment of conceptual change in mathematics – which seems essentially equivalent to Carnap's method of explication.

26 For a discussion of this point, see Nola (1980).

27 See, for example, Bishop (1967) and Heyting (1956).

28 This remark should be contrasted with Kuhn's position which argues against the cumulative model by claiming that successive scientific theories are incommensurable.

29 See Struik (1967), pp. 136–137.

30 Computers have recently been used in the proof of the famous four colour theorem (see Tymoczko, 1979; Teller, 1980; Defletsen and Luker, 1980).

Bibliography

Agassi, J., *Towards a Historiography of Science,* Mouton, The Hague, 1963.

Barwise, J., 'Axioms for Abstract Model Theory', *Annals of Mathematical Logic* 7 (1974), 221–265.

Benacerraf, P. and Putnam, H. (eds.), *Philosophy of Mathematics: Selected Readings,* Blackwell, Oxford, 1964.

Beth, E.W., *The Foundations of Mathematics,* North-Holland, Amsterdam, 1959.

Bishop, E., *Foundations of Constructive Analysis,* McGraw-Hill, New York, 1967.

Defletsen, M. and Luker, M., 'The Four-Color Theorem and Mathematical Proof', *The Journal of Philosophy* 77 (1980), 803–820.

Dummett, M., *Elements of Intuitionism,* Oxford University Press, Oxford, 1977.

Dummett, M., *Truth and Other Enigmas,* Duckworth, Oxford, 1978.

Feyerabend, P., *Against Method,* NLB, London, 1975.

Fine, A., 'Conceptual Change in Mathematics and Science: Lakatos' Stretching Refined', in P. Asquith and I. Hacking (eds.), *PSA 1978,* vol. 2, Philosophy of Science Association, East Lansing, 1981, pp. 328–341.

Hacking, I., *The Emergence of Probability,* Cambridge University Press, Cambridge, 1975.

Hadamard, J., *The Psychology of Invention in the Mathematical Field,* Dover, New York, 1954.

Hallett, M., 'Towards a Theory of Mathematical Research Programmes I–II', *The British Journal for the Philosophy of Science* **30** (1979), 1–25, 135–159.

Henkin, L. *et al.* (eds.), *Proceedings of the Tarski Symposium,* American Mathematical Society, Providence, 1974.

Heyting, A., *Intuitionism,* North-Holland, Amsterdam, 1956.

Hintikka, J., 'Surface Information and Depth Information', in J. Hintikka and P. Suppes (eds.), *Information and Inference,* D. Reidel, Dordrecht, 1970.

Hintikka, J. and Remes, U., *The Method of Analysis,* Reidel, Dordrecht, 1974.

Hintikka, J., Gruender, D. and Agazzi, E. (eds.), *Proceedings of the 1978 Pisa Conference on the History and Philosophy of Science,* vol. II, D. Reidel, Dordrecht, 1981.

Howson, C. (ed.), *Method and Appraisal in the Physical Sciences,* Cambridge University Press, Cambridge, 1976.

Howson, C., 'Methodology in Non-Empirical Disciplines', in G. Radnitzky and G. Andersson (eds.), *The Structure and Development of Science,* Reidel, Dordrecht, 1979, pp. 257–266.

Knorr, W., *The Evolution of the Euclidean Elements: A Study of the Theory of Incommensurable Magnitudes and Its Significance for Early Greek Geometry,* D. Reidel, Dordrecht, 1975.

Kuhn, T.S., *The Structure of Scientific Revolutions,* 2nd ed., University of Chicago Press, Chicago, 1970.

Kuhn, T.S., *The Essential Tension,* The University of Chicago Press, Chicago, 1977.

Lakatos, I., 'History of Science and Its Rational Reconstructions', in Howson (1976), pp. 1–39.

Lakatos, I., *Proofs and Refutations: The Logic of Mathematical Discovery,* Cambridge University Press, Cambridge, 1976. (Published first in 1963–64.)

Lakatos, I., *Philosophical Papers vol. 2: Mathematics, Science and Epistemology,* Cambridge University Press, Cambridge, 1978.

Lehti, R., 'Matemaattisen ajattelutavan kehitys', in *Logiikka ja matematiikka,* WSOY, Porvoo, 1971, pp. 75–97.

Lindström, P., 'On Extensions of Elementary Logic', *Theoria* **35** (1969), 1–11.

MacLane, S., *Categories for the Working Mathematician,* 1971.

Maistrov, L.E., *Probability Theory: A Historical Sketch,* Academic Press, New York, 1974.

Moulines, C.U., 'Theory-Nets and the Evolution of Theories: The Example of Newtonian Mechanics', *Synthese* **41** (1979), 417–439.

Musgrave, A., 'Logicism Revisited', *The British Journal for the Philosophy of Science* **28** (1977), 99–127.

Nola, R., '»Paradigms Lost, or the World Regained» – An Excursion into Realism and Idealism in Science', *Synthese* **45** (1980), 317–350.

Paris, J. and Harrington, L., 'A Mathematical Incompleteness in Peano arithmetic', in J. Barwise (ed.), *Handbook of Mathematical Logic,* North-Holland, Amsterdam, 1977, pp. 1133–1142.

Polya, G., *How to Solve It,* Princeton University Press, Princeton, 1945.

Popper, K.R., *Conjectures and Refutations,* Routledge and Kegan Paul, London, 1963.

Price, D. de Solla, *Little Science, Big Science,* Columbia University Press, New York, 1963.

Rescher, N., *Scientific Progress: A Philosophical Essay on the Economics of Research in Natural Science,* Blackwell, Oxford, 1978.

Sarton, G., *The Study of the History of Mathematics and the Study of the History of Science,* Dover, New York, 1957.

Simon, H., Langley, P. and Bradshaw, G., 'Scientific Discovery as Problem Solving', *Synthese* **47** (1981), 1–27.

Struik, D., *A Concise History of Mathematics,* Dover, New York, 1967.

Teller, P., 'Computer Proof', *The Journal of Philosophy* **77** (1980), 797–803.

Todhunter, I., *A History of the Theory of Probability from the Time of Pascal to that of Laplace,* Cambridge, 1865. (Chelsea, New York, 1965.)

Tymoczko, T., 'The Four-Color Problem and Its Philosophical Significance', *The Journal of Philosophy* **76** (1979), 57–83.

Weil, A., 'History of Mathematics: Why and How?', in O. Lehto (ed.), *Proceedings of the International Congress of Mathematicians, Helsinki, 1978,* Helsinki, 1980, pp. 227–236.

Wright, C., *Wittgenstein on the Foundations of Mathematics,* Duckworth, London, 1980.

CHAPTER 9

REALISM, WORLDMAKING, AND THE SOCIAL SCIENCES

The definition of scientific progress as increasing truthlikeness was elaborated and defended above in Chapters 5 and 7. Some opponents – who may feel that this defense at best is valid for the natural sciences – will deny its applicability to the human and social sciences. Generally speaking, there are two grounds on which such a denial might be supported. First, as the realm of cultural and social phenomena is ontologically different from nature, it has been argued that the basic categories of the realist semantics and epistemology (especially the correspondence-theoretical concept of truth) are not appropriate for it. Secondly, independently of the first argument, it has been claimed that the social sciences have a peculiar aim which is different from informative truth. Even though these two grounds raise interesting problems for an attempt to extend the basic ideas of scientific realism to the social sciences, it seems to me that they are mistaken. In this chapter, my reasons for thinking so are summarized. I shall start from problems of ontology and then proceed to epistemological and methodological issues.

1. ONTOLOGICAL PRELIMINARIES

By 'realism' philosophers usually mean the doctrine that there exists a mind-independent reality. According to 'scientific realism', science is the best – and also to some extent successful – vehicle for pursuing knowledge about this reality.[1] For example, Richard Boyd's (1981) recent exposition of scientific realism is based upon the assumption that »reality is prior to thought»: scientific theories refer to »the unobservable features of the world» and »provide increasingly accurate accounts of the causal structure of the world». Further,

> »not only scientific knowledge, but the language and methods of the sciences as well, represent hard-won victories in a continuing struggle to accomodate our intellectual practices to the structure of an independently existing world.» (Boyd, 1981, p. 613.)

Also some critics of realism presuppose a similar view: for example, Hilary Putnam (1982) regards the observation that there is no »ready-made world» as counting against the correspondence theory of truth and against certain forms of scientific realism.

This characterization of scientific realism is clearly oriented towards the natural sciences, and it leaves open the questions concerning the existence and the status of the cultural and social sciences. As culture and society are produced through the plans and activities of human beings, there is a straightforward sense in which they are not 'mind-independent', 'prior to thought' or 'ready-made'. Does it follow that the scientific study of culture and society, conforming to the standards of scientific realism, is impossible? To see why such a conclusion would be hasty and mistaken, a more careful analysis of the concept of reality is needed.

Let us follow Charles Peirce in saying that those things are *real* »whose characters are independent of what anybody may think them to be».[2] For example, stones are real, since it does not depend on anyone's thoughts whether an object is a stone or not. When I imagine a pink elephant, my mental state is real, but the elephant is fictional, since it has only such characters that my thought impresses upon it. Further, natural numbers are real, since a newly discovered number is even or odd independently of our opinions about that question. Peirce's criterion does not imply that a thing is real only if its existence is independent of minds – it is sufficient that it has 'objective' characters. For example, if I build a house and paint it red, this object and its colour is genetically mind-dependent, but when I have finished my work the house is a public and real thing in Peirce's sense. We may thus follow Karl Popper (1972) in dividing the *reality* into three domains: non-organic or organic physical objects and processes (*nature* or *world 1*); thoughts, mental states or consciousness of higher animals and of human beings (*world 2*); concrete and abstract products or public creations of human minds (*world 3*).

The relations between worlds 1, 2, and 3 have been understood in many different ways. Popper himself defends an interactionist solution to the mind-body problem – ability to exert a causal effect upon ordinary size world 1 objects is his criterion of reality (see Popper and Eccles, 1977). World 3 interacts causally with world 1 via world 2, but it is only »more or less independent» of the other two worlds, since it is man-made. Monistic ontologies regard one of these worlds as primary relative to the others. For the *materialists,* world 1 consists of the basic 'stuff' from which worlds 2 and 3 have developed through biological and cultural evolution. World 2 is the primary one for the *subjective idealists,* whereas the *objective idealists* regard world 3 as the original and independent reality which in some way or another generates everything else (e.g. Platonic ideas copy themselves in the sensible world, God creates the material world and the soul of man, Hegel's absolute spirit generates nature by a process of alienation).

The slogan 'reality is prior to thought' should, in my view, be understood as saying that nature exists independently of subjective minds. Thus, scientific realism is certainly incompatible with subjective idealism. Some of its supporters have been objective idealists (Peirce[3]) or dualists. Still, most scientific realists have been some sort of materialists who think that worlds 2 and 3 cannot exist without world 1.

It is important to emphasize that, in endorsing materialism, scientific realists need not be *reductionists* who treat human minds and their products merely as complex material things and processes – and thereby reduce worlds 2 and 3 to world 1. For such reductionists, there is no important ontological difference between nature, minds, and society, and therefore they may regard all sciences alike – no matter what aspect of reality they study. Ontological unity of the subject matter is here taken as a guarantee of the methodological unity of the sciences. On the other hand, anti-reductionist materialists find the attempted reductions of worlds 2 and 3 to world 1 philosophically confusing – they may admit the existence of 'emergent' properties and laws concerning the worlds of thought, culture, and society which are not reducible to the level of world 1.[4] (See Section 3 below.)

Given the division of reality into the three worlds, a corresponding classification of systematic disciplines studying reality is obtained: world 1 is investigated by the natural sciences, world 2 by psychology, and world 3 by mathematics, the cultural sciences, and the social sciences. In the following, my focus will be in the comparison between the sciences in the first and third groups.

2. Nature and Maker's Knowledge

In Kant's philosophy there is a gap between the objective realm of things-in-themselves (or noumena) and the world of phenomena. We can meaningfully speak of *Dinge an sich,* but besides their existence we know nothing about them. It is rather the world of phenomena that is the object of our knowledge: according to Kant's 'Copernican revolution' in epistemology, knowledge about phenomena is possible because their world is partly constituted by the specifically human cognitive faculty. Here Kant associates himself with the ancient tradition which regards 'maker's knowledge' as the genuine kind of knowledge.[5]

Various attempts have been made to overcome the Kantian agnosticism concerning the things-in-themselves. In the idealist tradition of Fichte and Hegel, the subject and the object of knowledge are identified: *Dinge an sich* are abolished, nature is viewed as alienated spirit, and knowledge about nature is merely a moment in the self-knowledge of the spirit. Young Georg Lukács, in his *Geschichte und Klassenbewusstsein* (1923), followed this tradition by replacing Hegel's spirit by the proletariat which becomes the »identical subject-object of history» after having gained its class consciousness. 44 years later Lukács argues that the »fundamental and grave error» of his early work was the confused identification of alienation and objectification: when alienation is overcome, objective reality – nature or the domain of objects – disappears.[6] Other idealistic trends include the phenomenalism of Mach and Carnap's *Aufbau,* where the whole world is built or constructed from sensuous phenomena, and Husserl's transcendental phenomenology, where the ultimate reality is the 'transcendental subject' which 'constitutes' the world.

While the idealists try to avoid Kantian agnosticism by eliminating the world of noumena, the realists in turn do this by rejecting the world of phenomena. The naive realists believe that we are able to cognize, in our ordinary life, the world as it 'really' is. The critical realists agree with Kant that our experience is always influenced by the accepted conceptual and theoretical frameworks.[7] This means that in studying the world the scientists have to 'conceptualize' it by some suitably chosen framework (cf. Chapter 5). The realists may also agree with 'praxists' (like later Husserl and Heidegger) that our experience is influenced by our needs and interests, i.e., by our practical orientation towards the world. In other words, human knowledge is both theory-laden and praxis-laden. However, in contrast with Kant's view, our conceptual frameworks and practical interests are not tied up with the permanent nature of the human species, but rather they are historically changing and changeable. For the critical scientific realists this means that human knowledge is not confined to the 'manifest image' (i.e., to the everyday or common sense view of the world): it is possible to use the methods of science to build increasingly accurate 'scientific images' of the world which go beyond the manifest image.[8]

From this perspective it appears that Kant's mistake lies in his attempt to place a separate world of phenomena *between* the object and the subject of knowledge:[9] our knowledge may be imcomplete and distorted, it may be influenced by our theoretical assumptions and practical interests, but it is nevertheless knowledge about the things-in-themselves. An admirable exposition of a realist epistemology was outlined by Peirce who explicitly rejected the assumption of incognizable things-in-themselves: objective reality causally influences scientific knowledge formation so that this same reality is obtained as the limit of the opinion of the scientific community (see Fig. 1). (Cf. also Chapter 3 above.)

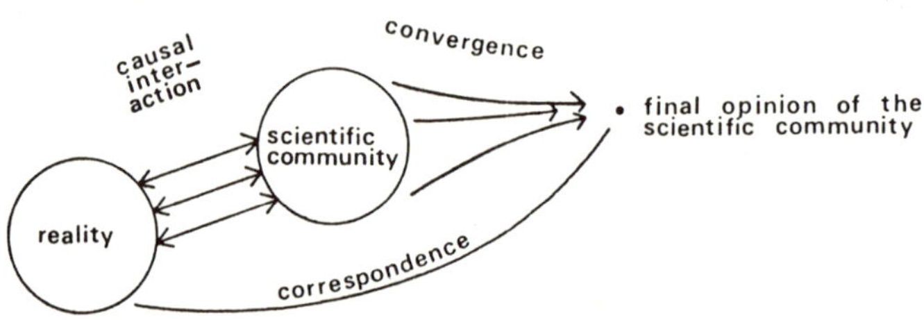

Fig. 1. Peirce's model of knowledge formation.

An interesting contrast can be made here between scientific realism and phenomenology. According to Sellars (1963), the theoretical entities of the most advanced scientific theories are ontologically primary – but not conceptually or epistemically primary – to the entities of the manifest image. Indeed, for a Sellarsian realist – who regards »science as the measure of all things» – the entities of the manifest image (such as stones, tables, etc.) do not 'really' exist but rather are composed of 'real' objects (such as atoms). Some philosophers from the phenomenological tradition agree with Sellars that natural science starts from the world of everyday experience – or 'life-world' (*Lebenswelt*) in Husserl's terminology.[10] However, they argue that the »resulting universe» obtained by the mathematical methods of »Galilean physics» is a human construction which should not be »mistaken for reality itself», since »reality is, and always remains, the life-world» (Gurwitsch, 1967).

In Peirce's realism, knowledge is produced through causal interaction with nature. Similarly, the concept of practical action – or work – has a great epistemological significance in Marxism. By work man continuously transforms nature and himself. The further this process advances, the more nature is *für uns* – and the more it is cognizable. Thus, Engels argues that Kantian agnosticism is refuted by »experiment and industry».

> »If we are able to prove the correctness of our conception of a natural process by making it ourselves, bringing it into being out of its conditions and making it serve our own purposes into the bargain, then there is an end to the Kantian ungraspable 'thing-in-itself'.» (Engels, 1946, p. 24.)

Engels and Lenin accept Hegel's dynamic conception of knowledge as a process which approaches to the 'absolute truth' through a sequence of 'relative truths'. Within this conception, truth is defined as correspondence with reality, and »practice is the criterion of truth» in the Baconian sense that our knowledge about nature and our ability to master nature in our practical activity develop together.

Besides this realist branch of the followers of Marx, there is a trend of Western Marxism, influenced by Lukács, which refuses to talk about nature as an ontological category. The representatives of this trend refer to Marx's own words from the *Economic and Philosophical Manuscripts* of 1844:

> »But *nature* too, taken abstractly, for itself, and fixed in its separation from man, is *nothing* for man.» (Marx, 1975, p. 398.)

Alfred Schmidt, a member of the Frankfurt school, argues that Marx's materialism is »non-ontological»: the concept of nature is »socially mediated», and one cannot speak of nature in abstraction of the level that man's mastery of nature has reached within the history of mankind (Schmidt, 1971, p. 167). Lenin's conception of 'reflexion' (between world and consciousness) is replaced by the concept of 'praxis' as the basic category of epistemology by the leaders of the Praxis group (Marković, 1968; Petrović, 1967).

These ideas of 'practical materialism' have a close connection to the pragmatist philosophical movement which tries to give a description of man's knowledge-seeking procedures without using 'metaphysical' concepts of reality and truth, i.e., concepts 'divorced' from man's practical activities.[11] In a sense, these philosophers replace the mind-independent nature with human praxis, and the idea of truth as correspondence with reality is replaced with the requirement that knowledge has to have a suitable kind of 'correspondence' with praxis. This position resembles that of phenomenology, since ordinary human activity takes place in the life-world – and this activity is here taken to be both epistemically and ontologically prior to the man-made worlds that human beings construct.[12]

In this section, we have considered various responses to Kant who postulated the existence of things-in-themselves but claimed that our cognition does not reach them. Idealism denies the existence of things-in-themselves; pragmatism finds it meaningless to speak of them; scientific realism admits their existence and argues that they can be known better and better through the systematic use of the scientific method. While the realists acknowledge the important epistemological function of human activity – both conceptualizing and practical activity – they think that scientific knowledge of nature is possible in spite of the fact that nature is not made by man.

3. World 3 and Our Knowledge of It

Without attempting at an exhaustive description, we may distinguish three kinds of entities among the man-made elements of world 3. First, there is the domain of *artefacts* or *cultural objects* (such as knives, tables, cars, houses, books, paintings, etc.), i.e., concrete physical things designed and created by human beings for specific purposes and uses. (Popper calls them »embodied» world 3 objects.) Artefacts exist as long as the underlying physical objects exist, but a description of them merely as physical things – i.e., as members of world 1 – is incomplete, since it does not express their historically developing role or function within human practice. For example, as a cultural object, a table has relational properties (such as its designer, owner, intended use, price, etc.) which presuppose other entities from worlds 2 and 3 – that is, properties that a mere physical object does not have.[13]

Secondly, world 3 contains *abstract entities* (such as numbers, geometrical figures, classes, concepts, propositions, conceptual systems, problems, theories, values, norms, etc.) which do not exist as physical objects at all but can be named or expressed by concrete (world 1) or mental (world 2) representations. For example, numbers can be denoted by numerals written on a paper or by ideas in the head of a mathematician, propositions can be expressed by sentences or by thoughts, etc. These entities resemble Plato's ideas: they are 'timeless' in the sense that the *same* entity can be reproduced arbitrarily many times.

Thirdly, world 3 contains more or less abstract *historical entities and processes*, from individual intentional acts to long-range historical events (such as the First World War), and from simple institutions to large-scale social systems and organizations (such as the University of Helsinki, the Lutheran religion, the state of France, capitalist economy, communism, the English language, and the legal order in Finland). These entities – which constitute human *history* and *society* – are created, reproduced, supported, transformed, and cancelled through the activity of human beings.[14] In most cases, social institutions are temporarily represented by certain persons (e.g., the contemporary administrators, teachers, and students in the University of Helsinki), but they are not reducible at any moment to these representatives.[15] On an institutional basis, certain groups of people constitute 'communities' – such as Finnish citizens, the linguistic community of the Finns, the legal community in Finland, and the scientific community of physicists. We may also say that most *societal facts* are 'institutional' in the sense that they are about social institutions ('The University of Helsinki was founded in Turku in 1640', 'Many universities in the Western countries are in an economic crisis') or presuppose the existence of institutions ('John payed his debt to the bank', 'Most residents of this area voted for the Conservatives').[16]

What we said of world 3 is already sufficient to make the following thesis plausible: it is possible to make such *statements about the elements of world 3 that are true or false in the correspondence-theoretical sense.* In other words, statements may quite well 'correspond' to a reality that is man-made rather than 'mind-independent'.

Let us give some illustrations of our thesis. Introductory courses of mathematical logic give a definition of the standard model **N** of arithmetic and of the Tarski-style truth conditions of arithmetical statements in **N**. If we pick up any numeral $\bar{n}$, then these definitions uniquely determine whether the statement '$\bar{n}$ is a prime number' is true or false: it is true if and only if the number in **N** denoted by $\bar{n}$ is not divisible by any smaller number other than 1. As this condition can be verified by purely conceptual operations, we have a priori knowledge about arithmetical truths. (More generally, every arithmetical statement will have a determinate truth value – true or false – in **N** independently of our knowledge of it. This view about arithmetic is thus 'realist' in the sense of Dummett (1982) without being Platonist.)

Suppose that an archeologist finds a strange wooden construction in Finland, and he claims that it is a primitive sledge which was built in order to help transportation on snow in winter time. This statement is true if and only if this object in fact was an artefact that the ancient Finns intended for the purpose of winter transportation. – In more general terms, what is called the interpretation or elucidation of the 'meaning' of cultural objects in hermeneutics gives results which are true or false (cf. Bleicher, 1980; Hookway and Pettit, 1978).

The statement 'Sibelius is the most popular composer in Finland' is a claim about a social state of affairs involving the musical taste of the Finns.

It is true if and only if the popularity of Sibelius, measured by the attitudes of the Finns, is greater than that of any other composer. Similarly, the statement 'Quantum mechanics belongs to present scientific knowledge' is a claim about the epistemic attitudes of the current scientific community of the physicists. It is true if and only if most members of this community accept QM as (approximately) true, believe that other members accept QM, and use QM in constructing explanations and predictions of microphysical phenomena.[17] A similar analysis can be given of the truth conditions of statements of the form 'Legal norm *n* is valid in Finland' and 'Linguistic norm *m* is valid in Finnish' – the former concerns the attitudes and the behaviour of the Finnish legal community and the latter the Finnish linguistic community.[18]

More generally, it may be said that typical descriptive statements about the domains of culture, history, and society express complex states of affairs. These states differ from those studied by the natural sciences through the fact that they involve the intentions, beliefs, desires or other attitudes of social agents (persons, groups, communities, institutions). In spite of this difference, it is correct to say that these statements are true if and only if the states of affairs they express actually obtain in reality, i.e., if and only if they correspond to facts.

This argument suggests that in studying culture and society, our basic descriptive statements will be true or false in the realist sense. The next step in developing a realist science of culture would be the establishment of – deterministic or probabilistic[19] – lawlike connections between the facts expressed by the basic statements.

The existence of laws about society has sometimes been denied on the ground that man has 'free will'. However, it seems more plausible to assume that the actions of all social agents are lawfully conditioned (but not necessarily determined) by their background and environment. Another strategy in arguing against laws of society might be called 'Lukácsian' (cf. Section 2): Simmel, young Lukács, and several Western Marxists have spoken of the capitalist society as 'second nature' which appears to us as objectively real (in the same sense as nature) but which in fact is man-made – and therefore can be abolished by men (cf. also Kosik, 1976). All this is compatible with our characterization of society as belonging to the realm of world 3. Moreover, if there are laws about capitalism, they do not disappear if the capitalist society is in some day abolished: it still remains true to say that whenever this kind of socio-economic structure exists in the world it must (by nomic necessity) have certain characteristic properties.

It follows from these observations that the distinction between the 'scientific image' and the 'manifest image' is relevant also in the social sciences. There is a temptation to think that, as society is man-made, there is nothing more in society than the 'life-world' constituted by men. Against this view, it can be argued that scientific knowledge about society should go beyond our ordinary conception of its basic features.

There is still one important problem for us to consider: What epistemological consequences, if any, does the ontological difference between worlds 1 and 3 have? Should we perhaps admit that we have maker's knowledge about world 3 which differs from our knowledge about nature? Indeed, already Hobbes argued that we have demonstrable knowledge both in geometry and »civil philosophy» (i.e., political science), and Vico based his famous idea of a »new science» upon the view that culture is made by man. It seems to me, however, that there is a great difference between mathematics and the other sciences of world 3. As the domain of mathematical entities can be produced and reproduced by each individual mathematician whenever he or she wants to, our mathematical knowledge is true a priori. On the other hand, facts about culture and society go beyond the limits of each individual investigator. In Durkheim's (1950) words, these facts are »external to the individual» and »endowed with a power of coercion by reason of which they control him». As social facts are never produced by an individual person, *no one can have complete maker's knowledge about them.*[20] Therefore, as Durkheim puts it, social phenomena »ought to be treated as things» within the social sciences. Therefore, outside mathematics, the truths about world 3 have to be established through 'objective' methods of investigation (such as the method of hypothesis[21]) which are familiar from the natural sciences.

4. Remarks on Methodological Monism

Methodological monism is the thesis that there is no significant methodological difference between the natural sciences and the human sciences. This thesis is sometimes called 'naturalism' (Bhaskar, 1979, p. 3), but this is slightly misleading: if it is claimed that the natural and the human sciences should be similar, this leaves still open the problem what kind of contemporary science, if any, they should resemble. Most methodological monists have been 'naturalists' in the sense that they have tried to show the reducibility of the cultural and social sciences to their own favourite model of the natural sciences (e.g. mechanical materialism, naive empiricism), and most opponents of monism have been 'anti-naturalists' who deny the validity of such reductions. However, for a scientific realist it is not very illuminating to know, e.g., that the social sciences do not fit the conception of science of the logical empiricists, since he does not believe that the natural sciences fit it either. On the other hand, the 'new philosophy of science' of the 60's has had the interesting consequence that the natural sciences and the social sciences have begun to look more and more similar to each other – as wittnessed by the numerous applications of Kuhn's and Lakatos's models of science to the human and social disciplines.[22]

The doctrine of methodological monism is of course relative to what is regarded as a 'significant methodological difference'.[23] In my view, monism is valid in the following sense: *in any field of science, progress can be defined primarily by means of increasing truthlikeness.* I shall conclude this paper by defending this thesis against some possible objections.

(a) Following Windelband and Rickert, it might be claimed that the historical and cultural sciences are 'idiographic' rather than 'nomothetic', i.e., they seek descriptions of singular events rather than general laws. Even if this claim were true (cf. Section 3, however), it would not contradict my thesis, since the concept of truthlikeness is equally well applicable to singular sentences as to generalizations.

(b) It may be claimed that social reality and laws are historically changing in a way which distinguishes them from nature. This is not quite true, since many natural sciences study historical entities (such as the universe, the earth, species of animals). Moreover, there is nothing in the theory of truthlikeness which would preclude its applicability to situations where the 'target' of our scientific statements is moving.[24]

(c) It may be pointed out that social reality is partly produced by the scientific theories that are accepted in the society, whereas nothing comparable can take place in the natural sciences. This situation may be illustrated by juridical theories about ownership: while the 'traditional theory' regarded ownership as a kind of substance belonging to one person at each moment of time, the 'analytical theory' defines ownership by the relations of being legally protected against certain other persons in certain respects (see Aarnio, 1981b). When the latter theory became dominant in Finland after the 50's, the institution of ownership itself changed. This example shows that a social scientist may be in the position to transform his society, not only in his role as a citizen, but also through his own scientific achievements. In contrast, the nature did not change, when Newton's theory was accepted. However, the juridical 'theory' of ownership is primarily a conceptual tool for handling certain cases – rather than a descriptive statement. An interesting observation concerning this example is Aarnio's suggestion that the 'analytical theory' of ownership is superior to the 'traditional theory' because its juridicial problem-solving ability is greater. This would mean that some 'theories' in the social sciences can be formalized in the 'structuralist' model of 'Kuhn-theories' (see Chapter 6 above): they would correspond to the conceptual 'core' of a Kuhn-theory, and, therefore, while they would not themselves be statements, they could be applied to juridical problems in a way which is similar to the role of 'theories' in Operations Research (see Chapter 11 below).

(d) Some philosophers may accept the existence of true or false statements about world 3, but still think that social scientists are not typically interested in such statements. For example, they may claim that the study of culture and society is not descriptive but rather ascribes meanings, evaluates cultural and social constructs, recommends actions, and transforms the society. This claim – which is related to the large problem area concerning the role of values in the social sciences – cannot be discussed in detail here, except for two general comments. First, one could argue that such non-descriptive activities belong to art and technology rather than science (cf. Chapter 12), and they also have to be based upon a descriptive – true or false – account of the 'meaning' of artefacts or of the actual state of society independently of the opinions of the investigator. Secondly, even when theories in the social sciences do not simply describe the existing society but instead outline desirable alternatives to it, they need not be construed as unconditional recommendations: rather they can be regarded as conditional recommendations ('technical norms' in von Wright's sense) of the form 'If you have these aims, then you should do so and so! '. The crucial difference between unconditional and conditional recommendations is that the latter, unlike the former, can be true or false. (A technical norm is true if the recommended action really is a means for attaining the goal.) Hence, the social sciences can be divided into *descriptive* and *design sciences.* The former (e.g. sociology, social history) tell what the man-made social world is like and how it has developed through times. The latter (e.g., social policy studies, economic planning, peace research) tell what the social world should be and what we should do in order to attain certain goals (e.g., social security and welfare, healthy economics, peace).[25] Both types of sciences produce knowledge – results which can be true or truthlike.

(e) It may be argued that different sciences are based upon different 'interests of knowledge', so that the notion of progress should also vary relative to these interests. Thus, Habermas (1972) claims that the natural sciences and some systematic social sciences are based upon a 'technical interest' of prediction and control, the humanities upon the 'practical' or 'hermeneutic interest' of communication, mediation of tradition, and increase of self-knowledge, and the critical social sciences and psychoanalysis upon an 'emancipatory interest' of criticism of ideology and abolition of reified social relations. In my view, this doctrine has two weak points. First, it presupposes the validity of an instrumentalist or technological conception of natural science[26] – which is denied in our realist conception where science is primarily regarded as pursuit of informative truth. Secondly, while scientific knowledge (in our realist sense) may in many cases serve precisely those 'interests' that Habermas mentions, the same 'interests' can be satisfied by several non-scientific activities: predictions are obtainable from chrystal balls, increased understanding of our own nature may be obtained by reading good novels, social structures can be transformed through direct political action. It therefore seems that the only specifically 'scientific' way of serving these 'interests' is to establish first highly truthlike statements and theories by ordinary scientific methods and then apply this knowledge for various purposes. The philosophical theory about 'interests of knowledge' is, therefore, based upon a confusion between scientific knowledge and its applications.

NOTES

1 Bas van Fraassen defines 'scientific realism' as follows: »Science aims to give us, in its theories, a literally true story of what the world is like; and acceptance of a scientific theory involves the belief that it is true.» (van Fraassen, 1980, p. 8.) The latter condition is too strong, however, since most contemporary realists would allow for the tentative acceptance of truthlike (or approximately true) theories. Cf. Chapters 5 and 7 above.

2 See Peirce, *CP* 5.405. Peirce proceeds to clarify this idea by applying his pragmatist theory of meaning to the concept of reality (cf. Chapter 3). Peirce's definition of 'the real' can be compared to Dummett's characterization of realism as requiring that »statements in the given class relate to some reality that exists independently of our knowledge of it» (Dummett, 1982, p. 55). If the well-known KK-thesis is valid for knowledge, it would be impossible to assume realism (in Dummett's sense) for knowledge statements of the form 'I know that p' (cf. Niiniluoto, 1981c, p. 186).

3 It should be added that Peirce's epistemology and philosophy of science can be largely read independently of his idealist metaphysics.

4 This is the position of Popper and Eccles (1977) and of Engels and Marxism-Leninism. Marxist arguments for the existence of 'the ideal' – corresponding to the Popperian world 3 – are presented by Ilyenkov (1977).

5 For the tradition of 'maker's knowledge', see Hintikka (1974), Ch. 4.

6 See Lukács (1971), p. xxiii. Chiodi (1976) argues that Sartre's *Critique de la Raison Dialectique* (cf. Sartre, 1976) is based upon the same error.

7 For an illuminating analysis of the 'conceptual components' in our everyday perceptual experience, see Kaila (1979), Ch. 4.

8 For the terms 'manifest image' and 'scientific image', see Sellars (1963).

9 This mistake is similar to the doctrine that we perceive sense data rather than physical objects. See Hintikka (1974), Ch. 11. The same error is made by those 'praxists', like Habermas (cf. Gutting, 1981; Hesse, 1982), who claim that the object domain of the natural sciences is 'constituted' by a 'framework of instrumental action'.

10 For Husserl's concept of life-world, see Husserl (1970), Luckmann (1978), Schutz (1962, 1967), Elliston and McCormick (1977). See also Föllesdal (1977). For a comparison between Husserl and Sellars, see Gutting (1978); between Habermas and Sellars, see Gutting (1981); and between phenomenology and Marxism, see Wartofsky (1978).

11 See also Chalmers (1978) who bases his 'radical instrumentalism' upon Althusser's materialism. For recent trends in pragmatism (in the large sense), see Rorty (1980), Putnam (1981), and Dummett (1982). Also Habermas' consensus theory of truth (cf. Hesse, 1981, 1982) represents a 'transcendental' version of pragmatism.

12 The British pragmatist F.S.C. Schiller argued, in his 'humanism', that the world is made by man. For an interesting analysis of 'worldmaking', see Goodman (1978). The radical view that worlds and their constituents (objects, facts, systems, etc.) are man-made can be viewed as an attempt to 'reduce' world 1 to world 3.

13 For this reason, a table as an artefact is not reducible to a collection of atoms. Cf. the discussion about 'Eddington's two tables' in Sellars (1963). The public character of artefacts is worth emphasising here. It does not matter if *I* happen to think that a certain table is not a table – it is a table nevertheless. But if you alone imagine that a certain stone is a divine thing to be worshipped, this 'private cultural object' is not real in Peirce's sense.

14 The phenomenologists have developed an elaborate account of the constitution of social reality: see Schutz (1962, 1972) and Luckmann (1978). In his »transformational model of social activity», Bhaskar argues against »reification» that society does not exist independently of human activity, and against »voluntarism» that men do not »make» or »create» society (Bhaskar, 1979, pp. 39–47).

15 The view of social institutions as man-made historical individuals is incompatible with some forms of 'methodological individualism'. Cf. Brodbeck (1968).

16 For discussion of institutional facts, see Searle (1969) and Taylor (1971).
17 Cf. the analysis of 'conventions' in Lewis (1969).
18 For a detailed defense of these suggestions, see Niiniluoto (1981a, b, c). For the alternative view that the truth of legal norm statements should be analysed by the coherence or consensus theory, see Aarnio (1981) and Aarnio *et al.* (1981). Against my view of languages as third-world entities, Platonists regard them as systems of »timeless abstract objects» (see Katz, 1982), mentalists (like Chomsky) as 'mental grammars' which are either elements of world 2 or reducible to world 1 (e.g. to neurophysiological states).
19 For a suggestion that some laws of society could be expressed through probabilistic propensities, see Niiniluoto (1976) and Urbach (1980).
20 This is true even of artefacts, such as works of art and novels, which are created by an individual person. It is not always the case that an artist or a writer has the best grasp of his own products. It is even more clearly true about institutional facts. Popper's proposal that the 'grasping' of a world 3 object means the 'making' or the 're-creation' of it applies thus only a very special class of abstract world 3 entities (cf. Popper and Eccles, 1977, p. 44).
21 Cf. Föllesdal (1979) who argues convincingly that the 'hermeneutical' method of interpretation follows the method of hypothesis.
22 For Kuhn, see the bibliography and the papers reprinted in Gutting (1980), Aarnio (1981b), and Thomas (1979). For Lakatos, see Latsis (1976). Progress in various social sciences is discussed in Littunen *et al.* (1980).
23 One methodological difference is worth noting: when a social scientist is a member of the culture that he studies, he has already learned quite a lot about his object of study. For example, a linguist can use his learned 'linguistic intuition' as evidence for his syntactic and semantic descriptions (cf. Niiniluoto, 1981c). It is a matter of dispute, however, how 'significant' difference this is.
24 This means, however, that the 'practically ideal' language relative to which the concept of progress is defined (cf. Chapter 5) is historically changing. For example, the price of a commodity may become dependent of some new factors which did not exist earlier.
25 For comments on the design sciences, see Chapter 12 below. Cf. also Chapter 10. Note that a single discipline in the social sciences often has both a descriptive and a designing branch. For example, legal dogmatics both describes the valid legal order in a society and presents conditional recommendations for extending or changing it. Engineering sciences, agricultural and forestry sciences are examples of design sciences among the natural sciences.
26 Hesse (1978) also claims that »one overriding value for natural science», but not necessarily for social science, is given by the »pragmatic criterion» of »increasingly successful prediction and control of the environment». For criticism of Habermas' view of the natural sciences, see also Gutting (1981) and Hesse (1982).

BIBLIOGRAPHY

Aarnio, A., 'On Truth and the Acceptability of Interpretative Propositions in Legal Dogmatics', in A. Aarnio *et al.* (eds.), *Methodologie und Erkenntnistheorie der juristischen Argumentation, Rechtstheorie, Beiheft* 2, Duncker and Humblot, Berlin, 1981, pp. 33–52. (a)

Aarnio, A., 'On the Paradigm Articulation in Legal Research', in I. Tammelo and A. Aarnio (eds.), *Zum Fortschritt von Theorie und Technik in Recht und Ethik, Rechtstheorie, Beiheft* 3, Duncker and Humblot, Berlin, 1981, pp. 45–56. (b)

Aarnio, A., Alexy, R. and Peczenik, A., 'The Foundation of Legal Reasoning', *Rechts-*

theorie **12** (1981), 133–158, 257–279, 423–448.
Asquith, P. and Hacking, I. (eds.), *PSA 1978,* vol. 2, Philosophy of Science Association, East Lansing, 1981.
Bhaskar, R., *The Possibility of Naturalism: A Philosophical Critique of the Contemporary Human Sciences,* Harvester Press, Brighton, 1979.
Bleicher, J., *Contemporary Hermeneutics,* Routledge and Kegan Paul, London, 1980.
Boyd, R., 'Scientific Realism and Naturalistic Epistemology', in P.D. Asquith and R.N. Giere (eds.), *PSA 1980,* vol. 2, Philosophy of Science Association, East Lansing, 1981, pp. 613–662.
Brodbeck, M. (ed.), *Readings in the Philosophy of the Social Sciences,* Macmillan, New York, 1968.
Chalmers, A.F., *What is this thing called Science?*, The Open University Press, Milton Keynes, 1978.
Chiodi, P., *Sartre and Marxism,* Harvester Press, Hassocks, 1976.
Dummett, M., 'Realism', *Synthese* **52** (1982), 55–112.
Durkheim, E., *The Rules of Sociological Method,* Free Press, New York, 1950. (Partly translated in Brodbeck, 1968.)
Elliston, F. and McCormick, P. (eds.), *Husserl: Expositions and Appraisals,* University of Notre Dame Press, Notre Dame, 1977.
Engels, F., *Ludwig Feuerbach and the End of Classical German Philosophy,* Progress, Moscow, 1946.
Föllesdal, D., 'Husserl and Heidegger on the Role of Actions in the Constitution of the World', in E. Saarinen *et al.* (eds.), *Essays in Honour of Jaakko Hintikka,* D. Reidel, Dordrecht, 1979, pp. 365–378. (a)
Föllesdal, D., 'Hermeneutics and the Hypothetico-Deductive Method', *Dialectica* **33** (1979), 319–336. (b)
Goodman, N., *Ways of Worldmaking,* Harvester Press, Hassocks, 1978.
Gurwitsch, A., 'Galilean Physics in the Light of Husserl's Phenomenology', in E. McMullin (ed.), *Galileo, Man of Science,* Basic Books, New York, 1967, pp. 388–401. (Reprinted in Luckmann (1978), pp. 71–89.)
Gutting, G., 'Husserl and Scientific Realism', *Philosophy and Phenomenological Research* **39** (1978), 42–56.
Gutting, G. (ed.), *Paradigm and Revolutions: Applications and Appraisals of Thomas Kuhn's Philosophy of Science,* University of Notre Dame Press, Notre Dame, 1980.
Gutting, G., 'Habermas and the Natural Sciences', in Asquith and Hacking (1981), pp. 424–437.
Habermas, J., *Knowledge and Human Interests,* Beacon Press, Boston, 1972.
Hesse, M., 'Theory and Value in the Social Sciences', in Hookway and Pettit (1978), pp. 1–16.
Hesse, M., 'Habermas' Consensus Theory of Truth', in Asquith and Hacking (1981), pp. 373–396.
Hesse, M., 'Science and Objectivity', in J.B. Thompson and David Held (eds.), *Habermas: Critical Debates,* The Macmillan Press, London, 1982, pp. 98–115.
Hookway, C. and Pettit, P. (eds.), *Action and Interpretation: Studies in the Philosophy of the Social Sciences,* Cambridge University Press, Cambridge, 1978.
Husserl, E., *The Crisis of European Sciences and Transcendental Phenomenology,* Northwestern University Press, Evanston, 1970.
Ilyenkov, E.V., 'The Concept of the Ideal', in *Philosophy in the USSR: Problems of Dialectical Materialism,* Progress, Moscow, 1977, pp. 71–99.
Kaila, E., *Reality and Experience,* D. Reidel, Dordrecht, 1979.
Katz, J., *Language and Other Abstract Objects,* Rowman and Littlefield, Totowa, 1982.
Kosik, K., *The Dialectics of the Concrete,* D. Reidel, Dordrecht and Boston, 1976.
Latsis, S. (ed.), *Method and Appraisal in Economics,* Cambridge University Press, Cam-

bridge, 1976.
Lewis, D., *Convention: A Philosophical Study,* Harvard University Press, Cambridge, Mass., 1969.
Littunen, Y., Pietilä, K., Pietilä, V., and Saarinen, A. (eds.), *Tieteen edistyminen ja yhteiskuntatieteet,* Yhteiskuntatieteiden tutkimuslaitos, Tampereen yliopisto, Tampere, 1980.
Luckmann, T. (ed.), *Phenomenology and Sociology,* Penguin Books, Harmondsworth, 1978.
Luckács, G., *History and Class Consciousness,* Merlin Press, London, 1972.
Marković, M., *Dialektik der Praxis,* Suhrkamp, Frankfurt am Main, 1968.
Marx, K., 'Economic and Philosophical Manuscripts', in L. Colletti (ed.), *Marx: Early Writings,* Penguin Books, Harmondsworth, 1975, pp. 279–400.
Niiniluoto, I., 'Inductive Explanation, Propensity, and Action', in J. Manninen and R. Tuomela (eds.), *Essays on Explanation and Understanding,* D. Reidel, Dordrecht, 1976, pp. 335–368.
Niiniluoto, I., 'On Truth and Argumentation in Legal Dogmatics', in A. Aarnio *et al.* (eds.), *Methodologie und Erkenntnistheorie der juristischen Argumentation, Rechtstheorie: Beiheft* 2, Duncker and Humblot, Berlin, 1981, pp. 53–76. (a)
Niiniluoto, I., 'On the Truth of Norm Propositions', in I. Tammelo and A. Aarnio (eds.), *Zum Fortschritt von Theorie und Technik in Recht und Ethik, Rechtstheorie: Beiheft* 3, Duncker and Humblot, Berlin, 1981, pp. 171–180. (b)
Niiniluoto, I., 'Language, Norms, and Truth', in I. Pörn (ed.), *Essays in Philosophical Analysis,* Acta Philosophica Fennica **32**, Helsinki, 1981, pp. 168–189. (c)
Petrović, G., *Marx in the Mid-Twentieth Century,* Doubleday, New York, 1967.
Popper, K.R., *Objective Knowledge,* Oxford University Press, Oxford, 1972.
Popper, K.R. and Eccles, J.C., *The Self and Its Brain,* Springer-International, Berlin, 1977.
Putnam, H., *Reason, Truth, and History,* Cambridge University Press, Cambridge, 1981.
Putnam, H., 'Why There Isn't a Ready-Made World', *Synthese* **51** (1982), 141–168.
Rorty, R., *Philosophy and the Mirror of Nature,* Blackwell, Oxford, 1980.
Sartre, J.P., *Critique of Dialectical Reason,* New Left Books, London, 1976.
Schmidt, A., *The Concept of Nature in Marx,* New Left Books, London, 1971.
Schutz, A., *The Problem of Social Reality* (vol. I of *Collected Papers,* ed. by M. Natanson), Nijhoff, The Hague, 1962.
Schutz, A., *The Phenomenology of the Social World,* Northwestern University Press, Evanston, 1967. (Heinemann, London, 1972.)
Searle, J., *Speech Acts. An Essay in the Philosophy of Language,* Cambridge University Press, Cambridge, 1969.
Sellars, W., *Science, Perception and Reality,* Routledge and Kegan Paul, London, 1963.
Taylor, C., 'Interpretation and the Sciences of Man', *Review of Metaphysics* **25** (1971), 1–51.
Thomas, D., *Naturalism and Social Science: A Post-Empiricist Philosophy of Social Science,* Cambridge University Press, Cambridge, 1979.
Urbach, P., 'Social Propensities', *The British Journal for the Philosophy of Science* **31** (1980), 317–328.
Wartofsky, M., 'Consciousness, Praxis, and Reality: Marxism vs. Phenomenology', in Elliston and McCormick (1977), pp. 304–313.

CHAPTER 10

FINALIZATION, APPLIED SCIENCE, AND SCIENCE POLICY

Science policy is a systematic effort for promoting and directing the development of science. But is it possible to plan and control science at all? Is a rational science policy possible? Are attempts to guide the development of science harmful to the autonomy of science? While questions like these continue to be heatedly disputed, it is natural to inquire whether the recent discussion about models of scientific change within philosophy of science can throw any light upon them.[1] This chapter deals with some aspects of this problem by giving special attention to the 'finalization thesis' of Böhme, van den Daele, and Krohn.

1. DISPUTES ABOUT SCIENCE POLICY

According to the most ambitious programmes for national science policy, the task of science policy is to identify problem areas within the progress of society, to express these social problems as the goals of scientific research and development, to find means for achieving these goals, to allocate resources for various branches of research, and to advance the utilization of the results of research.[2] On the basis of such programs, a demand is often made to the effect that the organization and administration of the national research system (universities, research institutes, etc.) has to be structured so as to make the general aims of science policy possible. Much of the recent empirical work within the 'science of science' tries to find organizational ways of increasing the effectivity of science as an instrument for solving social problems.

Some opponents of science policy claim that any attempt to direct science towards external goals is bound to distort the true nature of scientific activity and to arrest scientific progress. Science policy, they argue, is an attempt to subordinate science under political domination and thus violates the principles of autonomy and freedom of science – as expressed by Peirce's maxim: »Do not block the way of inquiry» (*CP*, 1.135).

A recent clash between these two positions occurred in West Germany as a result of the articles 'Alternativen in der Wissenschaft' (1972) and 'Finalizierung der Wissenschaft' (1973) by Gernot Böhme, Wolfgang van den Daele, and Wolfgang Krohn.[3] This group worked at the Max-Planck-Institut at Starnberg which, under the leadership of C.F. von Weizsäcker and Jürgen Habermas, investigated »living conditions in the industrial world». While the Starnberg group suggested that on some conditions it is possible to »plan science in accordance with political, economical and social goals», its opponents (among them many philosophers associated with German Popperianism) feared that the »finalization thesis» is a disguised Marxist attempt to put an end to »free research». A congress of antifinalists was organized in Munich in 1976, and many German newspapers paid attention to the »menace» of the »politicization» of science.[4]

The finalization debate will be discussed below in Sections 4–6, but first some general comments are made on the relevance of philosophy of science to issues of science policy (Sections 2 and 3).

2. The Aim of Science

In the Aristotelian conception, the aim of science is to 'inform' the soul – to make man good and happy through the possession of truth.[5] The method of science, which systematizes and organizes knowledge which already has been found, is thus not a tool of seeking new truths. Questions of science policy became important as soon as this Aristotelian conception was rejected by the spokesmen of new science. Galileo's trial showed the indispensability of the freedom of research for the search of new truths. Francis Bacon argued further that the proper use of the scientific method leads to new knowledge and to »inventions that may in some degree subdue and overcome the necessities and miseries of humanity». Bacon's *New Atlantis* (1624) contains already a proposal for a rational organization of scientific research in a community.

While Bacon and many of his followers in the Royal Society emphasized the practical utility of new knowledge, the progress of science was also given a theological or an eschatological justification: Bacon interpreted *Daniel 12:4* as indicating that »the thorough passage of the world» and »the advancement of the sciences» are »destined by the fate, that is, by Divine Providence, to meet in the same age».[6] Thus, science should be cultivated, but not because of the power that knowledge gives to man, but rather because the increase of knowledge is part of the plan that God has made for the history of mankind.

Auguste Comte's positivism holds that the study of nature furnishes us »with the true rational basis for acting upon nature». In 1830 he writes:

> »For it is only by knowing the laws of phenomena, and so foreseeing their occurrence, that we are able in active life to modify one another for our advantage ... We may sum up very exactly the general relation of science to art ... by the following very simple formula: from science comes prevision; from prevision comes action.» (Comte, 1970, p. 38.)

However, Comte immediately makes the following important remark:

> »But, in spite of the vital importance of this relation, which must never be ignored, we should form a very imperfect idea of the sciences if we were to regard them only as the bases of the arts, an error to which our age is, unhappily,

> too much inclined. Immerse as are the services rendered to industry by science, and although according to the striking aphorism of Bacon – knowledge is power – we must never forget that the sciences have a yet higher and more direct destination, that of satisfying the craving of our minds to know the laws of phenomena.»

In the same spirit, Peirce says that for »the scientific men» nature is »a cosmos, so admirable, that to penetrate to its ways seems to them the only thing that makes life worth living» (*CP*, 1.43). Science consists in

> »diligent inquiry into truth for truth's sake, without any sort of axe to grind, nor for the sake of the delight of contemplating it, but from an impulse to penetrate into the reason of things ... For it is not knowing, but the love of learning, that characterizes the scientific man.» (*CP*, 1.44.)

Peirce adds that if a man »occupies himself with investigating the truth of some question for some ulterior purpose», he may be »ever so much better than a scientific man, if you will», but »he is not a scientific man».

The sociologists of science have sometimes claimed that the idea of pure science as a pursuit of truth is just an *ideology* by which scientists have defended themselves against »external enemies» and protected their academic and professional interests.[7] Ravetz predicts that this ideology »will persist for some time to come, but its base in experience is eroding very rapidly» (Ravetz, 1973, p. 28), and it has to give way to a »technocratic conception of science» which views science as a basic factor in production. Johnston and Jagtenberg (1978), p. 30, go further and claim that this ideology has become dangerous:

> »In spite of the evidence that research has become a much more highly controlled and directed activity, academic scientists and in particular the elite who represent the scientific community in negotiation with the paymasters and political interests, continue to project the image of science as autonomous and apolitical, concerned solely with the objective pursuit of truth, i.e., as 'pure academic science' ... We would wish to argue that the continuation of this misleading view of science represents a dangerous mystification which can serve only to obscure the role of science in modern society, confuse the scientist and science administrator, and alienate the public.»

It has been estimated that about 25 per cent of all resources for research work is now directed to military technology. The majority of the community of researchers would not count as 'scientific men' in Peirce's sense. Science is not 'pure' any more, since it has sold its autonomy in order to get money from »dirty work», as Ravetz says. However, these calculations include everything that is classified as R&D (research and development), where *research* is defined as systematic pursuit of new scientific knowledge and *development* as non-routine activity aiming new products or new means of productions. Research is customarily divided into *basic research* (»the systematic pursuit of new scientific knowledge without the aim of specific practical application») and *applied research* (»the systematic, planned pursuit of scientific or technical knowledge with the aim of obtaining a specific goal.»).[8] Moreover, these calculations do not make a difference between science and technology, even if a large part of technology belongs to development rather than research (cf. Chapter 12 below).

From the existence of many different kinds of activities which factually are associated with science it does not follow that the ideal of science as the pursuit of informative truth has to be given up as a »mystification». Rather, one has to acknowledge the existence of these different activities, and then to examine the legitimacy of this ideal relative to each of them.[9]

Ravetz and Johnston are right in pointing out that conceptions of science may have an *ideological function* in society. But, of course, if the ideal of truth can serve as an ideology, the same is true likewise of alternative conceptions of science – such as the 'technocratic' view. Moreover, it does not follow that a philosophical view which functions as an ideology is also *false*[10] – or that the only way of arguing for or against such views is reduced to a power struggle between interest groups.

For example, even though Comte and Peirce understood very well the power that scientific knowledge can give to us – they were the founders of two 'utilitarian' philosophical movements: positivism and pragmatism, respectively – they also gave good arguments to show why it is vitally important to define science independently of its practical concerns. Comte points out that without the »strong natural craving» to know the »laws of phenomena»

> »the human intellect would be confined to researches of immediate practical utility, and, as Condorcet very justly remarked, would for that reason alone be completely arrested in its progress.» (Comte, 1970, p. 39.)
>
> »... if we hamper our search for truth with the extraneous condition that it shall have some immediate practical utility, it would be almost always impossible for us to succeed.» (*Ibid.*, p. 40.)

As an example of »purely theoretical labors» which »after a long series of generations» proved to be indispensable for the art of navigation, Comte mentions the work of Archimedes and Apollonius on conic sections. The same point is made by John Herschel (1831). He starts from an 'Aristotelian' remark:

> »The question '*cui bono*' to what practical end and advantage do your researches tend? is one which the speculative philosopher who loves knowledge for its own sake, and enjoys, as a rational being should enjoy, the mere contemplation of harmonious and mutually dependent truths, can seldom hear without a sense of humiliation. He feels that there is a lofty and distinterested pleasure in his speculations which ought to exempt them from such questioning.» (p. 10)

But, Herschel notes, if *this* is not alleged as a sufficient reply,

> »he has only to point to the history of all science, where speculations apparently the most unprofitable have almost invariably been those from which the greatest practical applications have emanated.» (p. 11)

The position of Comte and Herschel is not hostile to science policy:

they are in effect saying here that in the long run the most effective way of guaranteeing practical applications is to promote basic research. The later development of science has shown the validity of this idea: it has been estimated that about 75 per cent of modern technology is based upon innovations made within basic research.

3. Laws of the Development of Science

It was seen in Section 2 that philosophical conceptions of science may have an ideological function which makes them relevant to issues about science policy. It is clear, however, that such conceptions belong not only to the scientists in their research activity but also to all persons who are taking part in science policy: scientific members of advisory boards, administrators in research organizations, and politicians in the government. Decisions about science policy are after all made by human beings, and their behaviour is naturally influenced by their – explicitly or implicitly adopted – personal view about the nature and aims of science.

There is also another way in which philosophy of science is relevant to science policy. As Böhme (1977), p. 320, points out, »one must know the 'laws' of a field before one can reasonably attempt to plan its development». Hence, the planning of research becomes possible only if we know some laws concerning the ways in which science develops. The question about the existence and the nature of such laws has been a central topic within the study of the 'dynamics' of science in the 1960's and 70's. Philosophers and historians of science have proposed various 'models' of scientific change which try to give an overall picture of the factors that influence science and indicate the typical patterns which the development of science follows.[11]

Some models of scientific change entail that the planning of science is impossible. This will be the case, if the development of science is purely irregular or chaotic or if it follows 'automatically' a strictly determined 'inner logic' with which we cannot interfere at all. Some views entail that no planning is needed, since in science »anything goes» (Feyerabend). All these extreme positions are implausible, since in practice the scientists and their financiers have always made choices concerning the fields of study, research problems, and research strategies – and these choices have had an influence on the development of science.

On the other hand, the excessive faith in the possibility of science policy is usually based upon a naive belief that science possesses a mechanical method of problem solving. In this view, the only important step in research planning is the choice of the relevant problems. This choice can be done by any administrative board (e.g. a parliamentary committee or the Ministry of Education) on the basis of the social or political weight of these problems; then these problems are handed down to the scientists who solve them in a routine manner.[12] The trouble with this view is that there are no mechanical rules for transforming social problems to scientific problems. What is more, there are no mechanical methods of discovery in science[13]: the success of research depends in unpredictable ways on the creativity and luck ('serendipity') of the researches. To be sure, sociological generalizations can be established to the effect that certain kinds of circumstances tend to increase the productivity of research. Such generalizations suggest that it is possible to *stimulate* research in desired directions, but they do not *guarantee* success or progress.

The mechanical view of the scientific method ignores also the fact that the solution of a problem may presuppose an answer to other problems. There are systematic interconnections between scientific problems which any attempt at rational science policy has to take into account. In most cases, this means that a problem which is close to practice can be solved only if basic research has already provided a solution to many theoretical problems. For this reason, it is rational – even from the practical viewpoint – to let the scientists to pursue effectively their research front, i.e., those theoretical issues which have been generated by the inner dynamics of science, just as the Comte–Herschel position claims.

This view is still compatible with many different patterns which the development of science may follow. In Popper's scheme, attempts to solve scientific problems generate always new problems (cf. Chapter 3 above): the most effective way of progress consists in the proposal of bold hypotheses and their most severe tests. Lakatos (1971) admits no »instant rationality»: the basic unit of appraisal is a »research programme». Even though an individual scientist may always hope that he is able to turn a degenerative research programme into a progressive one, rational science policy should give support to progressive programs which have been – and still promise to be – successful.[13] Similar consequences may be based upon Laudan's (1977) account of the problem-solving ability of »research traditions».

Kuhn gives emphasis to the effectivity of tradition-bound »normal science» both in the solution of »puzzles» generated by an accepted »paradigm» and in the eventual »revolutionary» overthrow of such a research-guiding paradigm:

> »At least for the scientific community as a whole, work within a well-defined and deeply ingrained tradition seems more productive of tradition-shattering novelties than work in which no similarly convergent standards are involved.» (Kuhn, 1977, p. 234.)

Generally speaking, Kuhn's model of scientific change is a compromise between 'internalism' and 'externalism': he admits that both 'internal' and 'external' factors have an influence on science, but he suggests that their effects depend crucially on the stage which a scientific discipline has reached in its development. In his 1968 article on the history of sciences Kuhn describes his own views as follows:

»Early in the development of a new field, he suggests, social needs and values are a major determinant of the problems on which its practitioners concentrate. Also during this period, the concepts they deploy in solving problems are extensively conditioned by contemporary common sense, by a prevailing philosophical tradition, or by the most prestigious contemporary sciences. The new fields which emerged in the seventeenth century and a number of the modern social sciences provide examples. Kuhn argues, however, that the later evolution of a technical specialty is significantly different in ways at least foreshadowed by the development of the classical sciences during the Scientific Revolution. The practitioners of a mature science are men trained in a sophisticated body of traditional theory and of instrumental mathematical, and verbal technique. As a result they constitute a special subculture, one whose members are the exclusive audience for, and judges, of each other's work. The problems on which such specialists work are no longer presented by the external society but by an internal challenge to increase the scope and precision of the fit between existing theory and nature. And the concepts used to resolve these problems are normally close relatives of those supplied by prior training for the specialty. In short, compared with other professional and creative pursuits, the practitioners of a mature science are effectively insulated from the cultural milieu in which they live their extraprofessional lives.» (Kuhn, 1977, pp. 118–119.)

»... Nevertheless, the apparent autonomy of the internal approach is misleading in essentials, and the passion sometimes expended in its defense has obscured important problems. The insulation of a mature scientific community suggested by Kuhn's analysis is an insulation primarily with respect to concepts and secondarily with respect to problem structure. There are, however, other aspects of scientific advance, such as its timing. These do depend critically on the factors emphasized by the external approach to scientific development. Particularly when the sciences are viewed as an interacting group rather than as a collection of specialties, the cumulative effects of external factors can be decisive.

Both the attraction of science as a career and the differential appeal of different fields are, for example, significantly conditioned by factors external to science. Furthermore, since progress in one field is sometimes dependent on the prior development of another, differential growth rates may effect an entire evolutionary pattern. Similar considerations, as noted above, play a major role in the inauguration and initial form of new sciences. In addition, a new technology or some other change in the conditions of society may selectively alter the felt importance of a specialty's problems or even create new ones for it. By doing so they may sometimes accelerate the discovery of areas in which an established theory ought to work but does not, thereby hastening its rejection and replacement by a new one. Occasionally, they may shape the substance of that new theory by ensuring that the crisis to which it responds occurs in one problem area rather than another. Or again, through the crucial intermediary of institutional reform, external conditions may create new channels of communication between previously disparate specialties, thus fostering cross-fertilization which would otherwise have been absent or long delayed.»

4. The Finalization Thesis

We are now ready to consider the 'finalization thesis' of Böhme, van den Daele, and Krohn. According to Böhme *et al.*, there are always alternative ways in which science may develop. The actual choice between them is decided by the social environment of science. In the »Darwinian» phase, this selection process is spontane, not a result of conscious direction and planning. The finalization thesis claims that in the middle of the 19th century it became possible to overcome the »Darwinian» phase: in the new phase of finalization »it is possible to rationalize scientific progress by conscious planning» which is »based as much on the interests of science as on the needs of society» (Böhme *et al.*, 1976, p. 312). The term 'finalizierung' is derived from 'causa finalis': finalization is a »process through which external goals for science become the guidelines of the development of the scientific theory itself» (*ibid.*, p. 307). It refers to »those scientific developments which are characterized by their linkage with social, military, and economic purposes, but which are not adequately described by the traditional category of 'applied research'» (*ibid.*, p. 308). This »strategic theory development according to externally set goals» is different from »the mere application of theories to technical goals» (van den Daele and Weingart, 1976), since it influences the »cognitive structure» of theories (Böhme *et al.*, 1978, pp. 10–11).

To show the possibility of finalization, the Starnberg group derives mainly from Kuhn a three-stage model for the development of scientific disciplines:

(i) The *explorative phase* is characterized by the dominance of empirical and descriptive methods and by the lack of an accepted theoretical framework; it thus resembles Kuhn's account of 'pre-paradigmatic' sciences (e.g., chemistry before Lavoisier).

(ii) The *paradigmatic phase* (or the phase of theory dynamic) consists in the emergence of an organizing theoretical framework or paradigm and in its further development and articulation.

(iii) In the *post-paradigmatic phase* a »stable» paradigm is »specialized, differentiated, and modified in order to extend its range of applications» (Böhme *et al.*, 1976, p. 315).

In the explorative phase (i), it is possible to some extent to stimulate and to initiate research (Böhme *et al.*, 1977), but the attempts to use empirical research strategies for the solution of externally defined problems is severly limited in their success. In the paradigmatic phase (ii), the process of theory construction follows an inner logic, and scientific disciplines are resistant to external regulation. Through this development, some research fields reach an established or »closed» theory, so that the post-paradigmatic phase (iii) becomes possible. In this stage, external planning of science is feasible – social problems can be transformed to research strategies (cf. van den Daele *et al.*, 1977) and new scientific fields may emerge for problem areas with politically defined goals (van den Daele and Weingart, 1976). Thus, finalization is always based upon a »closed» or »mature» theory (Böhme *et al.*, 1976, pp. 316–319; Böhme, van den Daele, Hohlfeld, 1978), since »as long as science has not developed a fundamental theory ... attempts to link the advance of science to external mission-orientation offer little promise» (Böhme *et al.*, 1976, p. 315). As a »closed» theory has already solved its relevant theoretical problems, the inner logic guiding its development dissolves itself, and the question »where and with what intensity theory will further develop» (*ibid.*, p. 315) becomes dependent on the emergence of new practical problems.

Several questions may be raised about this approach.[15] I start with comments on the underlying philosophical framework; in the next section I discuss the relation between 'finalization' and 'applied science'; the concluding section considers the consequences of the finalization thesis for the autonomy of science.

It is a virtue of the analysis of Böhme *et al.* that they have tried to make the underlying methodological model explicit. In view of the many possibilities of reconstructing Kuhn's basic concepts and ideas, it seems that the three phase models would need much further articulation.[16] In particular, Kuhn's own account was restricted to the natural sciences, and it is still a much debated question whether the notion of 'paradigm' is applicable to the social sciences at all. While some philosophers think that the social sciences still are in the pre-paradigmatic phase, some others claim that in these sciences there are several co-existing rival paradigms which are linked with political ideologies (see, for example, von Wright, 1972). It remains as an open problem whether the finalization thesis can be applied to such disciplines as social policy studies or peace research – even though this should be a crucial question for anyone who tries to show that scientific methods can be consciously used for the solution of social and political problems. Moreover, the finalization thesis is intended to cover cases where natural sciences are oriented towards problems (such as noise) which embody elements of social valuation (cf. Böhme *et al.*, 1976, pp. 311–312).

The Starnberg group modifies Kuhn's model in one important respect: while Kuhn's normal science ends with a crisis and a revolutionary change of a paradigm, the paradigmatic phase of Böhme *et al.* typically ends with a theory which is *closed* in Heisenberg's sense, i.e., with a theory which solves adequately the fundamental problems of a discipline and »remains valid for all times». As examples of such 'closed' theories, Heisenberg and von Weizsäcker mention Classical Mechanics, Special Theory of Relativity, and Quantum Mechanics. This emphasis on »theoretical stability» (cf. van den Daele, 1977) against scientific revolutions is inconsistent not only with Kuhn's account of science but the more so with Popper's falsificationism (cf. Böhme, van den Daele, Hohlfeld, 1978). Indeed, it seems strange to say that Classical Mechanics is a 'closed' theory in spite of the fact two other 'closed' theories (Special Relativity and Quantum Theory) in fact contradict it. The fact that Newton's mechanics has still retained its paradigmatic status in many technological applications (Böhme, van den Daele, Hohlfeld, 1978, p. 206) can be explained by the assumption that Newton's laws are *approximately true* (cf. Chapter 5 above) – without claiming that this theory has already solved all its fundamental theoretical problems. It is doubtful that there are any physical theories which are 'closed' in the latter sense: the class of potential applications for a theory is open-ended, so that the set of theoretically interesting problems is non-exhaustible (cf. Chapters 5 and 6). The same is even more clearly true of mathematical theories (cf. Chapter 8). Moreover, even if some theory were 'closed' in the required sense, it is doubtful that we could ever know it.

The Heisenbergian notion of a 'closed' theory is not only controversial philosophically: it seems to be unnecessarily strong even for the purposes of the finalization thesis. Böhme, van den Daele, and Hohlfeld (1978), p. 219, in fact admit that organic chemistry was not a 'closed' theory in 1840 when agricultural chemistry was created. They try, therefore, to replace this notion by the concept of the *theoretical maturity* of a theory, but they have troubles in finding any practical criteria for 'maturity'. They also fail to explain how the concept of 'maturity' is connected with the existence or non-existence of an 'inner logic' determining the development of a theory.

In my view, these difficulties indicate that the whole three-phase model is in need of revision. As soon as the fundamental concepts and laws of a theory have been formulated sufficiently clearly, its theoretical development and its applications to areas close to practice can be pursued *at the same time.* The possibility of the latter activity depends on the degree of articulation of the theory, and its success depends on how close to the truth the fundamental principles of the theory are. A good example of this process is probability theory: as soon as its basic ideas had been formulated by Pascal, Fermat, and Huygens in the late 17th century, it simultaneously developed both theoretically and towards several areas of practical application (games of chance, annuities and insurance, testimony of witnesses, errors of measurements, etc.). The theoretical development led eventually, with the help of set theory and measure theory, to Kolmogorov's axiomatic treatment in the 1930's, and thereafter the theory has again been pursued in theoretical directions (e.g., probability on metric spaces) and towards subtheories relevant to practice (e.g., risk theory, mathematical statistics).

Hence, the Starnberg thesis that there is first a paradigmatic phase entirely governed by an 'inner logic' and then *after that* a post-paradigmatic phase with no 'inner logic' at all has to be rejected. It is possible to specialize even theoretically developing theories to practical problem areas. Moreover, it can be argued that the emerging new areas of 'applied science' (such as risk theory dealing with insurance companies) develop temporally through an 'inner logic' which partly depends on the development of the corresponding 'mother science'.

5. Applied Science

One of the aims of the Starnberg group has been to defend a thesis of 'social externalism' (Schäfer, 1978, p. 384) which claims that the development of science and the social history of society are in mutual interaction with each other. The influence of external social interests to science is indeed immediately visible at the level of research institutions and their disciplines. In view of the important role that agriculture and forests have played in the Finnish national economy, it is no accident that in 1898 a special section of Agricultural Economics was founded in the Faculty of Philosophy of the University of Helsinki. Later, in 1923, this section became an independent Faculty of Agriculture and Forestry which, among others, has the following departments: plant breeding, plant husbandry, plant pathology, agricultural chemistry, agricultural economics, dairy science, meat technology, food chemistry and technology, household economics, land use economics, limnology, horticulture, silviculture, peatland forestry, forest mensuration and management, logging and utilization of forest products, social economics of forestry, business economics of forestry, forest product marketing, and environmental conservation.

The various disciplines of Agriculture and Forestry represent 'applied sciences' which use information provided by basic sciences in other faculties (such as physics, chemistry, biology, botanics, zoology, physiology, economics). Some of these disciplines are comparable to the 'technological sciences' or 'engineering sciences' which in Finland are represented e.g. in the Helsinki University of Technology and in the State Center for Technological Research. (Cf. Chapter 12.) The birth of these special sciences was not a result of any 'inner necessity' within the basic sciences, but rather it was occasioned by the economic needs of the society.

More generally, it is a commonplace to claim that the birth of science in Egypt and Babylonia was a response to practical problems associated with agriculture and commerce. This development in turn stimulated the invention of basic research in Greece.

In addition to the 'social externalism', which concerns the genesis of scientific knowledge, the Starnberg group also defends a more interesting thesis of 'cognitive externalism'. Even if practical 'external' problems have in fact influenced the development of science from its infancy – where and with what intensity it is going – 'finalization' is intended to mean something more than this. When it is claimed that in finalization »external purposes become the guidelines not only for the 'off-springs' of scientific progress, i.e., the applications of research, but also for the underlying theory» (Böhme *et al.*, 1976, p. 312), this may be interpreted as a claim that the external goals not only guide and influence the choice of problems for theoretical study but they also modify and influence the general methodological criteria for theory formation. In other words, external goals do not only influence the choice of the problem areas which will be studied in some community, but they also give rise a new type of research with qualitatively new standards for the aims and the methods of theoretical work. Böhme *et al.* (1978), pp. 10–11, mention the following examples of areas of external cognitive influence: the choice of research problems, the definition of problems, the determination of explanatory ideals, and the rules for the limit of interesting research.

van den Daele (1977) in fact admits that the Starnberg group has not yet found sufficient evidence for the strong thesis of cognitive externalism, and he leaves it an open question whether it still holds in some area of science. This negative conclusion is also reflected in van den Daele's otherwise illuminating discussion of the structure of 'applied research' as progressive specialization of theories, since this structure seems to be entirely independent of the origin (i.e., theoretical vs. external) of the problem where the theory is applied.

Schäfer (1978), p. 385, suggests that social factors »influence also the truth of objective knowledge». While the claim that *truth* is relative to social interests leads to excessive relativism, it is a fact that the criteria of acceptability of scientific knowledge have changed historically – and even vary among the contemporary sciences. Theories within applied science and engineering science have to satisfy a condition of practical applicability, and depending on the area of intended application approximations and other smoothing techniques are used to quarantee this condition. Hence, the errors that are tolerated in these sciences are greater than in basic research – which also explains the relative robustness of some applied sciences relative to theoretical changes in their mother sciences.

Can these remarks be reconciled with the claims of Böhme *et al.* that finalization is something different from applied science and that finalization first occurred in the middle of the 19th century? In my view the answer to this question is negative: the use of the self-contradictory term 'oriented basic research' (Böhme, van den Daele, Hohlfeld, 1978, p. 233) shows that Böhme *et al.* have misinterpreted the content of the traditional category of 'applied science', and therefore they have to reinvent it and re-name it as 'finalization'.

The term 'applied science' is used in two senses. In the wide sense, it refers to all 'problem-oriented' or 'mission-oriented' research (i.e., research applied or applicable *to* something), and thus may cover both empirical pre-paradigmatic research and theory-based post-paradigmatic research. In the narrow sense, applied science means problem-oriented research which is based upon results of basic research (i.e., research which is an application *of* something).[17] This does not mean that applied science in the narrow sense is just an exercise in deduction: the standard OECD definition emphasizes that applied research is systematic and planned pursuit of new scientific knowledge. Since it uses information provided by basic sciences, applied science is based upon 'mature theories'. It is not basic research, however, because it is related to a specific external goal. In brief, finalization has all the characteristics of applied science in this sense.

The source of the terminological unclarity of Böhme *et al.* seems to be based on a possible confusion between *applied science* and *applications of science.* For example, 'applied mathematics' is a name of those branches of mathematics which construct mathematical theories – by using concepts from 'pure mathematics' – related to some area of application (e.g., numerical analysis, risk theory, queuing theory, etc.). It has to be distinguished from the actual applications of mathematical concepts and theories to practical problems. This difference may be obscured if the »traditional distinction of basic and applied research» is understood as the »distinction between theory construction and theory application» (Böhme, van den Daele, Hohlfeld, 1978, p. 210).

Applied sciences in the wide sense include both *predictive sciences* (e.g. applied astronomy, meteorology) and *design sciences* (natural and social 'engineering sciences').[18] The former attempt to establish causal laws of the form

(1) X causes A in circumstances B,

and use them to predict the occurrence of events of type A (e.g., eclipse, rain). The latter are based upon manipulable causal relations (1) where we can do or bring about X. In such cases, with some conditions about our preference system, (1) can be written as a so-called technical norm:

(2) If you want A, and if you believe that you are in a circumstance B, then you should do X!

The typical results of the engineering sciences, the agricultural and forestry sciences, medicine, nursing science, education, social policy studies, peace research, and military research have this form, where the 'external' goal A (e.g., safe aeroplanes, healthy forests, recovered patients, just society) is directly built into the technical norms as their antecedent conditions. Nevertheless, one can in principle establish statements of the form (2) by objective scientific methods – this can be done either by deriving the corresponding causal claim (1) from a more general theory (applied science in the narrow sense) or by defending (2) directly by optimazion methods. (Cf. Chapter 11 below.)

6. Autonomy of Science

The Starnberg group, while thinking that finalization binds the development of science together with social goals, emphasizes that »in the mission-oriented science practical valuations do not replace criteria of truth» (Böhme, van den Daele, Hohlfeld, 1978, p. 240). They argue, however, that finalization leads to a rejection of the traditional requirement of the *autonomy* of science: when scientific disciplines reach theoretical maturity and start the post-paradigmatic phase of development, the justification for the autonomy requirement disappears (Böhme *et al.*, 1978, p. 17).

I think that this conclusion is unwarranted. We have seen above that the core of the finalization thesis is given by the following claim: the planning of research towards practical problem areas becomes possible when and only when basic research has advanced to a sufficiently high stage. This widely accepted claim is in fact identical with the Comte-Herschel position of Section 3. It does not challenge the traditional view of the autonomy of science (cf. Chapter 1) – which includes the right of the scientists to formulate their research problems, to choose their basic concepts and methods, to accept or reject their hypotheses and to publish their results without 'external' interference. Finalization does not give any mechanical ways for transforming social problems or science policy programs into research programmes: the formulation of the detailed research problems is still the task of the scientists themselves, and only the scientists will be able to judge whether basic research has in fact advanced sufficiently to make the solution of practical problems possible.

We have also seen that typically the theoretical research front of a discipline and its problem-oriented specializations co-exist – and may mutually influence each other. The development of applied sciences receives support from their mother sciences, but otherwise it is plausible to assume that these areas of research grow according to some internal patterns[19] – and therefore the properly understood requirement of autonomy concerns applied sciences as well.[20]

No sciences have ever been absolutely autonomous in the society, since they need financial support from some 'external' source. It is therefore important to note that the descriptive sciences (i.e., basic research and the predictive applied sciences) and the design sciences have different relations to moral and political valuations. Both parties in the German debate on finalization failed to analyse the situation correctly. Even though it is possible to appraise in a neutral way technical norms of the form (2) – a scientist who does it need not endorse the goal A – they are not morally neutral. A successful defence of the conditional recommendation (2) gives us the power or the means to promote the goal A. The choice of this goal A may be a philosophical issue (e.g., the concept of health in medicine, the concepts of peace and social justice), but it also may be a political question (e.g., the avoidance of pollution or noise, the reduction of the risk of war, more powerful military technology). Thus, in spite of the fact that the research process of the design sciences is similar to other scientific disciplines, the existence of – or at least the public support for – these sciences depends on the question whether the promotion of their goal is found to be politically desirable in the society.

NOTES

1 For an attempt towards this direction, see Böhme (1977).

2 This kind of characterization of science policy is given e.g. in the *Programme of Science Policy* of the Central Board of Research Councils, The Academy of Finland (Helsinki, 1973).

3 For the work of the Starnberg group, see Böhme, van den Daele, and Krohn (1972, 1976, 1977, 1978), van den Daele (1975, 1977), van den Daele and Weingart (1976), van den Daele, Krohn, and Weingart (1977), Böhme, van den Daele, Hohlfeld, Krohn, Schäfer, and Spengler (1978). Recently a collection of their work was published in English (see Schäfer, 1983). Since 1977 Böhme has worked in Darmstadt, and the Max-Planck-Institute was closed in 1981.

4 Cf. the book *Die politische Herausforderung der Wissenschaft* (Hübner *et al.*, 1976) and the reviews by Pfetsch (1979) and Lenhardt (1981). One paradox of this debate is the fact that consistent Marxists in a capitalist society will defend the autonomy of science and be critical of its external control, since that control would be an instance of the domination of the Capital over men. Finalization is a social democratic rather than Marxist idea – Pfetsch (1979) sees it as representing a 'centre-left' position.

5 In St. Augustine's *Contra academicos* the question is raised whether a man who seeks the truth but has not yet reached it can be happy. Augustine's answer is negative. (Cf. Augustine, 1950.)

6 See Aphorism I, 93 of *Novum Organum.* Cf. also Ravetz (1975).

7 As Karl Popper and Isaac Levi have clearly demonstrated, the aim of science cannot simply be *truth,* because many truths are uninteresting, but rather *informative truth* (cf. Chapter 3 above). When I speak here of 'pursuit of truth', this should be understood in the sense of 'pursuit of informative truth'.

8 These standard OECD definitions are in many ways problematic: they do not tell what 'scientific knowledge' is supposed to be, and whose aims (the researcher's or the financier's) are relevant in classifying research as basic or applied. They are nevertheless used in many countries for the purpose of R&D statistics.

9 This question of legitimacy is not an empirical matter concerning the actual behaviour or attitudes of scientists. Rather, it is a philosophical question about the grounds which make certain kinds of activities rational.

10 Sometimes 'ideology' is defined as 'false consciousness', but in my view also true statements may have an 'ideological' function.

11 One problem here is that these models are often not intended as simple descriptive statements comparable to the laws of nature but rather as normative philosophical theories about what is rational and what is not rational in science. Such norms are relevant to the planning of science in the sense that they help us in our attempt to maintain the rationality of the scientific enterprise.

12 This idea could be called Baconian, since Bacon compared his inductive method with a »machine» which handles the »stuff» fed into it. Bacon had also a strong opinion on the proper order in which scientific problems have to be treated (i.e., from simple empirical situations to more general and abstract principles).

13 For an attempt to show how the »heuristic power» of a research programme can be objectively defined and how the »potential for future development» of a research programme can be assessed, see Urbach (1978).

14 The first examples of finalization are claimed to be found in agricultural chemistry (cf. Krohn and Schäfer, 1978). Other examples include the theory of fermentation (Böhme, 1978), aerodynamics (Böhme, 1978), noise research (Böhme *et al.,* 1976), and fusion research (van den Daele, 1977). For a discussion on cancer research, see Hohlfeld (1978). (See also Schäfer, 1983.)

15 Cf. also Johnston (1976), Symes (1976), Hieber (1975, 1976).

16 For steps towards this direction, see Böhme, van den Daele, Hohlfeld (1978), where the Sneed-Stegmüller reconstruction of Kuhn (cf. Chapter 5 above) is to some extent taken into account.

17 For example, Feibleman (1961) defines applied science as the »use of pure science for some practical human purpose». He distinguishes this from technology (trial-and-error or skilled approaches derived from concrete experience) and engineering (application of technology to particular cases).

18 Design sciences correspond to Simon's 'sciences of the artificial'.

19 For attempts to show that the development of technology follows an essentially Kuhnian pattern of change, see Johnston (1972) and Wojick (1979). See also Chapters 11 and 12 below.

20 It is a different matter that often applied research (e.g., in the engineering and military fields) is actually done under conditions which do not satisfy the traditional conditions of science (such as freedom in the choice of methods, publication of results). It is a controversial matter whether such use of scientific methods, i.e., without public results which can be critically discussed within the scientific community, deserves to be called 'science' at all.

Bibliography

Augustine, St., *Against the Academics,* The Newman Press, Westminster, 1950.

Böhme, G., 'Models for the Development of Science', in I. Spiegel-Rösing and D. de Solla Price (eds.), *Science, Technology, and Society,* Sage Publications, London, 1977, pp. 319–351.

Böhme, G., 'Autonomisierung und Finalisierung', in Böhme *et al.* (1978), pp. 69–130.

Böhme, G., 'On the Possibility of »Closed Theories»', *Studies in History and Philosophy of Science* **11** (1980), 163–172.

Böhme, G., Daele, W. van den, and Hohlfeld, R., 'Finalisierung revisited', in Böhme *et al.* (1978), pp. 195–250.

Böhme, G., Daele, W. van den, Hohlfeld, R., Krohn, W., Schäfer, W., Spengler, T., *Die gesellschaftliche Orientierung des wissenschaftlichen Fortschritts,* Suhrkamp, Frankfurt, 1978. (English translation in Schäfer, 1983.)

Böhme, G., Daele, W. van den, and Krohn, W., 'Alternativen in der Wissenschaft', *Zeitschrift für Soziologie* **1** (1972), 302–316.

Böhme, G., Daele, W. van den, and Krohn, W., 'Finalization in Science', *Social Science Information* **15** (1976), 307–330. (English version of the paper 'Finalisierung der Wissenschaft', *Zeitschrift für Soziologie* **2** (1973), 128–144.)

Böhme, G., Daele, W. van den, and Krohn, W., *Experimentelle Philosophie,* Suhrkamp, Frankfurt, 1977.

Böhme, G., Daele, W. van den, and Krohn, W., 'The »Scientification» of Technology', in Krohn *et al.* (1978), pp. 219–250.

Comte, A., *Introduction to Positive Philosophy,* Bobbs-Merrill, Indianapolis, 1970.

Daele, W. van den, 'Autonomie contra Planung: Scheingefect um die Grundlagenforschung?', *Wirtschaft und Wissenschaft* **2** (1975), 29–32.

Daele, W. van den, 'Scientific Development and External Goals', in P. Löppönen (ed.), *Proceedings of the International Seminar on Science Studies, Helsinki, 11–14 January 1977,* Reports of the Academy of Finland, Helsinki, 1977, pp. 139–165.

Daele, W. van den, Krohn, W. and Weingart, P., 'Political Direction of Scientific Development', in E. Mendelsohn, P. Weingart, and R.D. Whitley (eds.), *The Social Production of Scientific Knowledge,* D. Reidel, Dordrecht, 1977.

Daele, W. van den and Weingart, P., 'Resistance and Receptivity of Science – The Conditions for the Emergence of New Disciplines through Science Policy Control', in G. Lemaine *et al.* (eds.), *Perspectives on the Emergence of Scientific Disciplines,* Mouton, Paris, 1976. (English translation of the paper 'Resistenz und Rezeptivität der Wissenschaft', *Zeitschrift für Soziologie* **4** (1975), 146–164.)

Feibleman, J.K., 'Pure Science, Applied Science, Technology, Engineering: An Attempt at Definitions', *Technology and Culture* **2** (1961), 305–317.

Herschel, J., *A Preliminary Discourse on the Study of Natural Philosophy,* Longman,

Rees, Orme, Brown & Green and Taylor, London, 1831.
Hieber, L., 'Zum Konzept »Finalisierung der Wissenschaft»', *Leviathan* **3** (1975), 449–472.
Hieber, L., 'Gesellschaftliche Steuerung der Wissenschaft und das Konzept der Finalisierung', *Leviathan* **4** (1976), 284–289.
Hohlfeld, R., 'Praxisbezüge wissenschaftliche Disziplinen: Das Beispiel der Krebsforschung', in Böhme *et al.* (1978), pp. 131–193.
Hubig, C. and von Rahden, W. (eds.), *Konsequenzen kritischer Wissenschaftstheorie*, Walter de Gruyter, Berlin, 1977.
Hübner, K., Lobkovicz, N., Lübbe, H. and Radnitzky, G. (eds.), *Die politische Herausforderung der Wissenschaft: Gegen eine ideologische verplante Forschung*, Hoffman und Campe, Hamburg, 1976.
Johnston, R.D., 'The Internal Structure of Technology', *The Sociological Review, Monograph 18, Sociology of Science* (ed. P. Halmos), University of Keele, 1972.
Johnston, R., 'Finalization: A New Start for Science Policy?', *Social Science Information* **15** (1976), 331–336.
Johnston, R. and Jagtenberg, T., 'Goal Direction of Scientific Research', in Krohn *et al.* (1978), pp. 29–58.
Krohn, W., Layton, E.T. Jr. and Weingart, P. (eds.), *The Dynamics of Science and Technology*, D. Reidel, Dordrecht, 1978.
Krohn, W. and Schäfer, W., 'Ursprung and Struktur der Agrikulturchemic', in Böhme *et al.* (1978), pp. 23–68.
Kuhn, T., 'The History of Science', in *International Encyclopedia of the Social Sciences*, vol. **14**, Crowell Collier and Macmillan, New York, 1968, pp. 74–83. (Reprinted in Kuhn (1977), pp. 105–126.)
Kuhn, T., *The Structure of Scientific Revolutions*, 2nd ed., University of Chicago Press, Chicago, 1970.
Kuhn, T., *The Essential Tension*, The University of Chicago Press, Chicago, 1977.
Lakatos, I., 'History of Science and Its Rational Reconstructions', in R. Buck and R.S. Cohen (eds.), *Boston Studies in the Philosophy of Science*, vol. 8, D. Reidel, Dordrecht, 1971, pp. 91–136.
Laudan, L., *Progress and Its Problem*, Routledge and Kegan Paul, London, 1977.
Lenhardt, C., Review, *Philosophy of the Social Sciences* **11** (1981), 509–513.
Peirce, C.S., *Collected Papers*, Harvard University Press, Cambridge, Mass., 1931–35.
Pfetsch, F.R., 'The »Finalization» Debate in Germany: Some Comments and Explanations', *Social Studies of Science* **9** (1979), 115–124.
Ravetz, J.R., *Scientific Knowledge and Its Social Problems*, Penguin Books, Harmondsworth, 1973.
Schäfer, W., 'Normative Finalisierung: Eine Perspektive', in Böhme *et al.* (1978), pp. 377–415.
Schäfer, W. (ed.), *Finalization in Science: The Social Orientation of Scientific Progress*, D. Reidel, Dordrecht, 1983.
Simon, H., *The Sciences of the Artificial*, Second Edition, The MIT Press, Cambridge, Mass., 1982.
Spengler, T., '»Wissenschaft des Volkes». Politische Steuerung des Wissenschaftsbetriebes in China', in Böhme *et al.* (1978), pp. 251–337.
Symes, J., 'Policy and Maturity in Science', *Social Science Information* **15** (1976), 337–347.
Urbach, P., 'The Objective promise of a Research Programme', in G. Radnitzky and G. Andersson (eds.), *Progress and Rationality in Science*, D. Reidel, Dordrecht, 1978, pp. 99–113.
Weingart, P., 'The Relation between Science and Technology – A Sociological Explanation', in Krohn *et al.* (1978), pp. 251–286.
Wojick, D., 'The Structure of Technological Revolutions', in G. Bugliarello and D.B. Doner (eds.), *The History and Philosophy of Technology*, University of Illinois

Press, Urbana, 1979, pp. 238–261.

Wright, G.H. von, 'Wittgenstein on Certainty', in G.H. von Wright (ed.), *Problems in the Theory of Knowledge* (I.I.P., Entretiens in Helsinki 24–27 August 1970), Martinus Nijhoff, Haag, 1972.

CHAPTER 11

PARADIGMS AND PROBLEM-SOLVING IN OPERATIONS RESEARCH

Russell L. Ackoff, Professor of Systems Sciences in the University of Pennsylvania, published in 1979 in the *Journal of the Operational Research Society* two papers on the status of Operations Research (OR).[1] In the first paper, 'The Future of Operational Research is Past', Ackoff – who himself is one of the leading pioneers of OR – argues that »American Operations Research is dead even though it has yet to be buried». In the second article, 'Resurrecting the Future of Operational Research', Ackoff suggests that OR might be resurrected by adopting an »alternative paradigm to that currently used in OR».

While Ackoff already uses the term 'paradigm', which was made fashionable by Thomas Kuhn's book *The Structure of Scientific Revolutions* (1962)[2], Dando and Bennett (1981) have suggested that »OR is now in a period of crisis analogous to that described by Kuhn in the Natural Sciences». In this paper, I shall examine the question whether such Kuhnian notions as 'paradigm' and 'crisis' are applicable to the development of OR. I am not as such interested in evaluating any rival approaches to OR or making any suggestions for its future. Rather, I try to elaborate on the philosophical consequences that this case study may have concerning the ideas of Kuhn and others (such as Laudan) who treat science as a »problem-solving activity».

1. ACKOFF'S CRIRICISM OF OPERATIONS RESEARCH

The prehistory of Operations Research includes Frederick Taylor's ideas on scientific management in the end of the 19th century, A.K. Erlang's work on telephone traffic in 1908 (leading later to queuing theory), Frederick Lancaster's quantitative study of military strategies in 1916, Emile Borel's (1921) and John von Neumann's (1928) invention of the theory of games, and Abraham Wald's work on decision theory in 1939. The progress in probability theory, mathematical statistics, numerical methods, theory of automata, systems theory, and econometrics also contributed to the development of OR. It is usually said, however, that Operational Research was born in England during World War II when teams of scientists started to use mathematical methods for the planning of military operations. For example, one of the successful projects aimed at increasing the effectiveness of bombing missions. After the war OR was adopted in the civil sector as a tool of quantitative planning, and it was applied to various managerial problems in industry, business, and government.

Perhaps the most authoritative presentation of the aims and the methods of the new field was *Introduction to Operations Research* (1957) by C. West Churchman, Russell L. Ackoff, and E. Leonard Arnoff. (See also Ackoff, 1961, 1962; Ackoff and Sasieni, 1968.) They defined OR as follows:

> »O.R. is the application of scientific methods, techniques, and tools to problems involving the operations of a system so as to provide those in control of the system with optimum solutions to the problems.» (p. 18)

Churchmann *et al.* emphasized the application of OR methods to complex industrial organizations which face »executive-type problems» of combining the behaviour of its various departments so as to yield an optimum over-all result.

In his 1979 articles Ackoff complains that, after having gained a widespread acceptance by the mid 60's, OR has lost its »pioneering spirit» and »innovativeness». In his view, »academic OR and the relevant professional societies» are »primarily responsible for this decline». He argues that the current practice of OR has three major defects. First, the practitioners seek, select, and distort the problem situations so that »favoured techniques could be applied to them». Secondly, an increasing portion of OR, especially the teaching of mathematical methods, is »done by those who do not identify with the profession». Thirdly, when OR has become an isolated discipline, »the original interdisciplinarity of OR has completely disappeared».

Ackoff (1979a) gives a more detailed criticism of six points about OR. First, as the life time of solutions to many social and organizational problems is shorter than the time required to find them, there is a great need for »decision-making systems that can learn and adapt effectively». Secondly, OR's concept of optimality fails to take into account the aesthetic values associated with the intrinsic value of means and the extrinsic value of ends. Thirdly, managers are confronted with »messes», i.e., »complex systems of changing problems that interact with each other», and messes have to be treated holistically. Fourthly, OR's »paradigm» for problem-solving, which consists in the principle »predict the future and prepare for it», should be changed to »design a desirable future and invent ways of bringing it about». Fifthly, the treatment of messes requires an interdisciplinary approach. Sixthly, as the objectivity of science can be obtained only by taking »all possible values» into account, OR should involve »all those who can be affected by the output of decision making».

Ackoff's (1979b) »alternative paradigm» of »interactive planning» is based upon three operating principles. According to the »participative principle», »the principal benefit of planning comes from engaging in it». The »principle of continuity» requires that plans should be continuously revised in the light of their performance, unexpected problems and opportunities, and latest knowledge. The »holistic principle» states that »every part of a system and every level of it should be planned for simultaneously and interdependently» – instead of »top-down and bottom-up planning». Interactive planning should »formulate the mess», set the ideals of the organization by giving a continuously changing »idealized redesign of the system», and find the means and resources for pursuing these ideals. This new paradigm is teached in the programme of »Social Systems Sciences» that Ackoff has established at the University of Pennsylvania.

2. Kuhn's Account of Normal and Extraordinary Science

We may now ask whether Ackoff's description of the current state of OR bears any resemblance to Kuhn's account of a science in a crisis. To answer this question, let us first recall in this Section some of the basic ideas of Kuhn's analysis of science (cf. Figure 1).

Kuhn (1962) views scientific research as an activity of scientific communities. During the periods of *normal science*, this research is based upon past achievements or *paradigms* which the scientific community acknowl-

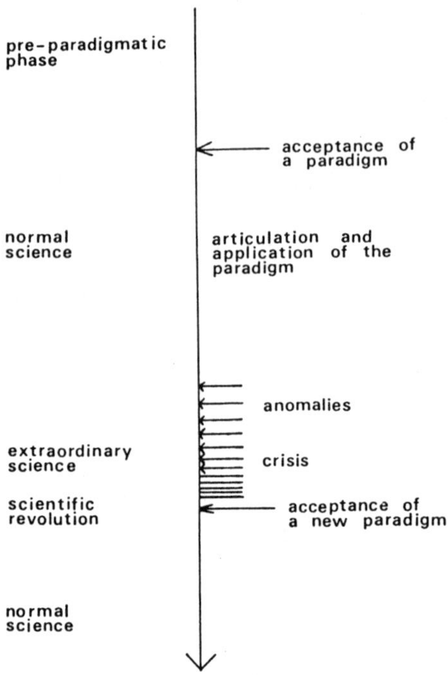

Fig. 1. Kuhn's model of scientific change.

edges as providing the rules and standards for its further practice. The community commits itself to certain methodological and theoretical assumptions without aiming at new sorts of phenomena or new theories. Thus, normal science consists in

> »extending the knowledge of those facts that the paradigm displays as particularly revealing, by increasing the extent of the match between those facts and the paradigm's predictions, and by further articulation of the paradigm itself». (p. 24)

To a great extent, normal-scientific research is »puzzle-solving», i.e., it uses skill and ingenuity to solve problems which are already guaranteed to have solutions by the paradigm.

Even though normal science is in a sense dogmatic, research under a paradigm is »a particularly effective way of inducing paradigm change» (*ibid.*, p. 52), since sooner or later nature will violate expectations based on the paradigm. Scientific theories can be to some extent adjusted to account for such anomalies, but a persistent breakdown of the normal-scientific puzzle-solving activity in important areas leads to a *crisis* (e.g., astronomy before Copernicus, chemistry before Lavoisier, physics before Einstein). A crisis is a »period of pronounced professional insecurity» (*ibid.*, p. 67) which begins with »the blurring of a paradigm» and with the »loosening of the rules for normal research» (*ibid.*, p. 84).

> »The proliferation of competing articulations, the willingness to try anything, the expression of explicit discontent, the recourse of philosophy and to debate over fundamentals, all these are symptoms of a transition from normal to extraordinary research.» (*ibid.*, p. 90)

The period of *extraordinary research* ends with the emergence (or recognition) of a new candidate for a paradigm and its acceptance in a *scientific revolution*. Such a paradigm change begins a new period of normal science.

Masterman (1970), who claims that Kuhn used the term 'paradigm' at least in twenty-one different senses in his book, makes a distinction between paradigms in three different senses: a *metaphysical paradigm* is a set of beliefs which serves as a *Weltanschauung* of the scientists; a *sociological paradigm* is a scientific achievement which is universally recognized in a community; an *artefact paradigm* is some concrete work. Metaphysical paradigms are larger entities than single scientific theories – they correspond closely to the »hard core» assumptions of Lakatos's (1970) *scientific research programmes* and Laudan's (1977) *research traditions*. The conceptual separation of metaphysical paradigms from sociological ones leaves open the possibility – important for many attempts to apply Kuhnian concepts outside natural science – that several 'paradigms' co-exist within the same community.

In his 'Postscript-1969' to the Second Edition of his book, Kuhn himself distinguishes between a *disciplinary matrix* (i.e., paradigm as a constellation of group commitments) and *exemplars* (i.e., paradigm as shared example). (See also Kuhn, 1977.) A disciplinary matrix consists of at least four components: (i) *symbolic generalizations* (i.e., laws), (ii) *models* (i.e., metaphysical beliefs, analogies, and metaphors), (iii) *values* (i.e., methodological standards and preferences), and (iv) *exemplars* (i.e., concrete problem solutions). In normal science, the members of the scientific community try to apply the symbolic generalizations to particular cases which are similar to the accepted exemplars. These cases are structured largely through the models of the matrix, and the success of the applications is evaluated relative to the values of the matrix. In this way, we may see how good a normal-scientific tradition has been in problem-solving – and whether it is *progressive* or *degenerative* in this respect (cf. Lakatos, 1970; Laudan, 1977). Then a crisis corresponds to the degenerative phase of a scientific tradition (cf. Kuhn, 1970, p. 256).

One way of reconstructing at least some aspects of Kuhnian normal science is to treat *Kuhn-theories* as triples $\langle K, I, I_0 \rangle$, where the *core* K consists essentially of a mathematical formalism (symbolic generalizations), I is the class of the *intended applications* of K, and $I_0 \subseteq I$ is the class of paradigmatic *exemplars*. The claim associated with theory $\langle K, I, I_0 \rangle$ is that K can be applied to the elements of I. Normal science consists in the gradual verification of this claim, and its crisis means the emergence of important anomalous cases $I' \subseteq I$ to which K cannot be successfully applied.[3]

3. Is Operations Research in a Kuhnian Crisis?

Dando and Bennett (1981) note that Ackoff is not the only person who has recently criticized OR: the optimism of the late 60's has in ten years changed to a widespread pessimism about the practical success of OR. It might be the case that this kind of feeling of insecurity is primarily a reflection of the decline in Western industry due to the oil crisis of the 70's. Such a professional crisis, generated by external reasons, should be distinguished from a crisis in Kuhn's sense, since the latter can be characterized by terms which are internal to a scientific tradition.[4]

Some of Ackoff's complaints about OR – especially his remarks against the character of OR as a »discipline», or against »academic OR» and »professional societies», and his claim that »favoured techniques» determine the problems that are studied – seem to be directed against the normal-scientific nature of current OR. But to be in a period of normal science does not constitute a »crisis» in Kuhn's sense. Thus, Ackoff's complaints might be interpreted as indicating that OR is not yet in a Kuhnian crisis – at most it may be just entering the phase of extraordinary research through the criticism of Ackoff and others.

As Kuhn has clearly shown the methodological rationale behind paradigm-based research, the criticism that all normal science is too 'dogmatic' is not justified. This criticism has nevertheless a point especially relative to those disciplines which serve as a basis for professional technologies (e.g., medicine, jurisprudence, mechanical engineering). These situations may bring about unhealthy dogmatism, since all challenges to the accepted theoretical background assumptions will be regarded as attacks against the social status of the whole profession. Perhaps Ackoff's message is that this has become the trouble with OR as well.

On the other hand, Ackoff's theses clearly imply also that in his view the current OR is in a degenerative phase, since it does not adequately treat certain important types of problems – which he calls »messes». Such messes are thus anomalies to the standard techniques of OR. This decline in the problem-solving ability of OR is due to changes in the social environment of OR[5] – the society is becoming increasingly complex as a collection of interrelated systems – and to the lack of sufficient feedback from the practice to the theory of OR.

If OR ever is to enter into a Kuhnian crisis, it first has to be based upon a Kuhnian paradigm or disciplinary matrix. Ackoff's discussion in this respect is not too clear, but in any case his own »paradigm» of »interactive planning» consists of a few general *norms* for the planning process. His criticism against OR also contains a somewhat implicit description of some general principles that he finds characteristic to the standard way of applying OR to planning problems. In other words, for Ackoff the old and new »paradigm» of OR are *theories of planning*. There are no a priori reasons why Kuhnian notions could not be applied to such theories – if they are formulated in a sufficiently explicit form.[6]

Dando and Bennett (1981, 1982) interpret the situation in a different way. They agree with Raitt (1979) that

> »OR has no distinctive subject matter in the way that physics has, and does not provide an accumulation of theoretical knowledge about the world.»
>
> »No-one expects the history of OR to show an accumulation of models of increasing power, precision or generality. A model is constructed for practical application in a particular situation. We do not ask if it is true, only if it works ...»

They come to the conclusion that, while OR does not generate its own »scientific-empirical» paradigms »in the original Kuhnian sense», OR does have a »methodological paradigm», since OR is committed to the scientific method. They view the Ackoff debate as a controversy between »official», »reformist», and »revolutionary» paradigms, which are related to positivist, interpretative, and critical methodologies in sociology.

In view of the recent developments within »soft systems theory», the idea of interpreting Ackoff's theory of interactive planning as a step from positivism towards hermeneutics is not so far-fetched on the whole as it may first seem: Ackoff's methods for dealing with »messes» are »softer» and more emphatic than the standard quantitative tools of OR. However, we have already seen above that there is no need to restrict the discussion of the paradigms of OR to general methodological issues, as Dando and Bennett suggest. Moreover – and this is for us the most interesting aspect of this paradigm debate – the discussion about OR has so far ignored the fact that Kuhn does not either regard scientific progress as »an accumulation of theoretical knowledge about the world». For Kuhn, science is a *problem-solving* rather than *truth-seeking* activity (cf. also Laudan, 1977). And what would be a more striking example of problem-solving activity than OR!

This observation shows that the question of the applicability of Kuhn's model of science to OR has to be reconsidered. We shall do this by looking into the nature of problem-solving in OR.

4. Problem-Solving in Operations Research

In their classical treatment[7], Churchman *et al.* (1957) outline the method of OR as consisting of the following steps:

»1. Formulating the problem.
2. Constructing a mathematical model to represent the system under study.
3. Deriving a solution from the model.
4. Testing the model and the solution derived from it.
5. Establishing controls over the solution.
6. Putting the solution to work: implementation.» (p. 13)

This view of scientific method as problem-solving is largely influenced by John Dewey's *Logic: The Theory of Inquiry* (1939). For Dewey, the locus of a problem concerns »what kind of responses the organism shall make» or »the interaction of organic responses and environing conditions in their movement toward an existential issue» (p. 107). For Churchman *et al.*, a problem involves four components: (i) a decision-maker (executive) who is »dissatisfied with some aspect of the state of affairs and consequently wants to make a decision with regard to altering it», (ii) the objectives (wants) of the decision-maker, (iii) the environment consisting of the resources available to the decision-maker, and (iv) alternative courses of action among which the decision-maker has to make his choice.

When a problem has been formulated, a mathematical model is constructed which expresses the »effectiveness» E of the whole system under study as a function of variables x_i subject to control by the decision-maker and variables y_j not subject to control:

(1) $E = f(x_i, y_j)$.

The problem is then reduced to the choice of the values of x_i which maximize (or minimize) the function (1), given the values of y_j and some restrictions on the permissible values of x_i. This optimatization problem is solved by analytical, numerical or Monte Carlo methods (simulation). The solution is then translated into a set of operating procedures, i.e., to recommendations of action for the decision-maker. This strategy is advertised as the general method of science in Ackoff (1962), p. 28.

A typical example is the *diet problem,* first studied by George Stigler in 1945. Assume that one unit of food x_i contains a_i units of vitamin A and b_i units of vitamin B, and that one unit of x_i costs c_i units of money

($i = 1,\ldots, n$). What is the minimum-cost diet which gives at least a units of vitamin A and b units of vitamin B? This problem can be solved by minimizing the cost function

$$(2) \quad c_1 x_1 + \ldots + c_n x_n$$

with the constraints

$$(3) \quad \begin{aligned} & x_i \geqslant 0 \quad (i = 1,\ldots, n) \\ & a_1 x_1 + \ldots + a_n x_n \geqslant a \\ & b_1 x_1 + \ldots + b_n x_n \geqslant b. \end{aligned}$$

This mathematical optimization problem can be handled with the techniques of *linear programming* (cf. Saaty, 1959). The same method is applicable e.g. to the *transportation problem*, formulated by F.L. Hitchcock in 1941: what is the most economical way of shipping a given amount of a product from m origins to n destinations.

Churchman *et al.* give a classification of the main types of problems to which OR is applicable:

a) inventory problems
b) allocation problems
c) waiting-time problems
d) replacement problems
e) competitive problems
f) combined problems.

Each type of problem is further divided into subtypes, and these are illustrated by specific applications. For example, the diet problem and the transportation problem are special cases of one important subtype of allocation problems. Some of the standard distinctions among competitive problems (where the effectivity of a decision depends upon the decision of another party with conflicting interests) are shown in Figure 2.

These classified problems are then correlated with mathematical models and techniques: for example, inventory models for inventory problems, queuing theory and sequencing models for waiting-time problems, game theory for competitive problems, linear and dynamic programming for several types problems, etc. The development of these models and techniques can be regarded as a branch of applied mathematics. In the applications of OR, they are used for finding solutions to particular decision-problems faced by various kinds of business, industrial, military, and administrative organizations.

On the basis of this survey, we can define the *disciplinary matrix* of OR as a quadruple $\langle T, M, P, I_0 \rangle$, where

T = a collection of mathematical *techniques* needed in the solution of optimization problems (e.g. linear programming, Monte Carlo methods)

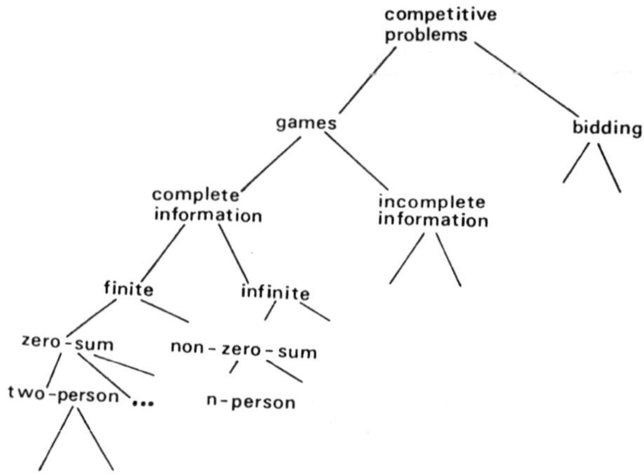

Fig. 2. Classification of competitive problems.

M = a collection of mathematical *models* which can represent the structure of decision-problem situations (e.g. queuing theory, game theory, decision theory, inventory models)

P = a hierarchical classification of various types of decision-problems

I_0 = a set of paradigmatic exemplars (applications of the techniques and models to particular problem situations; e.g. the diet problem).

A typical pattern of OR is to use the models in M to give structure to a particular decision-problem, to identify this problem in the hierarchy P and to apply the techniques of T to this problem by imitating the procedures used for the exemplars in I_0. A disciplinary matrix of OR may be combined also with a *theory of planning* which gives more general rules for the application process of OR methods (cf. Section 3).

This notion of disciplinary matrix is close to Kuhn's corresponding concept which includes symbolic generalizations, models, values, and exemplars as its components. It is also closely related to the notion of a Kuhn-theory in Section 2: T and M together correspond to the core K, P to the set I of intended applications, and I_0 to the set of exemplars.

Normal-scientific theoretical development of the matrix $\langle T, M, P, I_0 \rangle$ of OR means the invention of new or more effective techniques in T, the creation of new or more comprehensive models in M, and the refinement of the structure P of problems. Ackoff's »revolution» consists in his claim of the importance of a new class of problems – »messes» – and of the need to develop new kinds of methods for handling them.

5. Truth-Seeking and Problem-Solving

As a *problem-solving* activity, OR is oriented towards practice: it tries to use the methods of science to find optimal solutions to problems concerned with alternative courses of action. As these solutions are its primary aim, it is clear in which sense OR is not a *truth-seeking* activity: it is not a knowledge-seeking enterprise.

We have also seen that the Kuhnian conception of scientific change fits quite well to OR – even though OR is not a science with its own subject matter. As this fit is so good, one may start to wonder how Kuhn's account then can be applicable to sciences at all. Is it perhaps so that all sciences are comparable to OR as problem-solving activities? Some *instrumentalist* philosophies of science give a positive answer to this question. In particular, *behaviouralists* (to use L.J. Savage's term) think that the acceptance of a scientific hypothesis is always a decision to act upon that hypothesis. I shall argue for a negative answer by considering more closely the role of the notion of *truth* within problem-solving.

In the first place, it would be a mistake to conclude that questions about truth are irrelevant to the procedures of OR. The mathematical models that are used in OR are representations of the system under study. These models may be imperfect and idealized, but still the quality of the solutions that they yield crucially depends upon their closeness to reality in the relevant respects. Thus, Ackoff points out that »the optimal solution of a model is *not* an optimal solution of a problem unless the model is a perfect representation of the problem» (Ackoff, 1979a, p. 97). But since all models are »simplifications of reality», »it is critical to determine how well models represent reality». For this reason, the models of OR have to be tested and controlled before the implementation of the solutions (cf. Churchman *et al.*, 1957, Part IX). For example, in the diet problem it is assumed that the cost of a diet is measured by function (2). If the prices c_i of food change or if some additional costs have to be added (e.g., costs for preparing the meals), then the cost equation defined by (2) has to be changed as well. In this way, OR can be said to be a *truth-using* activity: even though it is not primarily interested in the models as such, since they serve as *instruments* for finding solutions to decision-problems, *the instrumental value of these models relative to a particular application depends partly upon their degree of truthlikeness.*[8] If the model is not sufficiently close to the truth, it gives non-optimal solutions which are disappointing, since they do not relieve the decision-maker from his initial dissatisfaction.[9]

Secondly, there is a crucial distinction to be made between *cognitive problems* (What to believe? What can be known?) and *problems of action* (What to do?). Cognitive problems involve hypotheses concerning alleged facts or regularities, past facts (postdiction), future facts (prediction), causes or reasons of facts (explanation). They can be expressed by WH-questions, why-questions, and how-questions – and in most cases they can be transformed to whether-questions of the following forms: Is h true or false? Which of the alternative hypotheses $h_1, \ldots, h_n$ is true? Which of the alternative hypotheses $h_1, \ldots, h_n$ is it rational to accept as true?[10] Relative to cognitive problems, one can make a distinction between *potential* answers and the *correct* answer, and between *partial* and *complete* answers. The aim of studying a cognitive problem is ultimately to find its complete correct answer – and thus to produce new knowledge. The outcome of such inquiry, viz. the acceptance of a hypothesis as true, is itself an act or a decision. Therefore, the procedure of solving cognitive problems can formally be represented as decision-making (see Levi, 1967; Niiniluoto, 1976). However, this activity is *truth-seeking* – and, hence, different from problem-solving with OR.

There is a class of cognitive decision-problems which are truth-using and truth-seeking at the same time: the employment of scientific theories to make predictions about particular empirical phenomena. In this case, the alternative hypotheses $h_1, \ldots, h_n$ are singular empirical statements with unknown truth-values. A theory which entails one of these hypotheses and refutes the other ones can be said to give a *potential solution* to this empirical problem.[11] But here, again, the quality of the solutions depends upon the nature of the theory: a potential solution must also be *correct* if the theory is true. If the degree of the truthlikeness of the theory is not sufficiently high, it may with good luck give a correct solution in some particular cases, but not consistently for an indefinitely large class of empirical problems. And if the theory were only a symbolic device which does not represent reality in any degree and does not have any truth-value, as some instrumentalists claim, its persistent ability to give correct solutions to new empirical problems would be – in Putnam's words – a miracle (cf. Chapter 5).[12]

These observations show that it is fully compatible with scientific realism to treat science as a problem-solving activity: if scientists seek answers to cognitive problems, then truth is both an aim of science and a factor which influences the instrumental value of theories. In contrast, Laudan's problem-solving model of science is based upon the idea that all issues about truth and falsity are left outside the account of the aims of science (cf. Laudan, 1977, 1981, 1982). We have seen one way in which such a project can be effectively pursued: as Ackoff (1962) does, one may accept the behaviouralist philosophy of OR and treat all scientific questions as problems of action. But even then one would have to acknowledge that science is a truth-using activity.

This is not Laudan's position, however. He endorses »semantic realism» (i.e., the claim that scientific theories have truth values), but denies that this is relevant to the goals of scientific inquiry. It remains a matter of controversy what he means by »problems» and their »solutions»[13], but recently he has claimed that »the *raison d'etre* of science» is to put us »in a position to anticipate nature» (Laudan, 1982, p. 80). We argued above, however, that problems of prediction are both truth-using and truth-seeking cognitive problems. It is not enough to characterize the task of scientific theories by saying that they are expected to »shape our expectations in various ways» (*ibid.*), since even a chrystal ball can do that. As soon as we admit that our theories are hoped to yield reasonably often »correct anticipations of nature», we have thereby acknowledged that science is a truth-seeking enterprise.[14]

NOTES

1 The term 'Operations Research' is used in America and 'Operational Research' in England. The term 'Management Science' is often used as another name for OR. I am grateful to Dr. Yrjö Seppälä for bringing the 'Ackoff debate' to my attention.

2 As the bibliography in Gutting (1980) shows, the applicability of Kuhn's account of scientific change has been explored in such fields as mathematics, physics, chemistry, astronomy, geology, biology, psychology, history, linguistics, theology, art and literature, economics, sociology, political science, and education. To this list we may add folklore, jurisprudence, technology, and Operations Research.

3 This is only a rough sketch of the formal treatment of »Kuhn-theories» and their development suggested by Sneed, Stegmüller, Balzer, and Moulines. Cf. Chapter 6.

4 Moreover, a Kuhnian crisis is a regular phenomenon which a discipline every now and then has to face. It is an essential element of the pattern by which scientific disciplines make progress – so that there are no negative overtones in this notion.

5 Kuhn notes the existence of this kind of phenomenon in Kuhn (1977), pp. 119–120.

6 Theories of planning which consist of *norms* (e.g. »take into account the values of all stakeholders of the organization») can be reformulated as *assertions* about rational or effective planning (e.g., »it is rational to take into account ...»).

7 The book of Churchman *et al.* has a 'paradigmatic' status in OR – in one sense of this word in Kuhn (1962), p. 10.

8 For the concept of truthlikeness, see Chapter 5 above.

9 It may of course happen that an inadequate model sometimes gives satisfactory results just *with good luck*, but such a case is exceptional. For a discussion of why »truth pays», see Loewer (1980). Take again the diet problem as an example: if the decision-maker has wrong information about the vitamins in different kinds of foods, his constraints (3) are mistaken, and the solution of the linear programming problem is not really optimal.

10 For an analysis of cognitive problems and their relation to questions, see Niiniluoto (1976).

11 As a weaker criterion for a solution, one might require that one and only one of the alternatives $h_1,...,h_n$ is inductively acceptable relative to the theory.

12 It may have been confusing in the dispute between scientific realists and instrumentalists that sometimes »mathematical models» *with free parameters* are called »theories». From a logical point of view, such »models» are open formulas and do not have a truth-value, but they become statements with a truth-value as soon as the values of the parameters are filled in relative to a particular application of this »model».

13 Laudan says that a theory solves a problem, if the »statement of the problem» can be deduced from the theory. It is difficult to see how problems of prediction (e.g. of the form 'p or non-p?') can be reduced to this idea. Laudan's account of problems

seems to be best applicable to why-questions of the form 'Why *p*?', where *p* is a statement which is believed to express a fact (i.e., *p* is believed to be true). He denies – in my view, for wrong reasons (cf. Chapter 5) – that problem-solving in his sense is the same as explanation. As one can distinguish between potential and correct explanations, the search of correct explanations is also a truth-seeking enterprise.

14 One reason for Laudan's opposition to truth as an aim of science is this complaint: the realists have failed to show that approximately true theories yield approximately true empirical predictions. An answer to this problem is given in Niiniluoto (1982). See also Chapter 7.

Bibliography

Ackoff, R.L., 'The Meaning, Scope, and Methods of Operations Research', in R.L. Ackoff (ed.), *Progress in Operations Research,* vol. I, John Wiley, New York, 1961, pp. 1–34.

Ackoff, R.L. (with the collaboration of S.K. Gupta and J.S. Minas), *Scientific Method: Optimizing Applied Research Decisions,* John Wiley, New York, 1962.

Ackoff, R.L., 'The Future of Operational Research is Past', *Journal of the Operational Research Society* **30** (1979), 93–104. (a)

Ackoff, R.L., 'Resurrecting the Future of Operational Research', *Journal of the Operational Research Society* **30** (1979), 189–199. (b)

Ackoff, R.L. and Sasieni, M.W., *Fundamentals of Operations Research,* John Wiley, New York, 1968.

Churchman, C.W., Ackoff, R.L. and Arnoff, E.L., *Introduction to Operations Research,* John Wiley and Sons, New York, 1957.

Dando, M.R. and Bennett, P.G., 'A Kuhnian Crisis in Management Science?', *Journal of the Operational Research Society* **32** (1981), 91–104.

Dando, M.R. and Bennett, P.G., 'Kuhnian Comments Squared? Replies to Gault and Williams', *Journal of the Operational Research Society* **33** (1982), 100–101.

Dewey, J., *Logic: The Theory of Inquiry,* Henry Holt and Company, New York, 1939.

Gault, A.R., 'A Metaphysical Debate rather than a Kuhnian Crisis', *Journal of the Operational Research Society* **33** (1982), 91–99.

Gutting, G. (ed.), *Paradigms & Revolutions: Applications and Appraisals of Thomas Kuhn's Philosophy of Science,* University of Notre Dame Press, Notre Dame, 1980.

Kuhn, T., *The Structure of Scientific Revolutions,* The University of Chicago Press, Chicago, 1962. 2nd enlarged edition, 1970.

Kuhn, T., 'Second Thoughts on Paradigms', in F. Suppe (ed.), *The Structure of Scientific Theories,* University of Illinois Press, Urbana, 1974, pp. 459–482.

Kuhn, T., *The Essential Tension,* The University of Chicago Press, Chicago, 1977.

Lakatos, I., 'Falsification and the Methodology of Scientific Research Programmes', in Lakatos and Musgrave (1970), pp. 91–196.

Lakatos, I. and Musgrave, A. (eds.), *Criticism and the Growth of Knowledge,* Cambridge University Press, Cambridge, 1970.

Laudan, L., *Progress and Its Problems,* Routledge and Kegan Paul, London, 1977.

Laudan, L., 'The Philosophy of *Progress* ...', in I. Hacking and P. Asquith (eds.), *PSA 1978,* vol. 2, Philosophy of Science Association, East Lansing, 1981, pp. 530–547.

Laudan, L., 'Problems, Truth, and Consistency', *Studies in History and Philosophy of Science* **13** (1982), 73–80.

Levi, I., *Gambling With Truth,* Alfred A. Knopf, New York, 1967.

Loewer, B., 'The Truth Pays', *Synthese* **43** (1980), 369–379.

Masterman, M., 'The Nature of Paradigm', in Lakatos and Musgrave (1970), pp. 59–90.

Niiniluoto, I., 'Inquiries, Problems, and Questions: Remarks on Local Induction', in R.J. Bogdan (ed.), *Local Induction,* D. Reidel, Dordrecht, 1976, pp. 263–296.

Niiniluoto, I., 'Truthlikeness for Qualitative Statements', in P.D. Asquith and P. Nickles (eds.), *PSA 1982,* vol. 1, Philosophy of Science Association, East Lansing, 1982, pp. 208–216.

Raitt, R.A., 'Viewpoints', *Journal of the Operational Research Society* **30** (1979), 835–836.

Saaty, T.L., *Mathematical Methods of Operations Research,* McGraw-Hill, New York, 1959.

CHAPTER 12

REMARKS ON TECHNOLOGICAL PROGRESS

Perhaps the most fundamental question of the new philosophy of technology is whether science and technology have to be conceptually distinguished from each other.[1] If technology were just a branch of science, then the philosophy of technology would likewise be just a branch of the philosophy of science. Therefore, the common definition of technology as 'applied science' has served as one of the reasons why the philosophy of technology has had difficulties in gaining the status of an independent branch of philosophy. On the other hand, some philosophers think that science is nothing but a tool of technology – and thus try to make the philosophy of science a subfield of the philosophy of technology. In this paper, I defend the view that science and technology are two historically and causally interconnected activities which have different aims and different patterns of progress. The connection between technological progress and social progress is also briefly discussed.

1. TECHNICS, TECHNOLOGY, AND SCIENCE

The words 'technics' and 'technology' are used in many different senses. They may refer to (a) the *tools* or *artifacts* that man has created for the interaction with nature,[2] (b) the *use* of such tools, (c) the *skills* (or *know how*) required in the use of these tools, (d) the *design* of these tools, (e) the *production* of these tools, or (f) the *knowledge* which is needed in the design and production of these tools.

Technics in the senses (a), (b), and (c) exists already among animals (e.g. spider's net); the primitive man is known to have made tools of stone almost three millions years ago (cf. Leakey and Levin, 1979). The creation of language and higher forms of thinking made man the only animal with technology in the senses (d), (e), and (f) and with technological progress – i.e., knowledge about technics and the corresponding ability to invent and design new tools. For this reason, man has been called a tool-making animal or *Homo faber* (Benjamin Franklin).

The primitive techniques presupposed only knowledge at the level of everyday experience. Some cultures developed a highly advanced technology without the help of science (e.g. China). In the Western countries, the progress of technology paved the way to the birth of science in Greece around 600 B.C. and also to its further development in the Middle Ages and in the Renaissance (cf. Stock, 1978).

Plato and Aristotle made a distinction between *science* which aims at knowledge (*episteme*) and productive *arts* (*techne*). However, Plato's model of knowledge was still tied with the technological idea of 'know how': genuine knowledge about shoes is possessed by a shoemaker who knows how to produce them.[3] According to Aristotle, a scientific account of things includes knowledge about their final causes – which in the case of artifacts is just their intended purpose of use. Still he distinguished theoretical and practical sciences (concerned with truth and action, respectively) from the productive arts. The 'mechanical arts' (distinguished from the seven 'liberal arts') received a new status in the 12th century, when Hugh of St. Victor classified the field of science into four independent groups: theoretical philosophy, practical philosophy, logic, and mechanical arts, where the last group included fabric-making, armament-making, commerce, agriculture, hunting, medicine, and theatrics (Weisheipl, 1978).

The word 'technology' was used in English in its modern sense first in the early 17th century – during the time when Francis Bacon gave an eloquent expression to the idea that scientific knowledge could serve as a ground for technological and social progress. Bacon's view – which was later emphasized by the fellows of the Royal Society, French Encyclopedists, and other philosophers of the Enlightenment – contains a threefold optimism (see Fig. 1). First, the employment of the correct method and a rational organization of research will ensure the growth of scientific knowledge; secondly, by scientific knowledge human life is »endowed with new discoveries and powers», since »human knowledge and human power meet in one»; thirdly, these new powers help »to subdue the necessities and miseries of human life».

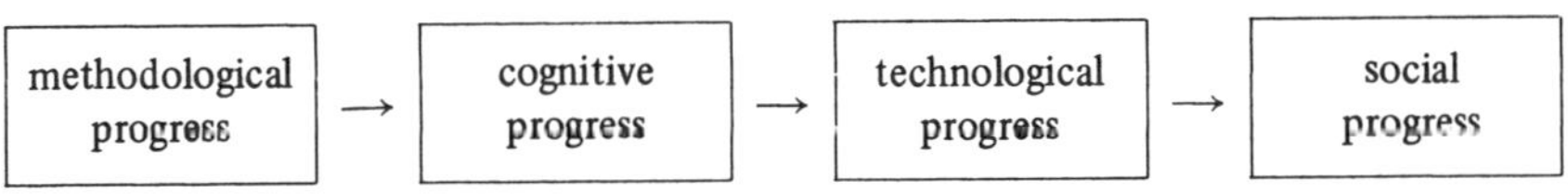

Fig. 1. Baconian optimism.

Bacon's programme started slowly but later accelerated rapidly: some historians of technology claim that its first real successes did not occur earlier than in the middle of the 19th century in the fields of chemical and agricultural industry.[4] While many areas of modern technology have later been successfully 'scientificized', in the sense that they rely on the results and methods of scientific research, there are still other areas of craft technologies which develop largely independently of science.

2. Scientific Progress vs. Technological Progress

The historical sketch of Section 1 already shows that technology and science are two different activities which were born at different times and with different aims. Skolimowski (1968) expresses this distinction in the following way: »Science aims at enlarging our knowledge by devising better and better theories. Technology aims at creating new artifacts through devising means of increasing effectiveness.» In other words, science is a systematic pursuit of new knowledge (i.e., well-grounded information about reality); technology is a systematic development of new tools and artifacts. This distinction, which views both science and technology as activities, is closely related to the standard definitions of 'research' and 'development': both *basic* (or fundamental) *research* and *applied research* aim at new knowledge, while *development* aims at new products and tools (cf. Chapter 9). (If science and technology are defined through their products, then the former consists of a body of knowledge and the latter of a collection of tools.)

Activities can be appraised by evaluating how 'good' results they produce. Therefore, *scientific progress* has to be defined by the increase that new theories contribute to human knowledge – how much new information they give and how close to the truth this information is.[5] *Technological progress* has to be defined by the ability of new tools to perform effectively their intended function of use. While scientific progress is measured by *epistemic utilities*[6] (such as truth, information content, truthlikeness, explanatory power, simplicity), technological progress is measured by *technological utilities* (effectiveness relative to a given practical purpose).

The theory of scientific progress has to identify the relevant epistemic utilities and to study their possible interrelations. Similarly, the relevant technological utilities have to be identified. Skolimowski (1966) has suggested that each major branch of technology is characterized by its own standard of efficiency – such as the accuracy of measurement in surveying, the durability of construction in civil engineering, or durability and aesthetics and utility in architecture – but the situation seems even more complex. Technological tools often serve many separate functions, and the relevant technological utilities and their weights vary from one type of tool to another. For example, maximum speed is very important for racing cars but less important for farm tractors. The effectivity of farm tractors is measured by such variables as 'fuel-consumption efficiency in horse-power-hours per gallon', 'ratio of drawbar horsepower to belt horsepower', 'belt horsepower per ballasted tractor weight', and 'price in relation to price of labor' (see Sahal, 1981). Beauty or artistic quality is an important utility in many fields of technology, but its measurement in quantitative terms is not easy. Moreover, technological utilities and their weights vary in different natural and cultural environments and in different periods of time. These problems have been discussed in the so-called Technology Assessment (TA) (see Durbin and Rapp, 1983).

It is interesting to compare these observations to Kuhn's remark that scientific communities may give different weights to epistemic utilities (Kuhn, 1977, pp. 320–339). Given a set of technological utilities and their weights, we may speak in an unambiguous sense of progress in the development of farm tractors, locomotives, semiconductors, computers, etc. (cf. Sahal, 1981; Schopman, 1981). However, when two groups of people disagree on these utilities (e.g., on the weight given to side-effects that are harmful to the natural and social environment) their evaluations become 'incommensurable'. Thus, the conflict between 'alternative technologies' can be reduced to the existence of rival frameworks or 'paradigms' in the Kuhnian sense.[7]

3. Technological Sciences

Philosophers who argue for the primacy of practice over theory tend to regard practical applications as the characteristic function of scientific knowledge. This *technological conception of science,* which views science only as a tool in man's interaction with his environment, values science for its practical utility – and thus establishes a *conceptual* link between scientific progress and technological progress. Different forms of this tendency are represented by John Dewey's instrumentalism, Martin Heidegger's philosophy of technology (cf. Heidegger, 1977; Ihde, 1979), Jürgen Habermas's claim that natural science is governed by a 'technical interest' for prediction and control, and Nicholas Rescher's (1978) methodological pragmatism.

The technological conception is in conflict with the *cognitivist* account which defines scientific progress in terms of epistemic utilities – and thus accepts only a *factual* connection from scientific progress to technological progress. There are genuine steps of scientific progress – discoveries of new theories which are more truthlike than their predecessors – which are not known to increase our technological capabilities and perhaps never will have any practical applications. Such steps are nevertheless valuable, since they help us to explain and to understand the reality better than earlier.

A partial compromise between instrumentalism and cognitivism can be given by the notion of *technological science* or *engineering science.* The design and production of new and better tools and artifacts requires knowledge – which we may call *technological knowledge.* Science may contribute to the development of technology by producing technological knowledge; the use of scientific methods for finding such knowledge may be called *technological research.*[8] The term 'la science des ingenieurs' occurs first in the early 18th century, and the oldest institutions for teaching such sciences (such as École polytechnic in Paris) were founded in the late 18th century. Several industrial laboratories for technological research (such as General Electric Laboratory) were founded in the first decade of this century (cf. Finch, 1961).

The mistake of instrumentalism is to think that all sciences basically are technological sciences. On the other hand, a purely cognitivist treatment of the engineering sciences is not adequate. These sciences have to be distinguished from technology, since they produce knowledge rather than artifacts. Their progress can be partly analysed by epistemic utilities, and their study may some times have the nature of 'basic research' (cf. Tondl, 1974) – in the sense that it is not yet known to which specific purposes the new knowledge can be used. But it is also the case that some steps of progress in the technological sciences do not exhibit scientific progress as increase in truthlikeness. This follows from the fact that there are special requirements for the results of the technological sciences which differ from ordinary sciences. Hence, the set of utilities for the evaluation of technological sciences includes, besides truth and information, also such 'practical' demands as *simplicity* and *manageability*. For example, to guarantee the applicability to engineering purposes, equations have to be solvable in some efficient and economical way, and this may be achieved by removing idealizations, introducing simplifications and approximations, etc.[9] Hence, sometimes a step from a theory T to its practically applicable variant T′ is progressive from the technological viewpoint even if T′ has less truthlikeness than T.

As the typical results of the engineering sciences are technical norms of the form 'If you want A, you should do X!' (see Chapter 10.5), we can also impose a requirement of *social relevance* to this group of sciences. This requirement – which has no analogue for the basic descriptive sciences – says that the goals A of the technical norms should be at least potentially acceptable to some 'auditory' in the society. The same requirement applies to other natural and social 'design sciences'. It is also clear that the assessment of social relevance involves social and political valuations.

4. Science, Technology and Social Progress

The humanist tradition in the Western countries has been divided into two parties concerning natural science and technology. The *Enlightenment* philosophers (such as Bacon, Comte, and Marx) have viewed science as a great liberating force in history which increases both the rationality of man (by making him free of religious or metaphysical superstition) and the value of his life (by giving him new powers and tools for realizing his potentials). The *Romantic* philosophers have criticized science – both because it challenges their own, often irrational, ways of thinking, and because they feel that science is either irrelevant or harmful to human life. In the 14th century, Petrarch raised the question whether knowledge about nature helps us toward a happy life (cf. Randall, 1940, p. 213), and his voice was echoed in Rousseau's famous claim, in 1750, that scientific progress has not advanced the happiness and virtues of man, and Husserl's complaint, in the 1930's, that the »European sciences» are in a »crisis» because they have lost their »meaning for life» (cf. Husserl, 1970).

Similarly, the philosophy of technology has been a battle field between cultural optimism and pessimism. The birth of *Philosophie der Technik* in the late 19th century was largely a reaction against the neglect of technology within the cultural and educational ideals of German idealism (cf. Mitcham, 1979; Koenne, 1979). In the other camp, Spengler argued in the 1920's that technology is a major factor which leads to the fall of the Western civilization. The Frankfurt School attacked, from 1930's onwards, the Enlightenment and the »technological reason» (cf. Horkheimer and Adorno, 1972) which leads to a suppressive »one-dimensional society» (cf. Marcuse, 1964). Heidegger (1977) sees the »danger» of technology in the possibility that the same instrumental attitude towards nature – which, he thinks, is characteristic to science – is directed to man himself. It has also been argued that man has lost his control over technology: like Frankenstein, technology has taken over the position of the real 'subject' of modern society and thereby made man its slave (cf. Rapp's, 1981, discussion).

No one can deny that the Industrial Revolution has brought about radical and positive changes in the living conditions of the humanity.[10] At he same time, our century – with its wars, armament race, exploitation of nature, pollution, and population explosion – has forced us to realize that technological progress has also had very undesirable effects which endanger our future. A naive technological optimism in Bacon's sense has proved to be mistaken: technological progress alone is not a sufficient condition of social progress.

It would be wrong to blame science for the problematic social effects of technology. As we argued above, science and technology have to be conceptually distinguished from each other. Technology was born millions of years before science – and we would have e.g. military technology today even if science had never come to existence.

If there is a »legitimation crisis» in the modern scientific-technological way of life[11], this is a crisis in *human values*. Technology develops only by decisions made by human beings. There is no inner necessity – or »technological imperative» – for realizing all technological possibilities. The development of technology and its social effects depend therefore upon the prevailing values in the society (cf. Rosenberg, 1979). Moreover, in contrast with the neutrality of scientific (cognitive) progress, what is taken to constitute technological progress also depends on moral values (cf. Bunge, 1979), since technological utilities and their weights reflect ethically relevant social valuations.

These problems were seen very clearly by Russell (1931), who argued that a »scientific civilization» will not be »good» unless increase in knowledge is accompanied with »increase in wisdom» concerning »the ends of life». Russell made it clear, however, that in his view scientific and technological progress are *necessary* conditions – even though they are not sufficient conditions – for social progress (see also Rescher, 1980, p. 20). This, I think, is the true kernel of Baconian optimism about science. It indicates why the simultaneous cultivation of scientific knowledge and of human values is a necessary task of mankind.

NOTES

1 For discussions of this issue, see the articles in *Technology and Culture* 2(1961) and 7(1966), Rapp (1974), Suppe and Asquith (1977), Bugliarello and Doner (1979). See also the monographs Ihde (1979) and Rapp (1981). Ihde classifies positions about science and technology in analogy with mind-body theories: (a) 'parallelism': science and technology are distinct but their development is correlated (e.g., Price's (1965) simile of »two dancers following the same tune»); (b) 'idealism': technology is reduced to science (e.g., the standard view of technology as applied science or as the application of science); (c) 'materialism': science is reduced to technology (e.g., Heidegger's view of technology as ontologically prior to science); (d) 'identity theory': science and technology are the same thing. The position that I defend in this paper represents 'dualist interactionism': science and technology causally influence each other in their developments and neither of them is *as such* the dominating party.

2 Interaction with nature should be understood in a broad sense – so that it includes also expressive activities, such as playing with toys, musical instruments, or sporting equipment. (Cf. Mumford, 1966.)

3 For an analysis of the idea of maker's knowledge, see Hintikka (1974).

4 For the history of technology, and its relation to the history of science, see e.g. Hall (1961), Finch (1961), Bugliarello and Doner (1979).

5 For an analysis of scientific progress in terms of the truthlikeness of theories, see Chapter 5 above. Truthlikeness is essentially a combination of truth and information. Sociologists of science often measure scientific progress in quantitative terms by the number of scientists, publications, journals, etc., but these factors are at best external indicators of real cognitive progress in the content of science. (The direct quantitative measurement of technological progress is usually easier, as we shall see below.) The most sophisticated work in the field of science indicators has been done by John Irvine and Ben Martin from the Science Policy Research Unit, University of Sussex (see Martin and Irvine, 1983). In evaluating the »relative contributions to scientific knowledge» made by different research groups or institutions, they regard peer-review, productivity (i.e., the number of publications), and citations as »partial indicators of scientific progress». They distinguish from each other the *quality* of a publication (»how well the research has been done», etc.), its *impact* (»its actual influence on surrounding research activities at a given time»), and its *importance* (its potential impact in an idealized scientific community with »perfect communication»). While these distinctions no doubt are useful, in my view it is not correct to claim that »it is the 'impact' of a publication that is most closely linked to the notion of scientific progress» (*ibid.*, p. 70). The impact of a publication tells how much attention or 'movement' it creates in the actual scientific community. If this movement happens to go the *wrong* direction, it is not clear that we have made progress – in other words, progress means impact to the right direction (or to one of the right directions). This is another way of saying that the concept of *truth* is needed in the definition of scientific progress, as argued in Chapter 5. The 'importance' of a paper, in Martin's and Irvine's sense, comes close to the pragmatist conception of truth (i.e., truth as the limiting opinion of an ideal scientific community). This notion may be a good indicator of truth, but still it is weaker than the concept of truth in the realist sense (cf. Chapter 5).

6 For a classical exposition of the notion of epistemic utility, see Levi (1967).

7 For attempts to discuss technology in Kuhn's terms, see Johnston (1972) and Wojick (1979). Besides Kuhn, one could try to apply other models of scientific change to the development of technology. For example, Sahal (1981) suggests that technological change is an evolutionary process (cf. Chapter 4 above). See also Elster (1983). The historical succession of certain kinds of artifacts (e.g., Fordson tractors, pistols, computers) could also be viewed as analogous to dynamically evolving 'research programmes' in Lakatos's sense.

8 Modern technological sciences are usually 'applied sciences' in the sense that they rely on theoretical results from the basic sciences (cf. Chapter 10 above), but they may also try to establish independently new theories by using the methods of science. For a discussion of two kinds of 'technological theories', see Bunge (1977, 1979).

9 Agassi (1966) proposes that science differs from technology by the feature that corroboration (or positive success in tests) is relevant only to the latter. This is problematic, since the standards of accuracy seem to be stricter in pure science than in the technological sciences. Newton's mechanics is still regarded as valid in many areas of technology, even if it has been refuted as a scientific theory. Agassi's argument for his view is partly based upon the assumption that the existence of a theory of confirmation ('inductive logic') would make the scientific method mechanical. This assumption is mistaken, since inductive probabilities for first-order theories cannot generally be effectively calculable.

10 The question whether men have also become happier is a tricky one, as Rescher (1980) points out. Many people think that their ancestors were happier than they themselves are, but still they would not like to change places with these ancestors. Rescher explains this 'paradox' by the assumption that the subjective feeling of happiness ('perceived happiness') depends upon the relation between their attainments and expectations. When technological progress has objectively increased the quality of life, men have escalated the level of their expectations.

11 An interesting defence of this thesis is given in von Wright (1981).

BIBLIOGRAPHY

Agassi, J., 'The Confusion between Science and Technology in the Standard Philosophies of Science', *Technology and Culture* 7 (1966), pp. 348–366. (Also in Rapp, 1974, pp. 40–59.)

Bugliarello, G. and Doner, D.B. (eds.), *The History and Philosophy of Technology*, University of Illinois Press, Urbana, 1979.

Bunge, M., 'The Philosophical Richness of Technology', in Suppe and Asquith (1977), pp. 153–172.

Bunge, M., 'Philosophical Inputs and Outputs of Technology', in Bugliarello and Doner (1979), pp. 262–281.

Durbin, P.T. and Rapp, F. (eds.), *Philosophy and Technology*, D. Reidel, Dordrecht, 1983.

Elster, J., *Explaining Technical Change*, Cambridge University Press, Cambridge, 1983.

Finch, J.K., 'Engineering and Science: A Historical Review and Appraisal', *Technology and Culture* 2 (1961), pp. 318–332.

Hall, A.R., 'Engineering and the Scientific Revolution', *Technology and Culture* 2 (1961), pp. 333–341.

Heidegger, M., *Die Technik und die Kehre*, Günther Neske, Pfullingen, 1962. English transl. *The Question Concerning Technology and Other Essays*, Harper & Row, New York, 1977.

Hintikka, J., *Knowledge and the Known*, D. Reidel, Dordrecht, 1974.

Horkheimer, M. and Adorno, T.W., *Dialektik der Aufklärung*, Querido Verlag, Amsterdam, 1947. English transl. *Dialectic of Enlightenment,* Herder and Herder, New York, 1972.

Husserl, E., *The Crisis of European Sciences and Transcendental Phenomenology,* Northwestern University Press, Evanston, 1970.

Ihde, D., *Technics and Praxis: A Philosophy of Technology,* D. Reidel, Dordrecht, 1979.

Johnston, R.D., 'The Internal Structure of Technology', *The Sociological Review, Monograph 18, Sociology of Science* (ed. P. Halmos), University of Keele, 1972.

Koenne, W., 'On the Relationship between Philosophy and Technology in the German Speaking Countries', in Bugliarello and Doner (1979), pp. 282–293.

Kuhn, T.S., *The Essential Tension,* The University of Chicago Press, Chicago, 1977.

Leakey, R.E. and Levin, R., *Origins,* 1977.

Levi, I., *Gambling with Truth,* Alfred A. Knopf, New York, 1967.

Lindberg, D.C., *Science in the Middle Ages,* The University of Chicago Press, Chicago, 1978.

Marcuse, H., *One-dimensional Man,* Beacon Press, Boston, 1964.

Martin, B. and Irvine, J., 'Assessing Basic Research: Some Partial Indicators of Scientific Progress in Radio Astronomy', *Research Policy* **12** (1983), 61–90.

Mitcham, C., 'Philosophy and the History of Technology', in Bugliarello and Doner (1979), pp. 163–201.

Mumford, L., 'Technics and the Nature of Man', *Technology and Culture* **7** (1966), pp. 303–317.

Price, D.J. de Solla, 'Is Technology Historically Independent of Science? A Study in Statistical Historiography', *Technology and Culture* **6** (1965), pp. 553–568.

Randall, J.H. Jr., *The Making of the Modern Mind,* Houghton Mifflin Company, New York, 1940.

Rapp, F. (ed.), *Contributions to a Philosophy of Technology: Studies in the Structure of Thinking in the Technological Sciences,* D. Reidel, Dordrecht, 1974.

Rapp, F., *Analytical Philosophy of Technology,* D. Reidel, Dordrecht, 1981.

Rescher, N., *Scientific Progress,* Blackwell, Oxford, 1978.

Rescher, N., *Unpopular Essays on Technological Progress,* University of Pittsburgh Press, Pittsburgh, 1980.

Rosenberg, N., 'Technology, Economy, and Values', in Bugliarello and Doner (1979), pp. 81–111.

Russell, B., *The Scientific Outlook,* London, 1931.

Sahal, D., *Patterns of Technological Innovation,* Addison-Wesley, Reading, 1981.

Schopman, J., 'The History of Semiconductor Electronics – A Kuhnian Story?', *Zeitschrift für allgemeine Wissenschaftstheorie* **12** (1981), 297–302.

Skolimowski, H., 'The Structure of Thinking in Technology', *Technology and Culture* **7** (1966), pp. 371–383. (Also in Rapp, 1974, pp. 72–85.)

Skolimowski, H., 'On the Concept of Truth in Science and in Technology', *Akten des XIV. Internationalen Kongresses für Philosophie, Wien: 2.–9. September 1968,* Verlag Herder, Wien, 1968, pp. 553–559.

Stock, B., 'Science, Technology, and Economic Progress in the Early Middle Ages', in Lindberg (1978), pp. 1–51.

Suppe, F. and Asquith, P.D. (eds.), *PSA 1976,* vol. 2, Philosophy of Science Association, East Lansing, 1977.

Tondl, L., 'On the Concepts of »Technology» and »Technological Sciences»', in Rapp (1974), pp. 1–18.

Weisheipl, J.A., 'The Nature, Scope, and Classification of the Sciences', in Lindberg (1978), pp. 401–482.

Wojick, D., 'The Structure of Technological Revolutions', in Bugliarello and Doner (1979), pp. 238–261.

von Wright, G.H., *Humanismi elämänasenteena,* Otava, Keuruu, 1981.

INDEX OF NAMES